THE INTERNATIONAL SERIES
IN
ELECTRICAL ENGINEERING

consulting editor *ROBERT F. LAMBERT*

Professor of Electrical Engineering
University of Minnesota

COMPUTER ANALYSIS OF CIRCUITS

COMPUTER ANALYSIS OF CIRCUITS

DAVID J. COMER

Director of Engineering
Chico State College

INTERNATIONAL TEXTBOOK COMPANY

College Division   Intext Educational Publishers

Scranton   Toronto   London

ISBN 0-7002-2292-8

LIBRARY OF CONGRESS CATALOG CARD NUMBER 70-151645

 Printed in the United States of America by The Haddon Craftsmen, Inc., Scranton, Pennsylvania.

TO

GAIL
SHAUNA
LOGAN
and
ANNA B.

PREFACE

It is difficult to recall a field that has grown as rapidly as that of computer-aided network design. In the last few years industries have not only accepted computer-aided design as a new and important field, but they have provided support for further development. There are always educational problems accompanying the rapid growth of a new area. University and college teachers have two problems with which to contend in this regard: (1) the limited number of student hours available over a normal undergraduate program, requiring that only topics of major importance be discussed; (2) self-education in a field that must eventually find its way into the curriculum. Generally by the time that the addition to the curriculum of a new topic is required, there is a gap between the teacher's knowledge of the topic and the literature of the day. Often the staff member chooses to become a "technician" in the new area and teach the subject matter on a superficial basis. This type of approach was popular in the late 1950's, when transistor courses were being added to electrical engineering programs. Staff members, thoroughly competent in vacuum-tube circuitry, did not have the necessary background to treat transistor circuitry. Rather than take the time to gain this background, many teachers taught only the "hows" of transistor circuitry, rather than the "hows" and "whys."

It would be easy to take the same approach to computer-aided network design. One can simply obtain a commercial program such as ECAP or SCEPTRE, read the instruction manual which tells how to enter an analysis program, and pass this information on to the student. The student gains very little from an exercise of this type, since he learns only "how" to run a program and nothing relating to the shortcomings of the program, other approaches that may be more efficient, or the direction that the entire field may take in coming years.

This text is a fundamental coverage of the more important concepts required in this field of computer-aided design. The necessary mathematics, topology theory, and network theory are provided in a single text. Superfluous theory is avoided in an effort to be concise. Mathematical techniques and background material are developed in the first four chapters in order to consider specific programs in later chapters. Although most chapters are general in nature, applications are often related to the two popular commercial programs, ECAP and SCEPTRE. The text is primarily for electrical engineers, but a student that has completed a basic Fortran programming course and an electronic circuits course should be adequately prepared. It is also helpful if the student has had

Laplace transforms. The text can be covered in one semester of a first-year graduate or advanced undergraduate course. Upon completion the student should know why certain approaches are more useful than others, he should understand some of the problems that arise in computer-aided design, and he should know how to effectively use ECAP.

The Introduction discusses the general concept of computer-aided design.

Chapter 1 is a review chapter on vector spaces, which may be deleted if the class has the necessary background.

Chapter 2 considers the solution of matrix equations with emphasis on the eigenvector equation. Functions of a matrix are also discussed here.

Chapter 3 develops some basic topological ideas and defines certain important circuit matrices. Analytical methods based on these topological matrices are derived in this chapter.

Chapter 4 introduces the state equation and outlines a method of formulation. This method is amenable to formulation using the digital computer. Selection of a proper tree by SCEPTRE is discussed here.

Chapter 5 covers several methods of solving the state equation. Specific programs implementing some of these methods are discussed. Some comparisons are made among the different methods of solution.

Modeling of devices is the topic of Chapter 6. This chapter is short and does not attempt to treat all possible models of a given device. It should give the student an appreciation for the advantages and shortcomings of modeling. Programs with stored models are discussed.

Chapter 7 discusses modeling of nonlinear networks using the ECAP approach and based on theory developed in Chapter 6. Analytical methods of nonlinear network analysis are also treated. Specific Newton-Raphson programs are reported in this chapter.

Chapter 8 covers ECAP and provides the necessary programming information to use the d-c, a-c, and transient analysis portions of the program. Significant features of SCEPTRE are also considered.

Several people have contributed to this textbook. I would like to thank the 1969 graduates in electrical engineering at The University of Calgary for their patience and fortitude in correcting the notes from which the manuscript was derived. The programming efforts of Professor Ray Anthes, John McDermid, and John Bancroft are also gratefully acknowledged.

I would like to thank the secretarial staffs at the University of Calgary and at Chico State College for their work relating to the typing of the manuscript and the distribution of class notes. I am particularly grateful to Miss Agnes Kabatoff for her excellent typing of the bulk of the manuscript.

DAVID J. COMER

Chico, California
April, 1971

CONTENTS

INTRODUCTION

Computer-aided design, as the name indicates, links the computer with the design procedure to result in a more efficient design process. The term computer-aided design should not be confused with "computer design," the latter being applicable only when the computer becomes capable of actual design and the former applying to the use of the computer in any of the design steps. Since "computer-aided design" applies to many possible situations, some clarification will be made in the following paragraphs regarding the usage of the term in this text.

Let us consider some possible steps in a typical network-design procedure. A set of requirements or specifications serves as the starting point for any design procedure. The next step is to choose some configuration that might reasonably be expected to satisfy the given specifications. Of course we may eventually find that another configuration is required or that any one of several possible configurations can be used. After selecting a configuration, a set of appropriate equations describing the network behavior must be formulated. Here again, several possible choices occur because of the existence of so many network-analysis methods. For a very simple network or very standard configurations, design equations may be available. In the general case, however, design is often based on analysis and trial-and-error techniques. That is, once the analysis equations are formulated, they are solved for the outputs or dependent variables of interest. This solution is based on the use of the proper independent variables as inputs and assumed values of elements. If the output quantities satisfy the given specifications, the design is complete. On the other hand, if the specifications are not met, key element values are modified and the analysis is repeated. Thus after each solution we must compare actual results to the desired results and decide whether to accept or reject the proposed (assumed) values and configuration. We might refer to these steps as *parameter modification and inspection* or *examination of results.* One might classify this part of the process as *optimization,* but the term is rather misleading when used in this context. In design there are a great many solutions to a given problem and each may offer both advantages and disadvantages when compared to other solutions. Referring to a given acceptable solution as the optimum has little meaning unless there is some additional criterion on which to base the choice of solutions. For example, after parameter modification and inspection have determined an acceptable solution, it may be desirable to minimize the spread of resistor or capacitor values. It may be desirable to minimize the absolute value of the largest capacitor of the circuit or the gain of the network may require maximization. Any number of possible constraints may be imposed on the network to require an optimized solution. In this case, optimization techniques can be applied as a final step in the design procedure.

A flow chart of the design procedure as outlined is shown in Fig. I-1. The dashed line to the box labeled configuration modification indicates that an

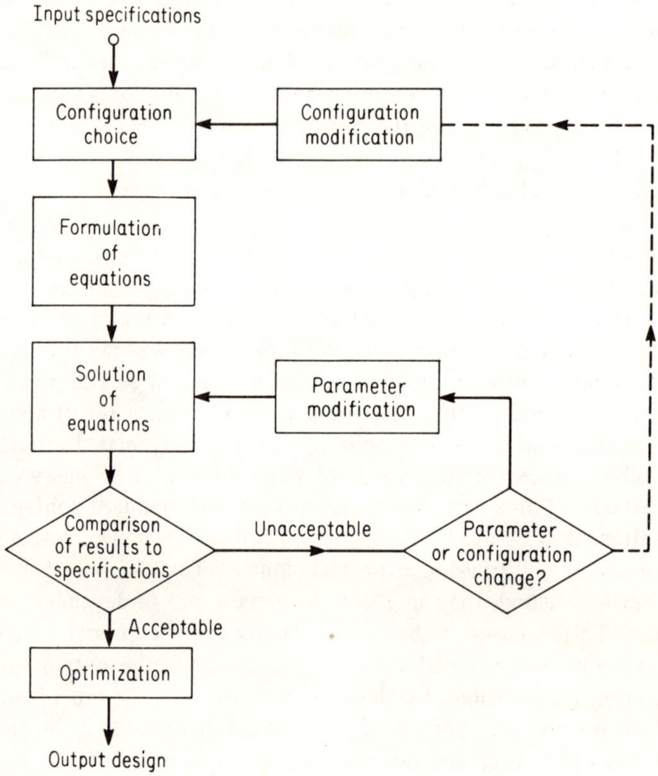

FIG. I-1. Flow chart of design process.

acceptable solution may never be obtained if an incorrect configuration is chosen; it may therefore be necessary to select a different configuration.

A. APPLICATION OF THE COMPUTER TO THE DESIGN PROCEDURE

The digital computer has long been looked upon as an extremely efficient calculator and has been applied to the solution of equations for some years. It has not been until recent years that algorithms for equation formulation and optimization have been developed. It is now possible to specify configuration and element values only, allowing the computer to formulate and solve the pertinent equations. Optimization can also be accomplished, although it is often difficult to guarantee that the absolute optimum conditions have been obtained.

A simple example can be considered to demonstrate manual design and examine the possibilities of applying the digital computer to each step of the procedure. A circuit is to be designed to have the transfer function magnitude indicated in Fig. I-2. The source resistance is 1 kΩ while the load resistance is 10 kΩ. The constant A must be greater than 0.1.

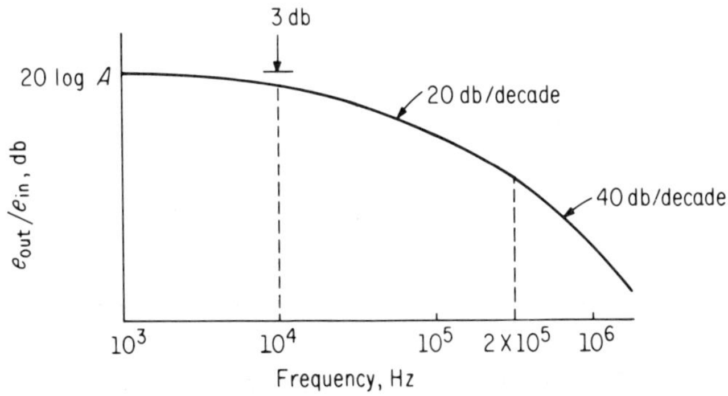

FIG. I-2. Asymptotic frequency response of transfer function.

Configuration

The choice of configuration is generally a manual one. While there is research work in the area of configuration selection by machine, general methods have not been developed to a practical extent. The computer, of course, can choose between a few simple configurations given a set of criteria, but for very complex networks a manual choice is necessary.

Modifications of configuration are possible in some optimization schemes. The original configuration is specified, but elements may be added or removed based on a prescribed error criterion applied to the output signal. As these types of programs are improved, configuration modification may develop to the point that configuration selection is possible.

For the example problem, two of the several possible configurations are shown in Fig. I-3. The selection of the configuration is based on several different considerations. The availability of one component over another might determine whether an *RC* or *RL* network is used. The ease with which the circuit can be analyzed or the cost of the circuit are other factors that may be important. Even the size of the components could be a determining factor especially if printed circuits are used. While other criteria may apply to a specific situation, we will assume that the capacitor filter is chosen to complete this example.

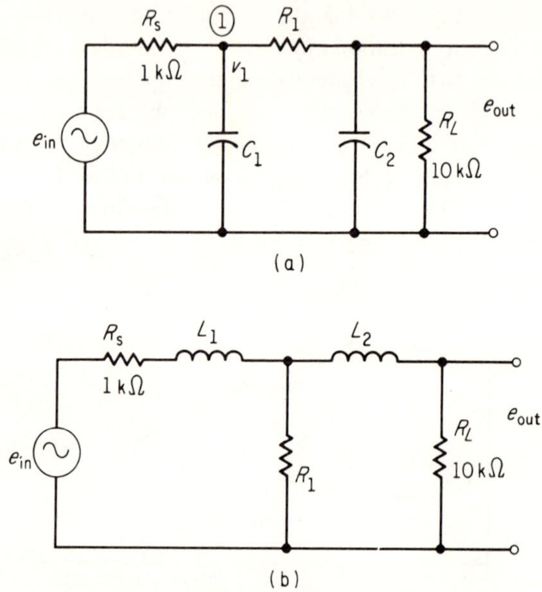

FIG. I-3. (a) Capacitor filter. (b) Inductor filter.

Formulation

Machine formulation of equations is an area that has developed very rapidly in the last ten years. This step must be concerned with two major requirements. The obvious function is that of formulating a set of analysis equations based on a minimal amount of information relating to configuration and element values. A second requirement is that the equations formulated must be amenable to efficient solution by the digital computer. The computer has been able to formulate equations for the last few years, but the efficiency of solution does not always justify the use of the computer to perform this step. Formulation of equations is a step that follows from orderly analysis procedures. For example, certain rules can be sequentially followed to write the node or loop equations for a network. Current and voltage vectors can be defined allowing the equations to be expressed as matrix equations, and the computer can easily manipulate matrices. However, the solution of the equations may require several matrix inversions, a time-consuming and therefore costly process for a digital computer. Some other formulation scheme might result in more efficient usage of the computer. It is easy to see then that there must be an interaction between the formulation of the equations and their solution. In the last five years rapid strides have been made in the development of more efficient methods of solution based on different sets of equations. The state-variable technique has become prominent during this period of time, and formulation procedures along with methods of solution based on the state variable are often more efficient than more classical approaches. A great deal of research continues in this area and more important developments can be expected in the next few years.

The circuit of Fig. I-3a can be analyzed manually by several methods. Using frequency-dependent impedances to represent the capacitors and applying the node method of analysis, we can formulate the following equations:

$$\frac{e_{\text{in}} - v_1}{R_S} = v_1 j\omega C_1 + \frac{(v_1 - e_{\text{out}})}{R_1}$$

$$\frac{v_1 - e_{\text{out}}}{R_1} = e_{\text{out}}\,(j\omega C_2 + 1/R_L)$$

We will see in Chapters 3 and 4 that the structure or topology of the graph and the element values are the data required by the computer to formulate the analysis equations. Most computer-aided design programs accept this information in a very simple form.

Solution

The efficiency of solution depends directly on the formulation method. Unfortunately no single method can be considered to be the most efficient for all types of analysis. The node method of analysis may have advantages over the state-variable approach in some instances, while other designs may result in advantages for the state-variable approach. To add to the complexity of the solution problem, we note that once an approach is determined by the formulation step, there can still be several methods of solution. As an example of this we can consider the analysis of a circuit using the state-variable approach. The formulation process results in a first-order differential equation in matrix form with the state variable as the dependent variable. If the time constants of the network are not widely separated, numerical integration may well be the most efficient method of machine solution. On the other hand, if widely separated time constants are present, some other method such as an eigenvector approach may be more desirable.

In considering the solution problem one must also consider consequences of parameter modification and recalculation of output quantities. Certain solution procedures might allow a parameter to be modified and the corresponding solution to be found without repeating the entire process of solution. An example of this occurs in the calculation of frequency response of a system. The node method may require matrix inversion at each frequency of interest to generate a frequency response. Another method, based on an eigenvector approach, requires only a single matrix inversion plus some relatively simple calculations at each frequency of interest. A second example occurs in the Electronic Circuit Analysis Program (ECAP) which performs sensitivity analysis without the necessity of resolving the original equations each time a resistor value is changed.

Returning to the example, the transfer function for the capacitor filter is found to be

$$\frac{e_{\text{out}}}{e_{\text{in}}} = \frac{1}{1+R_1/R_L+R_S/R_L+j\omega[C_1(R_S+R_SR_1/R_L)+C_2(R_S+R_1)]-\omega^2C_1C_2R_1R_S}$$

Using the specified values for R_S and R_L allows this function to be expressed as

$$\frac{e_{\text{out}}}{e_{\text{in}}} = \frac{1}{1.1+R_1/R_L+j\omega[C_1(10^3+10^{-1}R_1)+C_2(10^3+R_1)]-\omega^2 10^3C_1C_2R_1}$$

For this simple example we have only a single scalar equation. A more complex circuit analyzed by the computer will generally result in a matrix expression at this point.

We note that there are two specific constraints (required break frequencies) and one broad requirement on the low-frequency attenuation. Thus several combinations of element values can satisfy the design specifications.

It is at this step of the design procedure that the manual procedure often differs markedly from a general purpose computer-aided design program. The computer must base the element value selection on a trial-and-error process. Some sort of error criterion is applied to check each solution to determine if design specifications are met by the element values used for that solution. Element values are then changed and a new solution is found. Methods are applied which, hopefully, result in rapid convergence to an acceptable solution. However, efficient convergence methods are not readily available and this part of the design process often cannot be done effectively by the computer. In fact, commercial programs such as ECAP or SCEPTRE require that the error criterion be applied manually based on results of computer solutions obtained for several sets of element values.

The manual procedure would normally take one of three approaches. The first is to apply some well-known design method to the problem, provided such a method exists. If not, it may be possible to develop a method in which the necessary constraints can be considered in selecting element values. Another alternative is to simplify analysis equations by neglecting second-order effects until it becomes feasible to use trial-and-error methods with hand calculations.

For the filter under consideration, a design method can be developed. If we select $R_1 = 1.5\ \text{k}\Omega$ the d-c attenuation is 0.8, calculated from the transfer function by neglecting the frequency-dependent terms. The transfer function having the desired break points and a d-c attenuation of 0.8 is given by

$$\frac{e_{\text{out}}}{e_{\text{in}}} = \frac{1}{1.25+2.09\times10^{-5}j\omega-1.585\times10^{-11}\omega^2}$$

With the selection of $R_1 = 1.5\ \text{k}\Omega$, $C_1 = 16.9nF$, and $C_2 = 0.625nF$, the actual transfer function equals the desired one. Thus C_1 and C_2 can be calculated exactly by equating the two transfer functions.

When complex networks are designed it may occur that only trial-and-error methods can be applied. These are the cases in which the computer may be profitably utilized.

Comparison

Since the trial-and-error approach is so prominent in design problems, comparison methods are also important. Although the comparison of calculated to desired results can be straightforward for many networks, most large-scale programs presently available do not provide this feature. Parameter modification capabilities are generally provided, but comparison and conditional branching capabilities are not. The results of each trial solution must be examined visually to determine an appropriate set of parameter values.

A suitable error criterion for the capacitor filter could be specified in terms of attentuation tolerances at the break points or selected frequencies. For example, we might require that the actual response curve differ from the ideal curve by less than 1 db at all frequencies. Criterion such as this one can usually be expressed analytically for machine application.

While comparison is easy to perform, the appropriate course to follow after each comparison is not always prescribed. That is, the parameter modifications necessary to insure rapid convergence are not easy to calculate by machine.

Optimization

Since large-scale programs do not generally provide conditional branching capabilities, optimization is not possible. However, a fair amount of research has been devoted to this area resulting in proposals for various methods of optimization. Filter design is an area wherein machine optimization theory has been usefully applied. Two major problems exist in this area. The first is the inability of optimization schemes to guarantee absolute optimization. An appropriate error function may exhibit more than one local minimum; thus it is difficult to know if a given optimum design has been achieved. The second is the amount of time required by most present day schemes to reach an optimum condition.

There are several different variables with respect to which a network can be optimized. The most common requirement is to select element values of a network such that the response approximates a specified response. While many schemes can eventually lead to an acceptable network design, the efficiency with which this solution is reached by machine is of primary importance. Often a single trial solution can be expensive. If several solutions are required to reach the desired result, machine design can be far too costly to apply. For general network problems, acceptable optimization schemes are not usually available. It should also be mentioned that there are designs wherein many sets of element values will lead to an acceptable design. In these cases the word *optimization* may be a misnomer. As requirements become more stringent, optimization more accurately describes the process.

Another type of optimization consists of constraining element values to a given range. For example, the spread of capacitor values or actual values may be minimized. In the capacitor filter example, if R, is chosen to be smaller than 1.5 kΩ, the values of C_1 and C_2 are more nearly equal.

It is obvious that complete computer design of general networks is presently not possible. It is further obvious though, that computer design is a possibility that may be realized in the coming years. As computer-aided design techniques allow the machine to perform all steps in the design process, the term "computer-aided design" will become "computer design."

This text is concerned mainly with the formulation and solution of analysis equations. In addition to the fact that these steps are often the most time-consuming of the design process, they serve as the basis for development of other steps such as optimization. Furthermore, these areas have received more attention than others with the result that well-developed procedures for formulation and solution now exist.

B. COMMERCIAL CIRCUIT ANALYSIS PROGRAMS

There are several programs available at the present time for computer-aided design purposes. One of the earlier programs that remains popular is the Electronic Circuit Analysis Program (ECAP) previously mentioned, which can be obtained from IBM and other computer firms. ECAP is capable of formulating equations based on a knowledge of the network topology and element values. The equations are then solved and the results printed out. No automatic optimization is possible although parameter modification can be done giving a printout of results at each new set of parameter values. A more extensive discussion of ECAP is given in Chapter 8.

SCEPTRE is a program designed mainly for analysis of transient circuits, and while more limited than ECAP in overall flexibility it still offers several advantages. This program basically formulates and solves network equations. However, stored models of electronic devices can be employed without re-entering data for similar devices. The characteristics of a transistor can be stored and used several times as required by the circuit configuration. This feature is a real time-saver for networks containing several devices of the same type. SCEPTRE is also discussed in more detail in Chapter 8.

Most of the large-scale programs such as CIRCUS, PANE, NET-1, NET-2, CORNAP, POTTLE, etc., basically perform equation formulation and solution. Optimization is generally too complex and costly to use except with small, special-purpose programs.

While the material covered in this book consists mainly of general techniques, many of the specific applications cited are discussed in connection with ECAP and SCEPTRE. This allows the reader to become acquainted with these two useful programs while providing a vehicle that allows a more detailed discussion of the application of the mathematical methods.

chapter **1**

REVIEW OF MATRICES AND DETERMINANTS

A. INTRODUCTION

Electrical engineering is a relatively precise field which depends quite heavily on mathematics. While there may be disagreement about the type of mathematical approach to apply to a given problem, there is no doubt that the electrical engineer must be an applied mathematician in order to function effectively in his work. Often, however, there is a gap in the level of mathematics used by the practical designer and that developed by the engineering theorist. Many methods of solution appear to the designer to offer little more than elegance while at the same time complicating the design procedures. It must be admitted that certain theoretical procedures are more time-consuming than more practical approximate methods. These mathematical methods are so cumbersome as to discourage practical application in all but a few instances. Unfortunately, the simplifications or approximations necessary to apply practical methods rarely allow optimum design to be achieved.

In recent years the digital computer has necessitated a reevaluation of the appropriateness of certain theoretical methods. The efficiency and speed of the computer remove the major objections to the use of several theoretical approaches. Cut-and-try methods are still important in computer-aided design, but the process of approaching the required performance by parameter variation is governed by prescribed mathematical guidelines based on theoretical methods. It is often possible to attain optimum performance using the computer in conjunction with the appropriate design procedure. Accuracy and preciseness are, of course, two of the distinguishing features of the digital computer. A very precise set of input data and input instructions are prerequisite to the successful application of the computer. In manual design, much of the accuracy and preciseness is discarded in an effort to achieve usable design procedures.

Network graph theory is an example of an area which previously had not achieved widespread acceptance in most design fields (except for specialized areas such as communication networks). For most problems, manual analysis is performed more simply by applying specific techniques than using general graph theory. The order required by the computer has made graph theory a necessity in computer-aided design work. The practical designer must now familiarize himself to a degree with topological methods to understand the computer programs to aid in design.

State-variable analysis is another example of an approach long considered too elegant to usefully apply to circuit design which now appears indispensable to many computer-aided design procedures. It is now realized that the matrix theory, operator theory, and eigenvalue theory associated with the state variable is useful not only to physicists and mathematicians but to circuit and system designers as well.

The need for newer mathematical methods in engineering seems to be associated closely with the use of the computer. The state-variable approach was first applied in electrical engineering to control problems. Analog and digital computers are utilized for simulation and solution of control systems mainly due to the complexity of the problems encountered in this field. New techniques to utilize the potential of the computer have evolved and developed rapidly in the last few years. With computer-aided circuit design becoming more popular we might expect that the circuit designer now requires more mathematics. While it is difficult to predict the impact on circuit design that the computer will have, it is certain that the utility of the computer rises in proportion to the development and understanding of modern mathematical techniques.

In this textbook an attempt is made to present general techniques that can be applied to computer analysis in the fields of circuit and control-system design. However, emphasis is on computer-aided circuit design, since many mechanical and control systems can be described in terms of electrical analogs and treated by the methods considered.

B. *n*-DIMENSIONAL SPACE

Vectors

We are familiar with the concept of two-dimensional and three-dimensional spaces. We can often visualize a particular vector or vector variation occurring in two- or three-dimensional spaces. If we extend the discussion to spaces having four or more dimensions it is immediately impossible to associate the same physical visualization with this situation as with the lower-dimensional space. Certainly the physical space we live in can be described in terms of three mutually perpendicular (orthogonal) reference vectors; height, width, and depth, whereas four-dimensional space cannot. This does not mean that perhaps three of the four reference vectors of a four-dimensional space cannot be associated with physical dimensions of space, with the fourth reference vector associated with some other acceptable quantity such as time. However, it must be stated at the outset that the space referred to in n-dimensional space does not usually refer to or include the physical space described in terms of height, width, and depth. As a matter of fact, the coordinates in three-dimensional space often refer to other quantities than physical space, although we tend to visualize the functional variation of these quantities in terms of height, width, and depth.

A space containing n-dimensions is sometimes called an *abstract space*, further emphasizing the fact that physical space does not have a one-to-one correspondence with n-dimensional space. On the other hand, certain operations or concepts which are valid for any space, can be visualized in two- or three-dimensional space and then applied to higher-order spaces. Thus, it is often useful to work examples in three-dimensional space to understand underlying concepts.

At this point, one might justifiably ask just what the six dimensions in six-dimensional space actually represent. In general, this question must be answered in terms of specific problems. The dimensions can represent variables such as the voltages at six different nodes of a circuit, the currents through six different branches of a circuit, a combination of three node voltages and three branch currents, or any other set of variables that describe a circuit or system.

There is a very close correspondence between algebraic equations and vector spaces. A set of n simultaneous equations can be represented by a set of vectors in n-dimensional space provided the proper relationship between the n variables

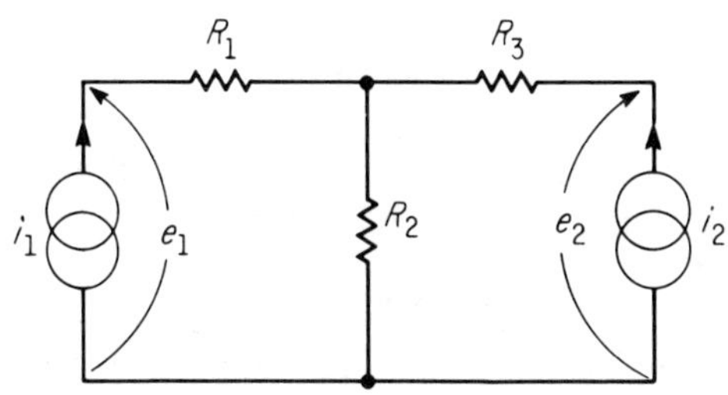

FIG. 1-1. Resistive network.

and the n dimensions is established. Consider the circuit of Fig. 1-1. The voltages e_1 and e_2 can be expressed in terms of the currents i_1 and i_2 by

$$e_1 = (R_1 + R_2)\, i_1 + R_2 i_2 \tag{1-1a}$$

$$e_2 = R_2 i_1 + (R_2 + R_3)\, i_2 \tag{1-1b}$$

The voltages e_1 and e_2 can be considered to be components of a voltage vector **E** while i_1 and i_2 are components of a current vector **I**. With these identifications, the two previous equations can be written simply as

$$\mathbf{E} = \mathbf{RI} \tag{1-2}$$

The term **R** represents a matrix which must multiply the current vector to obtain the voltage vector. Matrices will be discussed in greater detail in succeeding sections. Before proceeding further we should examine the motive for expressing a set of simultaneous equations as a vector equation. Two important reasons for making this change are (1) the resulting simplicity of notation, and (2) the resulting procedural simplicity in solving for the unknowns (or unknown vector). In attempting to solve a large number of simultaneous equations and/or in manipulating unknowns that are expressed in terms of equations, the use of vector equations can be far more efficient than other methods. This fact becomes extremely important when a digital computer, costing several hundred dollars per hour, is used to aid the circuit design. In order to represent a set of vectors, we choose reference vectors or basis vectors which can be used to describe any one of the vectors in the set. The vector **I** is expressed in terms of the components i_1 and i_2, but each of these components must be associated with a direction. Since two dimensions are required in this case, we can choose a coordinate system having two dimensions. For example, we could choose a set of perpendicular axes with the unit vector in the horizontal direction given by

$$\mathbf{u}_1 = \begin{bmatrix} 1 \\ 0 \end{bmatrix}$$

and the unit vector in the vertical direction given by

$$\mathbf{u}_2 = \begin{bmatrix} 0 \\ 1 \end{bmatrix}$$

If the direction of the i_1 component is associated with $\mathbf{u}_1$ and the direction of the i_2 component with $\mathbf{u}_2$, then **I** can be written as

$$\mathbf{I} = i_1 \mathbf{u}_1 + i_2 \mathbf{u}_2$$

The vectors $\mathbf{u}_1$ and $\mathbf{u}_2$ are called *basis vectors*, since all other vectors of the set (all two-dimensional vectors) can be expressed in terms of these vectors.

If we use a third current i_3, we can express Eq. 1-1 as

$$e_1 = (R_1 + R_2/2)i_1 + (R_2/2)i_2 + (R_2/2)i_3 \tag{1-3a}$$

$$e_2 = (R_2/2)i_1 + (R_2/2 + R_3)i_2 + (R_2/2)i_3 \tag{1-3b}$$

However, we now have two equations involving three unknowns, including the necessity of a third equation which is $i_3 = i_1 + i_2$. This step results in a more complex set of equations, and follows from the fact that **E** is expressed in terms of a vector that is not independent of the basis vectors. The component i_3 and the direction associated with i_3 are dependent on i_1, i_2 and $\mathbf{u}_1$ and $\mathbf{u}_2$. Therefore, a third component is not required to express any two-dimensional vector.

C. MATRICES

The relationship between two vectors can conveniently be expressed in terms of a matrix. Equation 1-1 can be represented by Eq. 1-2 if the correct resistance matrix **R** is chosen. By definition, a matrix is simply an array of numbers or elements arranged in square or rectangular fashion. Since this definition is so broad, perhaps it should be narrowed by requiring the array of elements to satisfy the rules of matrix algebra discussed next.

Before discussing these rules the standard matrix notation will be introduced. In the matrix below, the first subscript refers to the row and the second to the column in which a given element finds itself,

$$A = \begin{bmatrix} a_{11} & a_{12} & a_{13} & \cdots & a_{1n} \\ a_{21} & a_{22} & a_{23} & \cdots & a_{2n} \\ a_{31} & a_{32} & a_{33} & \cdots & a_{3n} \\ \cdot & \cdot & \cdot & \cdots & \cdot \\ \cdot & \cdot & \cdot & \cdots & \cdot \\ \cdot & \cdot & \cdot & \cdots & \cdot \\ a_{n1} & a_{n2} & a_{n3} & \cdots & a_{nn} \end{bmatrix} \tag{1-4}$$

This notation is convenient, since we now can discuss any individual element by specifying only the two associated subscripts.

We should note that the matrix itself has no single value. The matrix is defined only in terms of its elements.

Matrix Algebra

The sum of two matrices is formed by summing the two elements with corresponding subscripts (corresponding element position) of each matrix. The sum of **C** and **D** is

$$\begin{bmatrix} c_{11} & c_{12} & c_{13} \\ c_{21} & c_{22} & c_{23} \\ c_{31} & c_{32} & c_{33} \end{bmatrix} + \begin{bmatrix} d_{11} & d_{12} & d_{13} \\ d_{21} & d_{22} & d_{23} \\ d_{31} & d_{32} & d_{33} \end{bmatrix}$$

$$= \begin{bmatrix} c_{11}+d_{11} & c_{12}+d_{12} & c_{13}+d_{13} \\ c_{21}+d_{21} & c_{22}+d_{22} & c_{23}+d_{23} \\ c_{31}+d_{31} & c_{32}+d_{32} & c_{33}+d_{33} \end{bmatrix} \tag{1-5}$$

Addition of matrices is not defined in the strictest sense for matrices of different dimensions. In the practical case it is often possible and useful to add the necessary number of zero elements to the smaller matrix to equalize the dimensions of the two matrices. The word order is used to describe the dimensions of a matrix. For example, a 3×4 matrix refers to one with 3 rows and 4 columns. If the order of one matrix equals that of another matrix, both the number of rows and the number of columns are equal.

The product of two matrices **GH** is defined only if the number of columns of **G** equals the number of rows of **H**. If this condition is fulfilled then matrix multiplication is defined by

$$\mathbf{M} = \mathbf{GH}$$

where the elements of **M** are

$$m_{ik} = \sum_{j=1}^{n} g_{ij} h_{jk} \tag{1-6}$$

In this summation n refers to the number of columns of g or the number of rows of h. In words, Eq. 1-6 states that to form the element of the ith row and kth column of the product matrix, one must form a sum of the products of the elements of the ith row of **G** and the elements of the kth column of **H**. The product of a 3 X 3 matrix and a 3 X 2 matrix is

$$\mathbf{M} = \begin{bmatrix} g_{11} & g_{12} & g_{13} \\ g_{21} & g_{22} & g_{23} \\ g_{31} & g_{32} & g_{33} \end{bmatrix} \begin{bmatrix} h_{11} & h_{12} \\ h_{21} & h_{22} \\ h_{31} & h_{32} \end{bmatrix}$$

$$= \begin{bmatrix} g_{11}h_{11} + g_{12}h_{21} + g_{13}h_{31} & g_{11}h_{12} + g_{12}h_{22} + g_{13}h_{32} \\ g_{21}h_{11} + g_{22}h_{21} + g_{23}h_{31} & g_{21}h_{12} + g_{22}h_{22} + g_{23}h_{32} \\ g_{31}h_{11} + g_{32}h_{21} + g_{33}h_{31} & g_{31}h_{12} + g_{32}h_{22} + g_{33}h_{32} \end{bmatrix} \tag{1-7}$$

In light of this definition of matrix multiplication, the relationship to algebraic equations becomes obvious. Equation 1-1 is represented by Eq. 1-2, a matrix equation, and can be written as

$$\begin{bmatrix} e_1 \\ e_2 \end{bmatrix} = \begin{bmatrix} (R_1 + R_2) & R_2 \\ R_2 & (R_2 + R_3) \end{bmatrix} \begin{bmatrix} i_1 \\ i_2 \end{bmatrix} \tag{1-8}$$

In this equation the voltages and currents form vectors or column matrices to which the rule of matrix multiplication applies. When **I** is multiplied by **R** the result is an expression indicating the equality of two matrices. In order that two matrices be equal each element of one matrix must equal the corresponding element of the other matrix. In equating these elements, the two algebraic equations given by Eq. 1-1 result.

EXAMPLE 1-1. If

$$\mathbf{A} = \begin{bmatrix} 3 & 1 & 0 \\ -2 & 3 & 4 \\ 1 & 6 & 1 \end{bmatrix}, \quad \mathbf{B} = \begin{bmatrix} 2 & 2 & 1 \\ 0 & 1 & -1 \\ 3 & 1 & 2 \end{bmatrix}$$

and

$$\mathbf{I} = \begin{bmatrix} 1 \\ 4 \\ 2 \end{bmatrix}$$

find

(a) **A** + **B**
(b) **A** − **B**
(c) **A B**
(d) **B A**
(e) **A I**

Solution

(a) $$\begin{bmatrix} 3 & 1 & 0 \\ -2 & 3 & 4 \\ 1 & 6 & 1 \end{bmatrix} + \begin{bmatrix} 2 & 2 & 1 \\ 0 & 1 & -1 \\ 3 & 1 & 2 \end{bmatrix} = \begin{bmatrix} 5 & 3 & 1 \\ -2 & 4 & 3 \\ 4 & 7 & 3 \end{bmatrix}$$

(b) $$\begin{bmatrix} 3 & 1 & 0 \\ -2 & 3 & 4 \\ 1 & 6 & 1 \end{bmatrix} - \begin{bmatrix} 2 & 2 & 1 \\ 0 & 1 & -1 \\ 3 & 1 & 2 \end{bmatrix} = \begin{bmatrix} 1 & -1 & -1 \\ -2 & 2 & 5 \\ -2 & 5 & -1 \end{bmatrix}$$

(c) $$\begin{bmatrix} 3 & 1 & 0 \\ -2 & 3 & 4 \\ 1 & 6 & 1 \end{bmatrix} \begin{bmatrix} 2 & 2 & 1 \\ 0 & 1 & -1 \\ 3 & 1 & 2 \end{bmatrix} = \begin{bmatrix} 6 & 7 & 2 \\ 8 & 3 & 3 \\ 5 & 9 & -3 \end{bmatrix}$$

(d) $$\begin{bmatrix} 2 & 2 & 1 \\ 0 & 1 & -1 \\ 3 & 1 & 2 \end{bmatrix} \begin{bmatrix} 3 & 1 & 0 \\ -2 & 3 & 4 \\ 1 & 6 & 1 \end{bmatrix} = \begin{bmatrix} 3 & 14 & 9 \\ -3 & -3 & 3 \\ 9 & 18 & 6 \end{bmatrix}$$

Note: In general, **A B** ≠ **B A**.

(e) $$\begin{bmatrix} 3 & 1 & 0 \\ -2 & 3 & 4 \\ 1 & 6 & 1 \end{bmatrix} \begin{bmatrix} 1 \\ 4 \\ 2 \end{bmatrix} = \begin{bmatrix} 7 \\ 18 \\ 27 \end{bmatrix}$$

There are other important properties of matrices, many of which follow from the multiplication rule.

Scalar Multiplication

A matrix can be multiplied by a scalar by multiplying each element of the matrix by the scalar quantity. This is implied in Part (b) of Example 1-1 where the quantity **A** − **B** is desired. Since the rule for adding matrices is known, subtraction could be thought of as multiplying **B** by −1 and adding the result to **A**.

Associative and Commutative Laws

Matrices are associative and commutative with respect to addition. These laws are summarized as follows:

$$\mathbf{A} + (\mathbf{B} + \mathbf{C}) = (\mathbf{A} + \mathbf{B}) + \mathbf{C}$$

$$\mathbf{A} + \mathbf{B} + \mathbf{C} = \quad \mathbf{A} + \mathbf{C} + \mathbf{B} = \quad \mathbf{B} + \mathbf{C} + \mathbf{A} = \quad \mathbf{C} + \mathbf{A} + \mathbf{B}$$

Of course, the necessary constraint regarding matrix dimensions must be satisfied before performing the indicated operations. Matrices are associative with respect to multiplication, but are not commutative. Thus

$$\mathbf{A}(\mathbf{BC}) = (\mathbf{AB})\,\mathbf{C}$$

however, in general

$$\mathbf{A}\,\mathbf{B} \neq \mathbf{B}\,\mathbf{A}$$

We can also note that matrices are distributive, meaning that

$$\mathbf{A}\,(\mathbf{B} + \mathbf{C}) = \mathbf{AB} + \mathbf{AC}$$

Identity Matrix

The identity or unit matrix has the property that

$$\mathbf{UA} = \mathbf{AU} = \mathbf{A}$$

The unit matrix consequently has the same dimensions as the matrix **A** with diagonal elements of unity and zero elements elsewhere.

The Transpose of a Matrix

If **D** is defined as a matrix with elements d_{ij} then the transpose of **D**, designated $\mathbf{D}^t$, has elements d_{ji}. This operation amounts to replacing the columns by the rows. Consider a matrix given by

$$\mathbf{D} = \begin{bmatrix} 1 & 3 & 4 \\ 6 & 1 & 5 \\ 2 & 2 & 1 \end{bmatrix}$$

then

$$\mathbf{D}^t = \begin{bmatrix} 1 & 6 & 2 \\ 3 & 1 & 2 \\ 4 & 5 & 1 \end{bmatrix}$$

The Inverse of a Matrix

The inverse of a matrix is defined by the equation

$$\mathbf{A}^{-1}\,\mathbf{A} = \mathbf{AA}^{-1} = \mathbf{U} \tag{1-9}$$

The matrix $\mathbf{A}^{-1}$ is called the inverse matrix of **A** or simply the inverse of **A**. To find $\mathbf{A}^{-1}$, given the matrix **A**, we must find that matrix which is multiplied by **A** to give the unit matrix. The relationship between the elements of $\mathbf{A}^{-1}$ and those of **A** can be found in the general case, however it is more instructive to derive this relationship for a third order matrix and then extend to the case of an nth order matrix.

Let $\mathbf{A}$ be a known matrix with elements a_{ij} and let the unknown elements of $\mathbf{A}^{-1}$ be denoted by d_{ij}. Then

$$\mathbf{A}\,\mathbf{A}^{-1} = \begin{bmatrix} a_{11} & a_{12} & a_{13} \\ a_{21} & a_{22} & a_{23} \\ a_{31} & a_{32} & a_{33} \end{bmatrix} \begin{bmatrix} d_{11} & d_{12} & d_{13} \\ d_{21} & d_{22} & d_{23} \\ d_{31} & d_{32} & d_{33} \end{bmatrix}$$

$$= \begin{bmatrix} a_{11}d_{11} + a_{12}d_{21} + a_{13}d_{31} & a_{11}d_{12} + a_{12}d_{22} + a_{13}d_{32} & a_{11}d_{13} + a_{12}d_{23} + a_{13}d_{33} \\ a_{21}d_{11} + a_{22}d_{21} + a_{23}d_{31} & a_{21}d_{12} + a_{22}d_{22} + a_{23}d_{32} & a_{21}d_{13} + a_{22}d_{23} + a_{23}d_{33} \\ a_{31}d_{11} + a_{32}d_{21} + a_{33}d_{31} & a_{31}d_{12} + a_{32}d_{22} + a_{33}d_{32} & a_{31}d_{13} + a_{32}d_{23} + a_{33}d_{33} \end{bmatrix}$$

The result of multiplying $\mathbf{A}^{-1}$ by $\mathbf{A}$ must be the unit matrix to satisfy Eq. 1-9. Thus we can equate the elements of this third-order matrix to those of the third-order unit matrix

$$\mathbf{U} = \begin{bmatrix} 1 & 0 & 0 \\ 0 & 1 & 0 \\ 0 & 0 & 1 \end{bmatrix}$$

By properly grouping these nine equations we obtain three sets of three simultaneous equations as follows:

$$\left.\begin{aligned} a_{11}d_{11} + a_{12}d_{21} + a_{13}d_{31} &= 1 \\ a_{21}d_{11} + a_{22}d_{21} + a_{23}d_{31} &= 0 \\ a_{31}d_{11} + a_{32}d_{21} + a_{33}d_{31} &= 0 \end{aligned}\right\}$$

$$\left.\begin{aligned} a_{11}d_{12} + a_{12}d_{22} + a_{13}d_{32} &= 0 \\ a_{21}d_{12} + a_{22}d_{22} + a_{23}d_{32} &= 1 \\ a_{31}d_{12} + a_{32}d_{22} + a_{33}d_{32} &= 0 \end{aligned}\right\}$$

$$\left.\begin{aligned} a_{11}d_{13} + a_{12}d_{23} + a_{13}d_{33} &= 0 \\ a_{21}d_{13} + a_{22}d_{23} + a_{23}d_{33} &= 0 \\ a_{31}d_{13} + a_{32}d_{23} + a_{33}d_{33} &= 1 \end{aligned}\right\}$$

Note that the first set of equations contains the 3 unknowns d_{11}, d_{21}, and d_{31}. The second set contains only d_{12}, d_{22}, and d_{32}, while the third set contains d_{13}, d_{23}, and d_{33}.

Next let us consider solving the first set of equations for d_{11}, d_{21}, and d_{31} by using Cramer's rule (see Sec. D). In terms of determinants we find that

$$d_{11} = \frac{\begin{vmatrix} a_{22} & a_{23} \\ a_{32} & a_{33} \end{vmatrix}}{\begin{vmatrix} a_{11} & a_{12} & a_{13} \\ a_{21} & a_{22} & a_{23} \\ a_{31} & a_{32} & a_{33} \end{vmatrix}} = \frac{\text{cofactor } a_{11}}{\det \mathbf{A}}$$

$$d_{21} = \frac{-\begin{vmatrix} a_{21} & a_{23} \\ a_{31} & a_{33} \end{vmatrix}}{\begin{vmatrix} a_{11} & a_{12} & a_{13} \\ a_{21} & a_{22} & a_{23} \\ a_{31} & a_{32} & a_{33} \end{vmatrix}} = \frac{\text{cofactor } a_{12}}{\det \mathbf{A}}$$

and

$$d_{31} = \frac{\begin{vmatrix} a_{21} & a_{22} \\ a_{31} & a_{32} \end{vmatrix}}{\begin{vmatrix} a_{11} & a_{12} & a_{13} \\ a_{21} & a_{22} & a_{23} \\ a_{31} & a_{32} & a_{33} \end{vmatrix}} = \frac{\text{cofactor } a_{13}}{\det \mathbf{A}}$$

The other two sets of simultaneous equations can be solved to obtain similar relationships. The method of extension to an nth order matrix should be now obvious. The general expression for the elements of the inverse matrix is

$$d_{ij} = \frac{\text{cofactor } a_{ji}}{\det \mathbf{A}} \tag{1-10}$$

We have calculated the element values of $\mathbf{A}^{-1}$ to insure that $\mathbf{A}\,\mathbf{A}^{-1} = \mathbf{U}$. We have not shown that $\mathbf{A}^{-1}\,\mathbf{A} = \mathbf{U}$, however this is a trivial matter which is left to the student.

It should also be obvious that the inverse of a nonsquare matrix is undefined.

Matrix Partitioning

A matrix can be partitioned to create a set of submatrices within the original matrix. In performing certain algebraic operations with matrices, the effort required can occasionally be reduced considerably by partitioning a matrix or several matrices.

The partitioning of a matrix is arbitrary, but if this operation is to lead to simplified results, it must be done in a specific way. The 4 X 4 matrix **B** could be subdivided as

$$\mathbf{B} = \left[\begin{array}{ccc:c} b_{11} & b_{12} & b_{13} & b_{14} \\ b_{21} & b_{22} & b_{23} & b_{24} \\ \hdashline b_{31} & b_{32} & b_{33} & b_{34} \\ b_{41} & b_{42} & b_{43} & b_{44} \end{array}\right] = \begin{bmatrix} \mathbf{B}_{11} & \mathbf{B}_{12} \\ \mathbf{B}_{21} & \mathbf{B}_{22} \end{bmatrix}$$

where

$$\mathbf{B}_{11} = \begin{bmatrix} b_{11} & b_{12} & b_{13} \\ b_{21} & b_{22} & b_{23} \end{bmatrix}, \quad \mathbf{B}_{12} = \begin{bmatrix} b_{14} \\ b_{24} \end{bmatrix}$$

$$\mathbf{B}_{21} = \begin{bmatrix} b_{31} & b_{32} & b_{33} \\ b_{41} & b_{42} & b_{43} \end{bmatrix}, \quad \mathbf{B}_{22} = \begin{bmatrix} b_{34} \\ b_{44} \end{bmatrix}$$

On the other hand, if constraints are imposed on **B** due to a second matrix **C** which is to be added to or multiplied by **B**, a different partitioning scheme might be more reasonable. If two matrices are to be added or subtracted using submatrices, both matrices must be partitioned in the same way. For example, if a 4 X 4 matrix **C** is to be added to **B**, then **C** is partitioned as follows:

$$\mathbf{C} = \left[\begin{array}{ccc:c} c_{11} & c_{12} & c_{13} & c_{14} \\ c_{21} & c_{22} & c_{23} & c_{24} \\ \hdashline c_{31} & c_{32} & c_{33} & c_{34} \\ c_{41} & c_{42} & c_{43} & c_{44} \end{array}\right] = \begin{bmatrix} \mathbf{C}_{11} & \mathbf{C}_{12} \\ \mathbf{C}_{21} & \mathbf{C}_{22} \end{bmatrix}$$

where

$$\mathbf{C}_{11} = \begin{bmatrix} c_{11} & c_{12} & c_{13} \\ c_{21} & c_{22} & c_{23} \end{bmatrix}, \quad \mathbf{C}_{12} = \begin{bmatrix} c_{14} \\ c_{24} \end{bmatrix}$$

$$\mathbf{C}_{21} = \begin{bmatrix} c_{31} & c_{32} & c_{33} \\ c_{41} & c_{42} & c_{43} \end{bmatrix}, \quad \mathbf{C}_{22} = \begin{bmatrix} c_{34} \\ c_{44} \end{bmatrix}$$

The sum of **B** and **C** can be written as

$$\mathbf{B} + \mathbf{C} = \begin{bmatrix} \mathbf{B}_{11} + \mathbf{C}_{11} & \mathbf{B}_{12} + \mathbf{C}_{12} \\ \mathbf{B}_{21} + \mathbf{C}_{21} & \mathbf{B}_{22} + \mathbf{C}_{22} \end{bmatrix}$$

For multiplication employing submatrices, the partitioning of the rows of the first matrix and the columns of the second matrix is arbitrary. The columns of the first matrix and the rows of the second matrix must be partitioned in the same way to perform this operation. If we desire to form the product **BC** with matrix **B** partitioned as shown previously, then one possible partitioning scheme for **C** is

$$\mathbf{C} = \left[\begin{array}{cc:cc} c_{11} & c_{12} & c_{13} & c_{14} \\ c_{21} & c_{22} & c_{23} & c_{24} \\ c_{31} & c_{32} & c_{33} & c_{34} \\ \hdashline c_{41} & c_{42} & c_{43} & c_{44} \end{array}\right] = \begin{bmatrix} \mathbf{C}_{11} & \mathbf{C}_{12} \\ \mathbf{C}_{21} & \mathbf{C}_{22} \end{bmatrix}$$

where

$$\mathbf{C}_{11} = \begin{bmatrix} c_{11} & c_{12} \\ c_{21} & c_{22} \\ c_{31} & c_{32} \end{bmatrix}, \quad \mathbf{C}_{12} = \begin{bmatrix} c_{13} & c_{14} \\ c_{23} & c_{24} \\ c_{33} & c_{34} \end{bmatrix}$$

$$\mathbf{C}_{21} = [c_{41} \quad c_{42}], \quad \mathbf{C}_{22} = [c_{43} \quad c_{44}]$$

The product **BC** is expressible as

$$\mathbf{BC} = \begin{bmatrix} \mathbf{B}_{11}\mathbf{C}_{11} + \mathbf{B}_{12}\mathbf{C}_{21} & \mathbf{B}_{11}\mathbf{C}_{12} + \mathbf{B}_{12}\mathbf{C}_{22} \\ \mathbf{B}_{21}\mathbf{C}_{11} + \mathbf{B}_{21}\mathbf{C}_{21} & \mathbf{B}_{21}\mathbf{C}_{12} + \mathbf{B}_{22}\mathbf{C}_{22} \end{bmatrix}$$

It can easily be verified that the product **BC** calculated in the conventional way is equal to that given by the above expression. On this basis it is reasonable to inquire what is to be gained by using partitioning. While this question will subsequently be answered, a simple example can demonstrate one advantage.

EXAMPLE 1-2. If

$$\mathbf{A} = \begin{bmatrix} 1 & 0 & 0 \\ 2 & 0 & 0 \\ 1 & 3 & 2 \end{bmatrix} \quad \text{and} \quad \mathbf{B} = \begin{bmatrix} 1 & 2 & 0 \\ 2 & 1 & 2 \\ -1 & -1 & 1 \end{bmatrix}$$

find **AB** first by using conventional matrix multiplication and then by partitioning the matrices in such a way to minimize the number of operations required to obtain the product.

Solution. The product **AB** is

$$\mathbf{AB} = \begin{bmatrix} 1 & 0 & 0 \\ 2 & 0 & 0 \\ 1 & 3 & 2 \end{bmatrix} \begin{bmatrix} 1 & 2 & 0 \\ 2 & 1 & 2 \\ -1 & -1 & 1 \end{bmatrix} = \begin{bmatrix} 1 & 2 & 0 \\ 2 & 4 & 0 \\ 5 & 3 & 8 \end{bmatrix}$$

The number of multiplications of individual elements necessary to find this product is 27 or n^3, where n is the order of the square matrices. All three rows of **A** must be multiplied by the three columns of **B** and three element multiplications are required to form the product of a row and a column. Furthermore, 18 additions are required to complete the matrix multiplication.

If the matrices are partitioned such that

$$\mathbf{A} = \left[\begin{array}{c:cc} 1 & 0 & 0 \\ 2 & 0 & 0 \\ \hdashline 1 & 3 & 2 \end{array}\right] \quad \text{and} \quad \mathbf{B} = \left[\begin{array}{cc:c} 1 & 2 & 0 \\ \hdashline 2 & 1 & 2 \\ -1 & -1 & 1 \end{array}\right]$$

where

$$\mathbf{A}_{11} = \begin{bmatrix} 1 \\ 2 \end{bmatrix}, \quad \mathbf{A}_{12} = \begin{bmatrix} 0 & 0 \\ 0 & 0 \end{bmatrix}, \quad \mathbf{A}_{21} = [1], \quad \mathbf{A}_{22} = [3 \quad 2]$$

$$\mathbf{B}_{11} = [1 \quad 2], \quad \mathbf{B}_{12} = [0], \quad \mathbf{B}_{21} = \begin{bmatrix} 2 & 1 \\ -1 & -1 \end{bmatrix}, \quad \mathbf{B}_{22} = \begin{bmatrix} 2 \\ 1 \end{bmatrix}$$

the product **AB** can be written in terms of submatrices as

$$\mathbf{AB} = \begin{bmatrix} \mathbf{A}_{11}\mathbf{B}_{11} + \mathbf{A}_{12}\mathbf{B}_{21} & \mathbf{A}_{11}\mathbf{B}_{12} + \mathbf{A}_{12}\mathbf{B}_{22} \\ \mathbf{A}_{21}\mathbf{B}_{11} + \mathbf{A}_{22}\mathbf{B}_{21} & \mathbf{A}_{21}\mathbf{B}_{12} + \mathbf{A}_{22}\mathbf{B}_{22} \end{bmatrix}$$

The submatrix products are

$$\mathbf{A}_{11}\mathbf{B}_{11} = \begin{bmatrix} 1 \\ 2 \end{bmatrix} [1 \quad 2] = \begin{bmatrix} 1 & 2 \\ 2 & 4 \end{bmatrix}, \quad \begin{bmatrix} \text{4 multiplications} \\ \text{0 additions} \end{bmatrix}$$

$$\mathbf{A}_{12}\mathbf{B}_{21} = \begin{bmatrix} 0 & 0 \\ 0 & 0 \end{bmatrix} \begin{bmatrix} 2 & 1 \\ -1 & -1 \end{bmatrix} = \begin{bmatrix} 0 & 0 \\ 0 & 0 \end{bmatrix}, \quad \begin{bmatrix} \text{0 multiplications} \\ \text{0 additions} \end{bmatrix}$$

$$\mathbf{A}_{11}\mathbf{B}_{12} = \begin{bmatrix} 1 \\ 2 \end{bmatrix} [0] = \begin{bmatrix} 0 \\ 0 \end{bmatrix}, \quad \begin{bmatrix} \text{0 multiplications} \\ \text{0 additions} \end{bmatrix}$$

$$\mathbf{A}_{12}\mathbf{B}_{22} = \begin{bmatrix} 0 & 0 \\ 0 & 0 \end{bmatrix} \begin{bmatrix} 2 \\ 1 \end{bmatrix} = \begin{bmatrix} 0 \\ 0 \end{bmatrix}, \quad \begin{bmatrix} \text{0 multiplications} \\ \text{0 additions} \end{bmatrix}$$

$$\mathbf{A}_{21}\mathbf{B}_{11} = [1] \; [1 \quad 2] = [1 \quad 2], \quad \begin{bmatrix} \text{2 multiplications} \\ \text{0 additions} \end{bmatrix}$$

$$\mathbf{A}_{22}\mathbf{B}_{21} = [3 \quad 2] \begin{bmatrix} 2 & 1 \\ -1 & -1 \end{bmatrix} = [4 \quad 1], \quad \begin{bmatrix} \text{4 multiplications} \\ \text{2 additions} \end{bmatrix}$$

$$\mathbf{A}_{21}\mathbf{B}_{12} = [1] \; [0] = [0] \quad \begin{bmatrix} \text{0 multiplications} \\ \text{0 additions} \end{bmatrix}$$

$$\mathbf{A}_{22}\mathbf{B}_{22} = [3 \quad 2] \begin{bmatrix} 2 \\ 1 \end{bmatrix} = [8] \quad \begin{bmatrix} \text{2 multiplications} \\ \text{1 addition} \end{bmatrix}$$

The proper submatrix products can now be added (ignoring the zero products) to form the overall product, giving

$$\mathbf{A}_{21}\mathbf{B}_{11} + \mathbf{A}_{22}\mathbf{B}_{21} = [1 \quad 2] + [4 \quad 1] = [5 \quad 3] \qquad \begin{bmatrix} \text{0 multiplications} \\ \text{2 additions} \end{bmatrix}$$

The product **AB** is

$$\mathbf{AB} = \begin{bmatrix} 1 & 2 & 0 \\ 2 & 4 & 0 \\ 5 & 3 & 8 \end{bmatrix}$$

Using submatrices as chosen in this example, 12 multiplications and 5 additions are required to form the product.

The preceding example clearly demonstrates that, in certain cases, the number of arithmetic operations required to form a product can be greatly decreased by the appropriate partitioning of a matrix. In computer-aided design the saving in computer time can be a significant factor making partitioning a necessity. Of course, a certain amount of additional bookkeeping by the computer is required, but this can often be accounted for in the program. In large-matrix manipulation or in problems requiring the formation of many products, partitioning the zero elements into submatrices can result in substantial savings. In some network-analysis programs, the proper choice of variables will allow the arrangement of zero elements such that partitioning can be usefully applied.

In calculating the number of element multiplications required to form a matrix product, the dimensions of both matrices must be considered. The product of an $m \times n$ matrix and an $n \times p$ matrix requires a total number of

$$N_m = mnp \tag{1-11}$$

element multiplications. Also, required are

$$N_a = m(n - 1)p \tag{1-12}$$

additions. The number of operations required in the presence of partitioning depends on the number and size of the all zero element submatrices that can be formed.

D. DETERMINANTS

It is assumed in this section that the student is familiar with the basic rules relating to expansion and manipulation of determinants. While Cramer's rule is basic to the application of determinants in the solution of simultaneous equations, it will be reviewed here.

Cramer's Rule

Cramer's rule is based on the well-known Laplace expansion of a determinant. In terms of a formula, the Laplace expansion can be stated as

$$\det \mathbf{A} = \sum_{k=1}^{n} a_{ik} \text{ cofactor } a_{ik} \tag{1-13}$$

where i is an integer such that $1 \leqslant i \leqslant n$. Of course, the cofactor a_{ik} will be a subdeterminant of order $n - 1$ and will require further Laplace expansions to be applied. Eventually, the cofactors will consist of 2×2 determinants which can be evaluated by the usual formula. Equation 1-13 refers to an expansion along the ith row. The expansion along the jth column is given by

$$\det \mathbf{A} = \sum_{k=1}^{n} a_{kj} \text{ cofactor } a_{kj} \tag{1-14}$$

where j is an integer such that $1 \leqslant j \leqslant n$.

To derive Cramer's rule, we start by considering the n simultaneous equations

$$\begin{array}{c} a_{11}x_1 + a_{12}x_2 + \cdots + a_{1n}x_n = c_1 \\ a_{21}x_1 + a_{22}x_2 + \cdots + a_{2n}x_n = c_2 \\ \cdot \qquad \qquad \qquad \cdot \quad \cdot \\ \cdot \qquad \qquad \qquad \cdot \quad \cdot \\ \cdot \qquad \qquad \qquad \cdot \quad \cdot \\ a_{n1}x_1 + a_{n2}x_2 + \cdots + a_{nn}x_n = c_n \end{array} \tag{1-15}$$

This set of equations can be expressed in matrix form as

$$\begin{bmatrix} a_{11} & a_{12} & \cdots & a_{1n} \\ a_{21} & a_{22} & \cdots & a_{2n} \\ \cdot & \cdot & & \cdot \\ \cdot & \cdot & & \cdot \\ \cdot & \cdot & & \cdot \\ a_{n1} & a_{n2} & \cdots & a_{nn} \end{bmatrix} \begin{bmatrix} x_1 \\ x_2 \\ \cdot \\ \cdot \\ \cdot \\ x_n \end{bmatrix} = \begin{bmatrix} c_1 \\ c_2 \\ \cdot \\ \cdot \\ \cdot \\ c_n \end{bmatrix} \tag{1-16a}$$

or simply

$$\mathbf{A}\,\mathbf{X} = \mathbf{C} \tag{1-16b}$$

The column matrix $\mathbf{X}$ can be found by multiplying both sides of Eq. 1-16b by $\mathbf{A}^{-1}$, giving

$$\mathbf{A}^{-1}\,\mathbf{A}\,\mathbf{X} = \mathbf{X} = \mathbf{A}^{-1}\,\mathbf{C} \tag{1-17}$$

Individual elements of **X** can be expressed as

$$x_i = \sum_{j=1}^{n} d_{ij} c_j \tag{1-18}$$

where d_{ij} are the elements of $\mathbf{A}^{-1}$. Since these elements can be expressed as

$$d_{ij} = \frac{\text{cofactor } a_{ji}}{\det \mathbf{A}} \qquad [1\text{-}10]$$

Equation 1-18 becomes upon substitution for d_{ij}

$$x_i = \sum_{j=1}^{n} c_j \frac{\text{cofactor } a_{ji}}{\det \mathbf{A}}$$

$$= \frac{1}{\det \mathbf{A}} \sum_{j=1}^{n} c_j \text{ cofactor } a_{ji} \tag{1-19}$$

The similarity between the Laplace expansion along the ith column and Eq. 1-19 is obvious. Instead of element a_{ji} multiplying each cofactor, c_j assumes this role. Hence, if we desire to solve for x_i, we simply replace the ith column of **A** by the column matrix **C** and use the Laplace expansion along this column. This result is divided by the value of det **A** as shown by Eq. 1-19 to obtain x_i. Equation 1-19 is the mathematical statement of Cramer's rule.

Gauss Elimination Method

While this method does not require the use of determinants, it is discussed here in order to compare the efficiency in the solution of simultaneous equations to that of Cramer's rule.

In solving the set of equations given by Eq. 1-15, we can use the variable elimination method. We can add the ith equation to or subtract the ith equation from the jth equation and replace the jth equation by this result. Furthermore, we can multiply the ith equation by a constant before forming the sum or difference with the jth equation. If the proper constant multiplier is chosen, the equation that replaces the jth equation will have a zero coefficient associated with one of the variables. Thus only $n - 1$ variables appear in this equation. By applying this process of equation multiplication and addition or subtraction several times, all but one variable can be eliminated in one equation. The Gauss elimination method provides an orderly procedure to attain this goal.

Consider the set of equations given by Eq. 1-16a. It has been pointed out that any single equation can be multiplied by a constant and added to or subtracted from any other equation. In the matrix expression, multiplication of any equation by a constant amounts to multiplying a row by this constant. The element on the right-hand side of the equation corresponding to that same row must also be multiplied by the same constant. The first row of Eq. 1-16a (and c_1) can be multiplied by a constant such that when this row is subtracted from row 2, the result is a new row that has a zero in the element position where a_{21} previously appeared. A different constant multiplying the first row will allow row 1 to be subtracted from row 3 leaving a zero in the element position previously occupied by a_{31}. This same procedure can be followed until all elements in column 1 are zero, with the exception of a_{11}. The same procedure is repeated using row 2 to be multiplied by constants and subtracted from all rows except row 1 to create zeros in column 2 for all rows except rows 1 and 2. Continuing along these lines allows the equation to be expressed in upper triangular form as

$$\begin{bmatrix} b_{11} & b_{12} & b_{13} & \dots & b_{1n} \\ 0 & b_{22} & b_{23} & \dots & b_{2n} \\ 0 & 0 & b_{33} & \dots & b_{3n} \\ \cdot & \cdot & \cdot & & \cdot \\ \cdot & \cdot & \cdot & & \cdot \\ \cdot & \cdot & \cdot & & \cdot \\ 0 & 0 & 0 & \dots & b_{nn} \end{bmatrix} \begin{bmatrix} x_1 \\ x_2 \\ x_3 \\ \cdot \\ \cdot \\ \cdot \\ x_n \end{bmatrix} = \begin{bmatrix} e_1 \\ e_2 \\ e_3 \\ \cdot \\ \cdot \\ \cdot \\ e_n \end{bmatrix} \tag{1-20}$$

The last row corresponds to the equation

$$b_{nn} x_n = e_n$$

from which we immediately know x_n to be

$$x_n = e_n / b_{nn}$$

We can now substitute this result into the equation corresponding to the next to last row and solve for x_{n-1}. Using this back substitution $x_{n-2}, x_{n-3}, \dots x_1$ can be found. An orderly procedure (algorithm) can be followed to reduce the matrix to upper triangular form. This procedure always works from right to left by reducing all elements below the diagonal to zero. To reduce a given entry a_{ij} to zero we form the quotient

$$\alpha_{ij} = a_{ij} / a_{jj} \tag{1-21}$$

Multiply the jth row by α_{ij} and subtract from the ith row, replacing the ith row by this difference. An example will demonstrate the method suggested.

EXAMPLE 1-3. Solve the following equations for x_1, x_2, and x_3 first using Cramer's rule and again by the Gauss elimination method.

$$\begin{bmatrix} 2.00 & 1.00 & -1.00 \\ 1.00 & 3.00 & 1.50 \\ 2.50 & -2.00 & 4.00 \end{bmatrix} \begin{bmatrix} x_1 \\ x_2 \\ x_3 \end{bmatrix} = \begin{bmatrix} 6.50 \\ 9.75 \\ -4.25 \end{bmatrix}$$

Solution. Cramer's rule: Applying Eq. 1-19,

$$x_1 = \frac{1}{\det \mathbf{A}} \sum_{j=1}^{3} c_j \text{ cofactor } a_{ji}$$

$$= \frac{\begin{vmatrix} 6.50 & 1.00 & -1.00 \\ 9.75 & 3.00 & 1.50 \\ -4.25 & -2.00 & 4.00 \end{vmatrix}}{\begin{vmatrix} 2.00 & 1.00 & -1.00 \\ 1.00 & 3.00 & 1.50 \\ 2.50 & -2.00 & 4.00 \end{vmatrix}}$$

$$= \frac{6.50 \begin{vmatrix} 3.00 & 1.50 \\ -2.00 & 4.00 \end{vmatrix} -9.75 \begin{vmatrix} 1.00 & -1.00 \\ -2.00 & 4.00 \end{vmatrix} -4.25 \begin{vmatrix} 1.00 & -1.00 \\ 3.00 & 1.50 \end{vmatrix}}{2.00 \begin{vmatrix} 3.00 & 1.50 \\ -2.00 & 4.00 \end{vmatrix} -1.00 \begin{vmatrix} 1.00 & -1.00 \\ -2.00 & 4.00 \end{vmatrix} +2.50 \begin{vmatrix} 1.00 & -1.00 \\ 3.00 & 1.50 \end{vmatrix}}$$

$$= 1.50$$

Similarly, x_2 and x_3 can be found as

$$x_2 = 3.00 \quad \text{and} \quad x_3 = -0.50$$

Gauss reduction method: Starting with the matrix equation

$$\begin{bmatrix} 2.00 & 1.00 & -1.00 \\ 1.00 & 3.00 & 1.50 \\ 2.50 & -2.00 & 4.00 \end{bmatrix} \begin{bmatrix} x_1 \\ x_2 \\ x_3 \end{bmatrix} = \begin{bmatrix} 6.50 \\ 9.75 \\ -4.25 \end{bmatrix}$$

the multiplier α_{21} can be found from Eq. 1-21 to be

$$\alpha_{21} = \frac{a_{21}}{a_{11}} = \frac{1.00}{2.00} = 0.50$$

The value of α_{31} is

$$\alpha_{31} = \frac{a_{31}}{a_{11}} = \frac{2.50}{2.00} = 1.25$$

Row 1 is first multiplied by α_{21} and subtracted from row 2. Then row 1 is multiplied by α_{31} and subtracted from row 3. Carrying out these steps results in

$$\begin{bmatrix} 2.00 & 1.00 & -1.00 \\ 0 & 2.50 & 2.00 \\ 0 & -3.25 & 5.25 \end{bmatrix} \begin{bmatrix} x_1 \\ x_2 \\ x_3 \end{bmatrix} = \begin{bmatrix} 6.50 \\ 6.50 \\ -12.375 \end{bmatrix}$$

The multiplier α_{32} is now found to be

$$\alpha_{32} = \frac{-3.25}{2.50} = -1.30$$

Multiplying row 2 by α_{32} and subtracting from row 3 gives

$$\begin{bmatrix} 2.00 & 1.00 & -1.00 \\ 0 & 2.50 & 2.00 \\ 0 & 0 & 7.85 \end{bmatrix} \begin{bmatrix} x_1 \\ x_2 \\ x_3 \end{bmatrix} = \begin{bmatrix} 6.50 \\ 6.50 \\ -3.925 \end{bmatrix}$$

The variable x_3 is obviously

$$x_3 = \frac{-3.925}{7.85} = -0.50$$

Substituting this value into the equation corresponding to row 2 results in

$$2.50x_2 + 2.00(-0.50) = 6.50$$

and

$$x_2 = 3.00$$

The values of x_3 and x_2 are used in the first equation (row 1) to give

$$x_1 = 1.50$$

The same results were obtained by two different methods of seemingly equal computational difficulties in the above example. In fact, as the number of variables and simultaneous equations increase, the Gauss elimination method becomes more efficient than the use of Cramer's rule. In solving a problem of simultaneous equations in n variables with a digital computer, computing time is minimized with the Gauss elimination method, provided that n is larger than 3.

Because computing time is a major factor in choosing a method of analysis, let us consider the number of algebraic operations required by each method. To evaluate each variable using Cramer's rule requires that the determinant of **A** be evaluated along with the determinant formed by replacing one column of **A** by the known column matrix. Of course, the determinant of **A** must be evaluated only once. Thus, for n variables, $n + 1$ determinants, each of dimension $n \times n$, must be evaluated. Using the Laplace expansion requires $(n - 1)(n!)$ multiplications plus several additions to evaluate a given determinant. This can be shown by considering the evaluation of various determinants starting with a 2×2. To evaluate

$$\det \mathbf{A} = \begin{vmatrix} a_{11} & a_{12} \\ a_{21} & a_{22} \end{vmatrix} = a_{11}a_{22} - a_{12}a_{21}$$

requires 2 multiplications and one addition (subtraction). A 3 X 3 determinant

$$\det \mathbf{A} = \begin{vmatrix} a_{11} & a_{12} & a_{13} \\ a_{21} & a_{22} & a_{23} \\ a_{31} & a_{32} & a_{33} \end{vmatrix}$$

$$= a_{11}a_{22}a_{33} - a_{11}a_{32}a_{23} - a_{21}a_{12}a_{33} + a_{21}a_{32}a_{13} + a_{31}a_{12}a_{23} - a_{31}a_{22}a_{13}$$

requires 12 multiplications and 5 additions. One can continue in this manner to show that $(n - 1)(n!)$ multiplications are required. In general, since multiplication is done by repeated additions on a computer, multiplication time is far greater than addition time. Therefore an approximate estimate of computer time required for this problem can be obtained by considering only the total number of multiplications (or divisions) required. To determine n variables would apparently require

$$N_m = (n + 1)(n - 1)(n!) \tag{1-22}$$

multiplications. In practice, a well-written program can decrease this number by taking advantage of the recurrence of certain products. However, for purposes of discussion we will consider Eq. 1-22 to be a measure of computer time in applying Cramer's rule to the solution of n simultaneous equations in n variables.

The Gauss elimination method requires considerably fewer operations when n is large. To apply this method, the first row must be multiplied by a constant multiplier and subtracted from row 2. This requires $n + 2$ multiplications, including the determination of the multiplier. This process must be repeated $n - 1$ times to create zeros in the first column of all rows except row 1. Thus a total of $(n - 1)(n + 2)$ multiplications are required before the usefulness of row 1 is exhausted. Next, row 2 is used to create zeros in the second column of all rows except rows 1 and 2. This requires $(n - 2)(n + 1)$ operations, since $(n - 2)$ other rows are involved and the zero element of row 2 need not be multiplied. Proceeding in this way we find that the number of multiplications required is

$$N_m = (1)(4) + (2)(5) + \cdots + (n - 1)(n + 2)$$

$$= \sum_{p=1}^{n-1} p(p + 3) \tag{1-23}$$

The number N_m can be found in more convenient terms by considering the difference in N_m for n equations and N_m for $n - 1$ equations. From Eq. 1-23 this difference is

$$N_m(n) - N_m(n-1) = \sum_{p=1}^{n-1} p(p+3) - \sum_{p=1}^{n-2} p(p+3)$$

$$= (n-1)(n+2) = n^2 + n - 2$$

We now must find the function $N_m(n)$ such that the preceding equation is satisfied. If we assume a solution of the form

$$N_m(n) = a_0 + a_1 n + a_2 n^2 + a_3 n^3$$

then

$$N_m(n-1) = a_0 + a_1(n-1) + a_2(n-1)^2 + a_3(n-1)^3$$

Expanding the latter expression for $N_m(n-1)$ and subtracting from $N_m(n)$ gives

$$N_m(n) - N_m(n-1) = (a_1 - a_2 + a_3) + (2a_2 - 3a_3)n + 3a_3 n^2$$

which can be equated to $n^2 + n - 2$. Solving for a_3, a_2, and a_1 yields

$$a_3 = 1/3, \quad a_2 = 1 \quad \text{and} \quad a_1 = -4/3$$

The coefficient a_0 can take on any value, thus we will assume a zero value for simplicity. The value of $N_m(n)$ is then

$$N_m(n) = (1/3)n^3 + n^2 - (4/3)n \tag{1-24}$$

Equation 1-24 expresses the number of multiplications required to convert the matrix **A** into upper triangular form. Additional operations are required to complete the back substitutions and solve for the variables. It is easy to show that this number is equal to

$$N_{mbs} = (1/2)n^2 + (1/2)n \tag{1-25}$$

giving a total number of

$$N_{mt} = (1/3)n^3 + (3/2)n^2 - (5/6)n \tag{1-26}$$

multiplications to complete the problem. In comparing this expression to Eq. 1-22, it is obvious that for large n ($n > 3$), the Gauss elimination method is more efficient than Cramer's rule. For $n = 6$, Eq. 1-22 predicts 25,200 multiplications, while Eq. 1-26 predicts only 121.

Before leaving the numerical aspects of solving simultaneous equations, we should comment on the fact that the accuracy of the Gauss elimination method is often poor. When one row is multiplied and subtracted from another to form a zero at the proper location, round-off error can lead to an element that differs slightly from zero. If this new row is then multiplied by a large multiplier as the process is continued, the element that differs from zero can lead to further error in other rows as subtraction takes place. If the multipliers are minimized the error tends to be minimal [1].* During the Gauss elimination procedure by machine, rows or columns can be interchanged so that each multiplier is minimized.

In solving equations whose coefficients are determined by circuit parameters it is desirable that the spread of coefficients be small. If the larger coefficients are several factors greater than the smaller coefficients, the accuracy of the procedure will be poor. Some computer-aided design programs such as ECAP specify a maximum spread on element values to minimize the accuracy problem.

There are other methods of evaluating determinants or inverting matrices that are more efficient than Cramer's rule. Usually some variant of the Gauss procedure is used for computer analysis.

BIBLIOGRAPHY

1. Calahan, D. A., *Computer-Aided Network Design*, preliminary edition, McGraw-Hill Book Company, New York, 1968, Chap. 1.
2. Cruz, J. B., and M. E. Van Valkenburg, *Introductory Signals and Circuits*, Blaisdell Publishing Company, Waltham, Mass., 1967, Chap. 15.
3. DeRusso, P. M., R. J. Roy, and C. M. Close, *State Variables for Engineers*, John Wiley & Sons, Inc., New York, 1965, Chap. 4.
4. Friedman, B., *Principles and Techniques of Applied Mathematics*, John Wiley & Sons, Inc., New York, 1956.
5. Kuo, B. C., *Linear Networks and Systems*, McGraw-Hill Book Company, New York, 1967, Chap. 3.
6. Pipes, L. A., and S. A. Hovanessian, *Matrix Computer Methods in Engineering*, John Wiley & Sons, Inc., New York, 1969, Chaps. 1 and 2.
7. Skilling, H. H., *Electrical Engineering Circuits,* John Wiley & Sons, Inc., New York, 1957, Chap. 10.
8. Weiss, M. J., *Higher Algebra for the Undergraduate*, John Wiley & Sons, Inc., New York, 1949, Chap. 6.

PROBLEMS

1-1 Express the voltages at nodes 1, 2, 3, and 4 as a function of the loop currents i_1, i_2, i_3, and i_4.

(a) Write this relationship as a vector equation.

(b) Write the vector equation expressing the voltages across R_5, R_6, R_7, and R_8 as a function of the loop currents.

1-2 (a) Write the vector equation expressing the voltages e_1, e_2, e_3, and e_4 as a function of the loop currents i_1, i_2, i_3, and i_4.

(b) Express the voltages between ground and the points A, B, C, and D in terms of the loop currents.

*Numbers in brackets refer to bibliography at end of chapter.

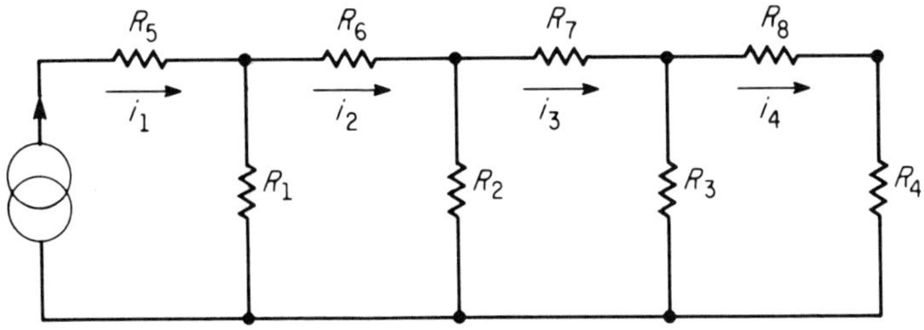

PROB. 1-1

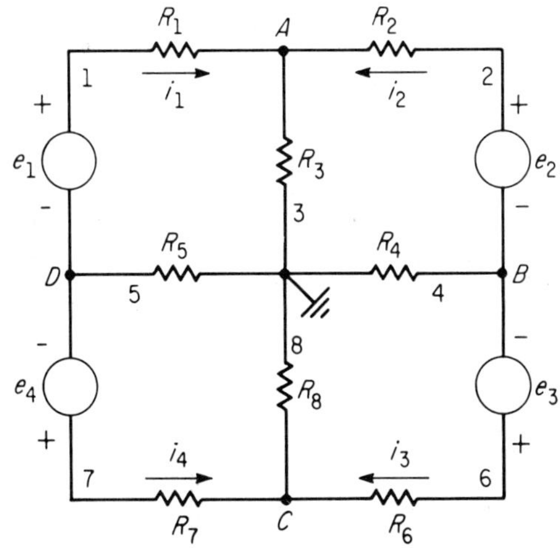

PROB. 1-2

1-3 Write the vector equation expressing the currents i_1, i_2, and i_3 as a function of the voltages e_1, e_2, and e_3.

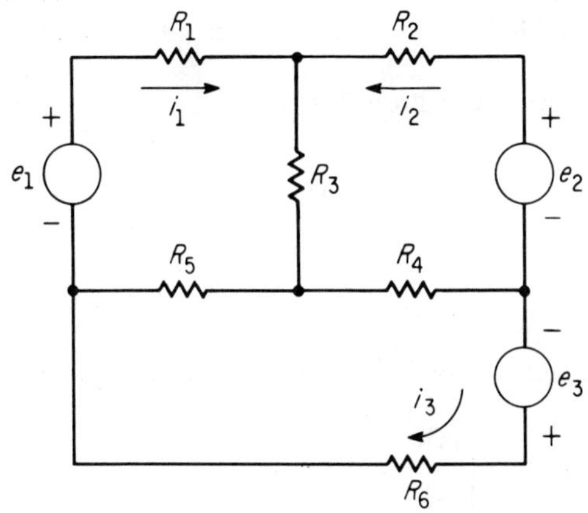

PROB. 1-3

1-4 If

$$\mathbf{A} = \begin{bmatrix} 4 & 1 & 2 & -1 \\ 3 & 0 & 0 & 2 \\ -1 & -1 & 2 & 2 \\ 0 & 2 & 0 & 4 \end{bmatrix}, \quad \mathbf{B} = \begin{bmatrix} 1 & 2 \\ -1 & 0 \\ 3 & 2 \\ -2 & -1 \end{bmatrix}, \quad \mathbf{C} = \begin{bmatrix} 4 & 3 \\ 2 & 1 \end{bmatrix}$$

calculate

(a) **A B**

(b) **B C**

(c) **A B C**

Does **A B** = **B A**? Does **B C** = **C B**? Does **A**(**B C**) = (**A B**)**C**? Can the matrix **A** + **B** + **C** be formed in this case?

1-5 If

$$\mathbf{A} = \begin{bmatrix} 3 & 2 & 0 \\ 1 & -1 & 1 \\ 2 & 1 & 3 \end{bmatrix}, \quad \mathbf{B} = \begin{bmatrix} 1 & 5 & 1 \\ 3 & 4 & 4 \\ 6 & 0 & -2 \end{bmatrix}, \quad \mathbf{C} = \begin{bmatrix} 0 & -1 & -2 \\ 2 & -3 & -1 \\ 2 & 2 & 1 \end{bmatrix}$$

calculate

(a) **A B**

(b) **A C**

(c) **B** − **C**

(d) **A**(**B** − **C**)

Demonstrate the distributive law by showing that

$$\mathbf{A\,B} - \mathbf{A\,C} = \mathbf{A}(\mathbf{B} - \mathbf{C})$$

1-6 Invert the matrix **A** of Prob. 1-5.

1-7 Invert the matrix **B** of Prob. 1-5.

1-8 Given two nonsingular matrices **A** and **B** of equal dimension. Does $(\mathbf{A\,B})^{-1} = \mathbf{A}^{-1}\mathbf{B}^{-1}$ or does $(\mathbf{A\,B})^{-1} = \mathbf{B}^{-1}\mathbf{A}^{-1}$? Prove that your choice is correct.

1-9 Invert the matrix **A** of Prob. 1-4.

1-10 If

$$\begin{bmatrix} x_1 \\ x_2 \\ y_1 \end{bmatrix} = \begin{bmatrix} 1 & 2 & 3 \\ 4 & 5 & 6 \\ 7 & 8 & 9 \end{bmatrix} \begin{bmatrix} u_1 \\ v_1 \\ v_2 \end{bmatrix}$$

write the equations for $\begin{bmatrix} x_1 \\ x_2 \end{bmatrix}$ and for $[y_1]$ in terms of submatrices and the variables $[u_1]$ and $\begin{bmatrix} v_1 \\ v_2 \end{bmatrix}$. That is, write

$$\begin{bmatrix} x_1 \\ x_2 \end{bmatrix} = \mathbf{A}[u_1] + \mathbf{B}\begin{bmatrix} v_1 \\ v_2 \end{bmatrix}$$

and

$$[y_1] = \mathbf{C}[u_1] + \mathbf{D}\begin{bmatrix} v_1 \\ v_2 \end{bmatrix}$$

then evaluate the matrices **A, B, C, D.**

1-11 If

$$\begin{bmatrix} x_1 \\ x_2 \\ y_1 \\ y_2 \end{bmatrix} = \begin{bmatrix} 1 & 0 & 4 \\ 3 & 1 & 6 \\ 2 & 1 & 2 \\ 3 & 1 & 6 \end{bmatrix} \begin{bmatrix} u_1 \\ u_2 \\ v_1 \end{bmatrix} + \begin{bmatrix} 1 & 0 \\ 2 & 1 \\ 3 & 4 \\ 2 & -1 \end{bmatrix} \begin{bmatrix} r \\ s \end{bmatrix}$$

write $\begin{bmatrix} x_1 \\ x_2 \end{bmatrix}$ and $\begin{bmatrix} y_1 \\ y_2 \end{bmatrix}$ in terms of $\begin{bmatrix} u_1 \\ u_2 \end{bmatrix}$, $[v_1]$, $[r]$, and $[s]$.

1-12 If

$$[x_1] = [1 \quad 2] \begin{bmatrix} u_1 \\ u_2 \end{bmatrix} + 4[v_1] + [r]$$

and

$$\begin{bmatrix} y_1 \\ y_2 \end{bmatrix} = \begin{bmatrix} 2 & -1 \\ 3 & 0 \end{bmatrix} \begin{bmatrix} u_1 \\ u_2 \end{bmatrix} + \begin{bmatrix} 0 \\ 1 \end{bmatrix} [v_1] + \begin{bmatrix} 2 \\ 0 \end{bmatrix} [r] + \begin{bmatrix} 4 \\ 0 \end{bmatrix} [s]$$

express $\begin{bmatrix} x_1 \\ y_1 \\ y_2 \end{bmatrix}$ in terms of $\begin{bmatrix} u_1 \\ u_2 \\ v_1 \end{bmatrix}$ and $\begin{bmatrix} r \\ s \end{bmatrix}$

1-13 Write the equations

$$x_1 + x_2 + 10x_3 = y_1$$
$$-2x_1 + x_2 + 20x_3 = y_2$$
$$40x_1 - 20x_2 + 40x_3 = y_3$$

in vector form. If $y_1 = 7.5$, $y_2 = 2.5$, and $y_3 = -6$, find **x** by Cramer's rule.

1-14 Find **x** in Prob. 1-13 by inverting the coefficient matrix.

1-15 Find **x** in Prob. 1-13 by using Gauss reduction.

1-16 Write the equations

$$10x_1 + 100x_2 + 100x_3 = y_1$$
$$x_1 + 100x_2 - 100x_3 = y_2$$
$$2x_1 - 1000x_2 + 100x_3 = y_3$$

in vector form. If $y_1 = 10.0$, $y_2 = 0.8$, and $y_3 = 3.1$, find **x** by Cramer's rule.

1-17 Find **x** in Prob. 1-16 by inverting the coefficient matrix.

1-18 Find **x** in Prob. 1-16 by using Gauss reduction.

1-19 Write a program in Fortran IV to invert a matrix of variable dimension up to 5×5. Use the Gauss reduction method.

chapter **2**

MATRIX EQUATIONS AND MATRIX FUNCTIONS

A knowledge of matrix theory is prerequisite to the application of state-variable techniques. Until recently network designers have not chosen to apply state-variable methods to any extent. However, the computer can often solve network equations more efficiently by using the state-vector concept rather than other methods. Matrix theory then becomes necessary in order to fully utilize the capabilities of the digital computer. Even in conventional analysis methods, matrices represent an efficient method of communicating with the computer.

The goal of this chapter is to develop matrix theory to the point that we can work with functions of matrices. For linear, time-invariant networks the state equation is a first-order differential matrix equation the solution of which involves an exponential function of a matrix. In solving a scalar differential equation, the end result is the solution expressed by an analytical function such as an exponential. On the other hand, the solution to a matrix differential equation must be expanded in order to obtain usable results. Often this expansion is the most difficult step of the entire procedure.

The theory of eigenvectors and eigenvalues is important in dealing with matrix functions. Evaluation of such a function can generally be expedited when eigenvector theory is applied.

The motivation for studying matrices and eigenvector theory is provided by the nature of the digital computer. The information necessary to formulate analysis equations can be stored in the computer and manipulated most efficiently by employing matrix methods. Chapter 3 demonstrates that even the structure or topology of a network can be described to the computer in terms of a special matrix. Chapter 5 discusses the machine solution of the matrix equations constructed by the computer. Many of the solution methods require the use of eigenvector theory and matrix functions. Thus in order to effectively link the digital computer to the network design process we must apply the theory appropriate to the computer rather than the conventional theory applied in manual design.

Some preliminary theory regarding linear independence and basis vectors is presented in this chapter prior to considering eigenvectors.

A. LINEAR INDEPENDENCE

In order to solve for n variables, a set of n linearly independent equations are required. If n equations are given, but one or more is linearly dependent on others of the set, all variables cannot be explicitly found. In this case, $n - r$ variables can be expressed as functions of r variables where r is the number of dependent equations of the set.

A linearly dependent equation is one that can be formed by linearly combining any number of independent equations. That is, if each independent equation is multiplied by a constant value and a sum of all such equations is formed, the result is a dependent equation. Given a set of homogeneous equations, one can determine the independence of these equations by means of a simple test. The equations written as

$$
\begin{aligned}
a_{11}x_1 + a_{12}x_2 + \cdots + a_{1n}x_n &= 0 \\
a_{21}x_1 + a_{22}x_2 + \cdots + a_{2n}x_n &= 0 \\
\vdots \qquad\qquad &\quad \vdots \\
a_{n1}x_1 + a_{n2}x_2 + \cdots + a_{nn}x_n &= 0
\end{aligned}
$$

can be represented by the matrix expression

$$\mathbf{A}\,\mathbf{X} = \mathbf{0} \tag{2-1}$$

where $\mathbf{0}$ is a zero vector of appropriate dimension.

By definition, if the n equations are linearly independent, equating a linear combination of the rows to zero, resulting in

$$C_1 a_{1j} + C_2 a_{2j} + \cdots + C_n a_{nj} = 0, \qquad j = 1, 2, \ldots, n \tag{2-2}$$

must imply that

$$C_1 = C_2 = \cdots = C_n = 0 \tag{2-3}$$

If Eq. 2-2 can be satisfied without requiring that all $C_i = 0$, then the homogeneous equations are not linearly independent. Equation 2-2 is expressed in expanded form as

$$
\begin{aligned}
C_1 a_{11} + C_2 a_{21} + \cdots + C_n a_{n1} &= 0 \\
C_1 a_{12} + C_2 a_{22} + \cdots + C_n a_{n2} &= 0 \\
\vdots \qquad\qquad &\quad \vdots \\
C_1 a_{1n} + C_2 a_{2n} + \cdots + C_n a_{nn} &= 0
\end{aligned}
$$

We must now check to see if the above set of equations implies that $C_1 = C_2 = \cdots = C_n = 0$. If we solve for any C_i using Cramer's rule, we replace the ith column with zeros and find that

$$C_i = \frac{\begin{vmatrix} a_{11} & a_{21} & \cdots & 0 & \cdots & a_{n1} \\ a_{12} & a_{22} & \cdots & 0 & \cdots & a_{n2} \\ \vdots & & & \vdots & & \vdots \\ a_{1n} & a_{2n} & \cdots & 0 & \cdots & a_{nn} \end{vmatrix}}{\begin{vmatrix} a_{11} & a_{21} & \cdots & a_{i1} & \cdots & a_{n1} \\ a_{12} & a_{22} & \cdots & a_{i2} & \cdots & a_{n2} \\ \vdots & & & \vdots & & \vdots \\ a_{1n} & a_{2n} & \cdots & a_{in} & \cdots & a_{nn} \end{vmatrix}} = \frac{0}{\det \mathbf{A}}$$

That the numerator is zero can be seen by expanding along the ith column. If det **A** has a nonzero value, then all C_i will equal zero. If det **A** is zero, then C_i can have an indeterminate but finite value. Thus, linear independence is verified by finding a nonzero value for det **A**. We can note above that the transpose of **A** appears rather than **A** in the expression for C_i, but it is trivial to show that $\det \mathbf{A} = \det \mathbf{A}^t$. Once it has been determined that a set of homogeneous equations is linearly independent, it follows that the only solution is the trivial one:

$$x_1 = x_2 = \cdots = x_n = 0$$

This can be shown by solving for any x_i by Cramer's rule. The numerator will always be zero, and if det **A** is nonzero, all x_i must be zero. Thus the equations only have a nontrivial solution if det **A** = 0, or if at least one equation is linearly dependent on others of the set. The presence of one dependent equation means that there are only $n - 1$ independent equations in n unknowns, therefore, $n - 1$ unknowns can be found in terms of 1 variable. In general, if q equations are dependent on the remaining $n - q$ equations, then $n - q$ unknowns can be expressed in terms of q variables.

EXAMPLE 2-1. Determine if the sets of equations are linearly independent.

(a)

$$\begin{aligned} x_1 + x_2 + 3x_3 &= 0 \\ x_1 - x_2 + x_3 &= 0 \\ -x_1 - 4x_2 + 6x_3 &= 0 \end{aligned}$$

(b)

$$x_1 + x_2 + 3x_3 = 0$$

$$x_1 - x_2 + x_3 = 0$$

$$x_1 + 5x_2 + 7x_3 = 0$$

Solution. (a) The matrix of coefficients is

$$\mathbf{A} = \begin{bmatrix} 1 & 1 & 3 \\ 1 & -1 & 1 \\ -1 & -4 & 6 \end{bmatrix}$$

which has a determinant value of -24. Thus the equations are linearly independent. The only values satisfying the three equations simultaneously are $x_1 = x_2 = x_3 = 0$.

(b) The matrix of coefficients is

$$\mathbf{A} = \begin{bmatrix} 1 & 1 & 3 \\ 1 & -1 & 1 \\ 1 & 5 & 7 \end{bmatrix}$$

In this case det $\mathbf{A} = 0$ and at least one of the equations is a linear combination of the remaining two. The third equation can be shown to be equal to three times the first minus two times the second equation. Two of the variables can be found in terms of the third. For example, $x_1 = -2x_3$ and $x_2 = -x_3$.

For the nonhomogeneous case, it can easily be proven that the equations are linearly independent if the matrix **A** is again nonsingular. The nonhomogeneous matrix equation

$$\mathbf{AX} = \mathbf{C}$$

can be solved for **X** giving

$$\mathbf{X} = \mathbf{A}^{-1}\,\mathbf{C}$$

The inverse of **A** will exist only if det **A** has a nonzero value. It is also an easy matter to show that the determinant of a matrix including at least one row that is a linear combination of other rows is zero. When the equations are linearly independent in the nonhomogeneous case, the *n* variables can be found to have nonzero values unlike the homogeneous case.

The problem of linear independence is an important one in network analysis. Writing loop or node equations in a disorderly manner sometimes results in equations that are dependent.

As an example of the occurrence of linearly dependent equations in network analysis, we can consider the circuit of Fig. 2-1. The loop equations for the cir-

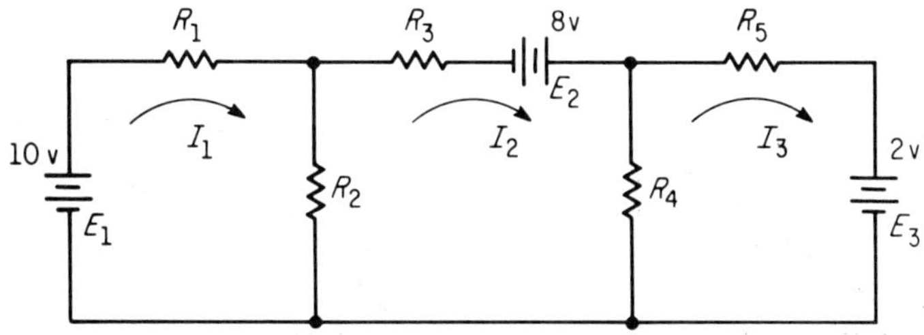

FIG. 2-1. A three-loop circuit.

cuit can be generated by summing voltages around the various loops. If we choose loop 1 to contain E_1, R_1, and R_2, loop 2 to contain E_2, R_2, R_3, and R_4, and loop 3 to contain E_3, R_4, and R_5 we can generate three independent equations. These equations are

$$
\begin{aligned}
(R_1 + R_2)I_1 - R_2 I_2 &= 10 \\
-R_2 I_1 + (R_2 + R_3 + R_4) I_2 - R_4 I_3 &= 8 \\
-R_4 I_2 + (R_4 + R_5) I_3 &= -2
\end{aligned}
$$

Suppose that rather than selecting loop 3 as above, it was selected to contain E_1, E_2, R_1, R_3, and R_4. In this case the third equation would be

$$R_1 I_1 + (R_3 + R_4) I_2 - R_4 I_3 = 18$$

which is linearly dependent on the equations written for loops 1 and 2. For this circuit it is obvious that we selected a dependent loop, but for complex networks it is not always obvious when a given loop is dependent. There are well-developed methods that guarantee the formulation of independent network equations. Network topology theory, as discussed in Chapter 3, not only provides the basis for independent equation formulation, but is very effectively utilized by the digital computer.

B. BASIS VECTORS

A vector with two components can be described in terms of basis vectors. In Fig. 2-2 the vector $\mathbf{x}$ can be related to the unit vectors in the $\mathbf{x}_1$ and $\mathbf{x}_2$ directions, $\mathbf{e}_1$ and $\mathbf{e}_2$ respectively.

The basis in (a) is called orthogonal for reasons we will consider in more detail shortly. For two dimensions, an orthogonal basis is one having perpendicular components. The basis in (b) is nonorthogonal, but has components which are independent. The set of reference vectors in (c), $\mathbf{x}_1$, $\mathbf{x}_2$, and $\mathbf{x}_3$ are not independent since $\mathbf{x}_3$ can be expressed as a linear combination of $\mathbf{x}_1$ and $\mathbf{x}_2$; hence, this set of vectors does not form a basis. By definition, a set of vectors forms a basis for a given space if the vectors span the space and also are linearly independent. Note that $\mathbf{x}$ can be expressed as a linear combination of the unit

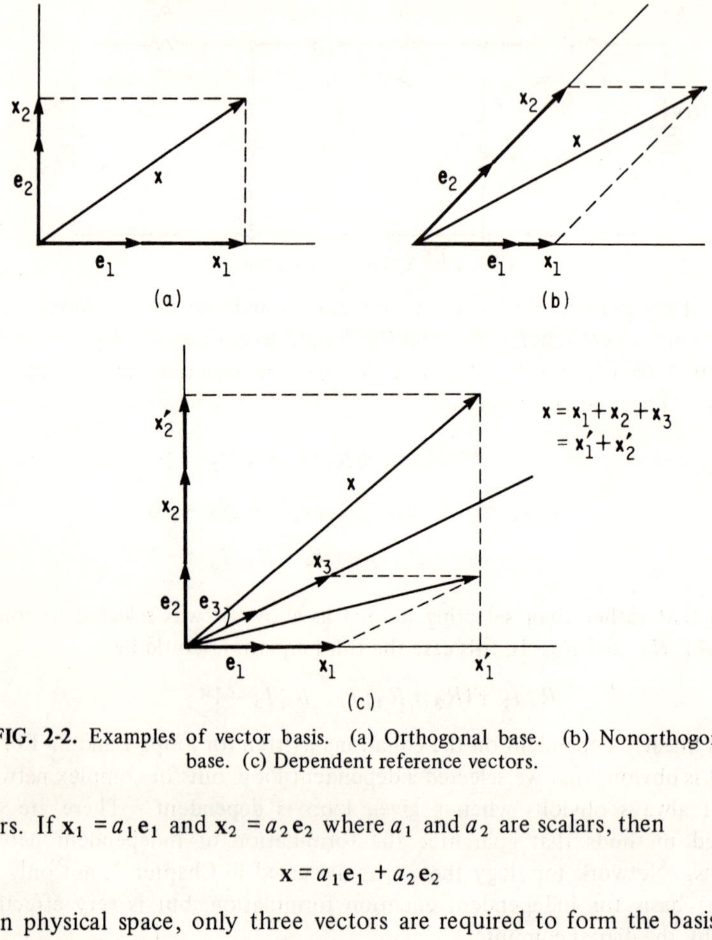

FIG. 2-2. Examples of vector basis. (a) Orthogonal base. (b) Nonorthogonal base. (c) Dependent reference vectors.

vectors. If $\mathbf{x}_1 = a_1\mathbf{e}_1$ and $\mathbf{x}_2 = a_2\mathbf{e}_2$ where a_1 and a_2 are scalars, then

$$\mathbf{x} = a_1\mathbf{e}_1 + a_2\mathbf{e}_2$$

In physical space, only three vectors are required to form the basis; additional reference vectors are dependent. However, in abstract space where variables are allowed to represent components of a vector, the basis may require a large number of reference vectors.

The scalar product of two vectors, $\mathbf{x}$ and $\mathbf{y}$, written $\langle\mathbf{x},\mathbf{y}\rangle$, is defined to be

$$\langle\mathbf{x},\mathbf{y}\rangle = a_1b_1 + a_2b_2 + a_3b_3 + \cdots + a_nb_n \tag{2-4}$$

where

$$\mathbf{x} = a_1\mathbf{e}_1 + a_2\mathbf{e}_2 + \cdots + a_n\mathbf{e}_n \tag{2-5}$$

and

$$\mathbf{y} = b_1\mathbf{e}_1 + b_2\mathbf{e}_2 + \cdots + b_n\mathbf{e}_n \tag{2-6}$$

In two or three dimensions, the length of a vector expressed in terms of an orthogonal basis is given by

$$|\mathbf{x}| = +\sqrt{\langle\mathbf{x},\mathbf{x}\rangle} \tag{2-7}$$

The positive square root of the scalar product of an n-dimensional vector $\mathbf{x}$ with itself is also referred to as the length or magnitude of the vector. If the scalar product of two vectors is zero, the vectors are said to be orthogonal or perpendicular. That is, if

$$\langle \mathbf{x}, \mathbf{y} \rangle = 0$$

the vectors are orthogonal to each other. In two or three dimensions, orthogonality of two vectors corresponds to a right angle between the vector directions. In abstract space of higher-dimension interpretation of orthogonality cannot be visualized, and the scalar product must be formed to test this property.

A set of basis vectors $\mathbf{e}_i$ form an orthonormal set if

$$\langle \mathbf{e}_i, \mathbf{e}_j \rangle = \delta_{ij} \tag{2-8}$$

where δ_{ij}, the Kronecker delta, is defined as

$$\begin{aligned} \delta_{ij} &= 0, \quad i \neq j \\ \delta_{ij} &= 1, \quad i = j \end{aligned} \tag{2-9}$$

Orthonormality implies orthogonality and normalization.

Often a set of variables will be chosen to correspond to basis vectors which are not orthogonal. Since orthogonality or orthonormality leads to ease of manipulation, it is sometimes convenient to construct an orthonormal basis spanning the given space. Before considering an orthonormalization process, the projection of one vector onto another must be discussed. Given two vectors, $\mathbf{x}_1$ and $\mathbf{x}_2$, the projection of $\mathbf{x}_2$ onto $\mathbf{x}_1$ equals the component vector of $\mathbf{x}_2$ having the same direction as $\mathbf{x}_1$. The vectors in Fig. 2-3 demonstrate the definition of projection on a graphical basis.

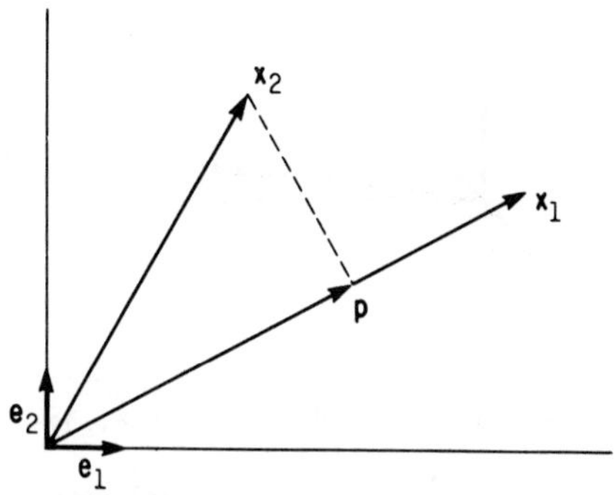

FIG. 2-3. The projection of x_2 on x_1.

The vector $\mathbf{p}$ is the projection vector. Given the vectors $\mathbf{x}_1$ and $\mathbf{x}_2$, $\mathbf{p}$ can be found by noting that the difference $\mathbf{x}_1 - \mathbf{p}$ is orthogonal to $\mathbf{x}_1$. We can form the scalar product

$$\langle \mathbf{x}_2 - \mathbf{p}, \mathbf{x}_1 \rangle$$

and choose $\mathbf{p}$ to cause this product to equal zero.

EXAMPLE 2-2. If $\mathbf{x}_1 = 10\mathbf{e}_1 + 2\mathbf{e}_2$ and $\mathbf{x}_2 = 6\mathbf{e}_1 + 8\mathbf{e}_2$, find the projection of $\mathbf{x}_2$ on $\mathbf{x}_1$ and the projection of $\mathbf{x}_1$ on $\mathbf{x}_2$.

Solution. Projection of $\mathbf{x}_2$ on $\mathbf{x}_1$:

We first note that $\mathbf{p}$ must lie along $\mathbf{x}_1$, hence $\mathbf{p}$ can be written as $\mathbf{p} = 10k\,\mathbf{e}_1 + 2k\,\mathbf{e}_2$ where k is a constant. The scalar product $\langle \mathbf{x}_2 - \mathbf{p}, \mathbf{x}_1 \rangle$ is

$$60 - 100k + 16 - 4k = 76 - 104k$$

Equating this product to zero results in $k = 19/26$. The vector $\mathbf{p}$ is then

$$\mathbf{p} = (95/13)\,\mathbf{e}_1 + (19/13)\,\mathbf{e}_2$$

Projection of $\mathbf{x}_1$ on $\mathbf{x}_2$:

If we want to find $\mathbf{p}$ in the direction of $\mathbf{x}_2$, we write $\mathbf{p} = 6k\,\mathbf{e}_1 + 8k\,\mathbf{e}_2$. Equating the scalar product $\langle \mathbf{x}_1 - \mathbf{p}, \mathbf{x}_2 \rangle$ to zero results in a value of $k = 38/7$. The vector $\mathbf{p}$ is

$$\mathbf{p} = (228/7)\,\mathbf{e}_1 + (304/7)\,\mathbf{e}_2$$

The projection of an n-dimensional vector can be found in the same manner as the two dimensional case. We can note that a vector orthogonal to a second vector can easily be found. In two dimensions, the difference vector $\mathbf{x}_2 - \mathbf{p}$ is always perpendicular to $\mathbf{x}_1$. In three dimensions the situation is slightly more complicated.

Consider three linearly independent vectors as shown in Fig. 2-4. The

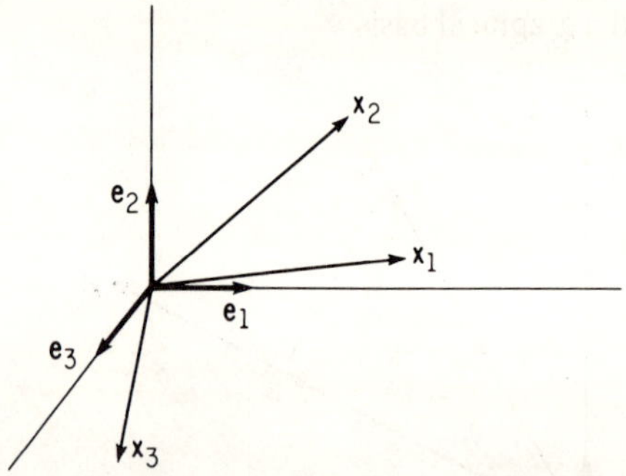

FIG. 2-4. Nonorthogonal basis vectors $\mathbf{x}_1, \mathbf{x}_2, \mathbf{x}_3$.

vectors, $\mathbf{x}_1$, $\mathbf{x}_2$, and $\mathbf{x}_3$ span the three-dimensional space, and thus can form a basis even though it is a nonorthogonal one. To construct an orthogonal basis, two vectors $\mathbf{y}_2$ and $\mathbf{y}_3$ can be constructed such that $\mathbf{x}_1$, $\mathbf{y}_2$, and $\mathbf{y}_3$ are mutually perpendicular. If $\mathbf{y}_2$ is linearly dependent on $\mathbf{x}_2$ and $\mathbf{y}_3$ is linearly dependent on $\mathbf{x}_3$, the three new basis vectors $\mathbf{x}_1$, $\mathbf{y}_2$, and $\mathbf{y}_3$ will obviously span the three-dimensional space, and hence form an orthogonal basis.

The fundamental process of establishing an orthogonal basis is easily visualized in three dimensions. Any one of the three original vectors can be chosen as the first basis vector for the new orthogonal space. In the case under discussion, we have chosen $\mathbf{x}_1$ as the first basis vector. Now $\mathbf{x}_2$ can be expressed in terms of two components, one parallel to $\mathbf{x}_1$ and one perpendicular to $\mathbf{x}_1$. The parallel vector is the projection of $\mathbf{x}_2$ on $\mathbf{x}_1$ ($\mathbf{p}_{21}$) and the perpendicular vector is the difference between $\mathbf{x}_2$ and the projection. It is obvious that subtraction of the component parallel to $\mathbf{x}_1$ from $\mathbf{x}_2$ leaves a vector orthogonal to $\mathbf{x}_1$. Thus the second basis vector is

$$\mathbf{y}_2 = \mathbf{x}_2 - \mathbf{p}_{21}$$

The remaining vector $\mathbf{x}_3$ can be resolved into three components; one parallel to $\mathbf{x}_1$, one parallel to $\mathbf{y}_2$, and one perpendicular to the plane containing $\mathbf{x}_1$ and $\mathbf{y}_2$. The component of $\mathbf{x}_3$ along $\mathbf{x}_1$ is equal to the projection of $\mathbf{x}_3$ on $\mathbf{x}_1$ ($\mathbf{p}_{31}$) and the component of $\mathbf{x}_3$ along $\mathbf{y}_2$ is equal to the projection of $\mathbf{x}_3$ on $\mathbf{y}_2$ ($\mathbf{p}_{32}$). Thus, the vector orthogonal to the plane containing $\mathbf{x}_1$ and $\mathbf{y}_2$ is

$$\mathbf{y}_3 = \mathbf{x}_3 - \mathbf{p}_{31} - \mathbf{p}_{32}$$

This argument can be extended to n-dimensional abstract space, where n orthogonal basis vectors are formed.

Before considering the mechanics of constructing an orthogonal basis, let us return to the projection vector. If $\mathbf{x}_2$ is projected on $\mathbf{x}_1$, we know two facts that allow us to relate $\mathbf{p}_{21}$ to $\mathbf{x}_2$ and $\mathbf{x}_1$. First of all, $\mathbf{p}_{21}$ is a scalar multiple of $\mathbf{x}_1$, that is

$$\mathbf{p}_{21} = k\,\mathbf{x}_1$$

Secondly, we know that the difference vector $\mathbf{x}_2 - \mathbf{p}_{21}$ is orthogonal to $\mathbf{x}_1$; therefore

$$\langle \mathbf{x}_2 - k\,\mathbf{x}_1, \mathbf{x}_1 \rangle = 0$$

Since the scalar product as defined previously is distributive, the above equation becomes

$$\langle \mathbf{x}_2, \mathbf{x}_1 \rangle - k \langle \mathbf{x}_1, \mathbf{x}_1 \rangle = 0$$

Solving for k gives

$$k = \frac{\langle \mathbf{x}_2, \mathbf{x}_1 \rangle}{\langle \mathbf{x}_1, \mathbf{x}_1 \rangle}$$

The projection vector is then expressible as

$$\mathbf{p}_{21} = \frac{\langle \mathbf{x}_2, \mathbf{x}_1 \rangle}{\langle \mathbf{x}_1, \mathbf{x}_1 \rangle}\,\mathbf{x}_1 \qquad (2\text{-}10)$$

Construction of a three-dimensional orthogonal basis then consists of finding the three vectors

$$\mathbf{x}_1$$

$$\mathbf{y}_2 = \mathbf{x}_2 - \frac{\langle \mathbf{x}_2, \mathbf{x}_1 \rangle}{\langle \mathbf{x}_1, \mathbf{x}_1 \rangle} \mathbf{x}_1$$

$$\mathbf{y}_3 = \mathbf{x}_3 - \frac{\langle \mathbf{x}_3, \mathbf{x}_1 \rangle}{\langle \mathbf{x}_1, \mathbf{x}_1 \rangle} \mathbf{x}_1 - \frac{\langle \mathbf{x}_3, \mathbf{y}_2 \rangle}{\langle \mathbf{y}_2, \mathbf{y}_2 \rangle} \mathbf{y}_2$$

It is often convenient to normalize the new basis at the same time it is orthogonalized. One such process is referred to as Gram-Schmidt orthonormalization and is outlined below.

Given three nonorthogonal vectors $\mathbf{x}_1$, $\mathbf{x}_2$, and $\mathbf{x}_3$. Let $\mathbf{y}_1$, $\mathbf{y}_2$, and $\mathbf{y}_3$ be the desired orthonormal basis. Then

$$\mathbf{y}_1 = \frac{\mathbf{x}_1}{|\mathbf{x}_1|} \quad \text{(normalization)}$$

$$\mathbf{y}_2^1 = \mathbf{x}_2 - \frac{\langle \mathbf{x}_2, \mathbf{y}_1 \rangle}{\langle \mathbf{y}_1, \mathbf{y}_1 \rangle} \mathbf{y}_1 = \mathbf{x}_2 - \langle \mathbf{x}_2, \mathbf{y}_1 \rangle \mathbf{y}_1 \qquad \text{(orthogonalization)}$$

$$\mathbf{y}_2 = \frac{\mathbf{y}_2^1}{|\mathbf{y}_2^1|} \quad \text{(normalization)}$$

$$\begin{aligned} \mathbf{y}_3^1 &= \mathbf{x}_3 - \frac{\langle \mathbf{x}_3, \mathbf{y}_1 \rangle}{\langle \mathbf{y}_1, \mathbf{y}_1 \rangle} \mathbf{y}_1 - \frac{\langle \mathbf{x}_3, \mathbf{y}_2 \rangle}{\langle \mathbf{y}_2, \mathbf{y}_2 \rangle} \mathbf{y}_2 \\ &= \mathbf{x}_3 - \langle \mathbf{x}_3, \mathbf{y}_1 \rangle \mathbf{y}_1 - \langle \mathbf{x}_3, \mathbf{y}_2 \rangle \mathbf{y}_2 \qquad \text{(orthogonalization)} \end{aligned}$$

$$\mathbf{y}_3 = \frac{\mathbf{y}_3^1}{|\mathbf{y}_3^1|} \quad \text{(normalization)}$$

Orthonormalization is usually undertaken in order to simplify succeeding mathematical manipulations. Scalar products, for example, are easier to work with when the vectors involved are expressed in terms of an orthonormal basis.

EXAMPLE 2-3. The three vectors

$$\mathbf{x}_1 = \begin{bmatrix} 1 \\ 1 \\ 0 \end{bmatrix}, \qquad \mathbf{x}_2 = \begin{bmatrix} 3 \\ 1 \\ 0 \end{bmatrix}, \qquad \mathbf{x}_3 = \begin{bmatrix} 2 \\ 1 \\ 2 \end{bmatrix}$$

are expressed in terms of the orthogonal basis (1,0,0), (0,1,0), and (0,0,1). Use the Gram-Schmidt procedure to form an orthonormal basis from $\mathbf{x}_1$, $\mathbf{x}_2$, and $\mathbf{x}_3$.

Solution. The first basis vector of the new basis is

$$\mathbf{y}_1 = \frac{\mathbf{x}_1}{|\mathbf{x}_1|} = \frac{\mathbf{x}_1}{\langle \mathbf{x}_1, \mathbf{x}_1 \rangle^{1/2}}$$

$$= \frac{1}{\sqrt{2}} \begin{bmatrix} 1 \\ 1 \\ 0 \end{bmatrix}$$

An orthogonal vector $\mathbf{y}_2^1$ can be formed as

$$\mathbf{y}_2^1 = \mathbf{x}_2 - \langle \mathbf{x}_2, \mathbf{y}_1 \rangle \mathbf{y}_1$$

$$= \begin{bmatrix} 3 \\ 1 \\ 0 \end{bmatrix} - 2 \begin{bmatrix} 1 \\ 1 \\ 0 \end{bmatrix} = \begin{bmatrix} 1 \\ -1 \\ 0 \end{bmatrix}$$

This vector is normalized to be

$$\mathbf{y}_2 = \frac{\mathbf{y}_2^1}{|\mathbf{y}_2^1|} = \frac{1}{\sqrt{2}} \begin{bmatrix} 1 \\ -1 \\ 0 \end{bmatrix}$$

The last orthogonal basis vector is

$$\mathbf{y}_3^1 = \mathbf{x}_3 - \langle \mathbf{x}_3, \mathbf{y}_1 \rangle \mathbf{y}_1 \langle \mathbf{x}_3, \mathbf{y}_2 \rangle \mathbf{y}_2$$

$$= \begin{bmatrix} 2 \\ 1 \\ 2 \end{bmatrix} - \frac{3}{2} \begin{bmatrix} 1 \\ 1 \\ 0 \end{bmatrix} - \frac{1}{2} \begin{bmatrix} 1 \\ -1 \\ 0 \end{bmatrix} = \begin{bmatrix} 0 \\ 0 \\ 2 \end{bmatrix}$$

The normalized vector is

$$\mathbf{y}_3 = \begin{bmatrix} 0 \\ 0 \\ 1 \end{bmatrix}$$

The orthonormal basis vectors are

$$\mathbf{y}_1 = \begin{bmatrix} \frac{1}{\sqrt{2}} \\ \frac{1}{\sqrt{2}} \\ 0 \end{bmatrix}, \quad \mathbf{y}_2 = \begin{bmatrix} \frac{1}{\sqrt{2}} \\ -\frac{1}{\sqrt{2}} \\ 0 \end{bmatrix}, \quad \mathbf{y}_3 = \begin{bmatrix} 0 \\ 0 \\ 1 \end{bmatrix}$$

We should note that if a given vector had been expressed in terms of the original basis vectors $\mathbf{x}_1$, $\mathbf{x}_2$, and $\mathbf{x}_3$, for example,

$$\mathbf{z} = a_1 \mathbf{x}_1 + a_2 \mathbf{x}_2 + a_3 \mathbf{x}_3$$

a new representation for $\mathbf{z}$ must be found in terms of the orthonormal basis. The determination of this new representation will be covered in a succeeding section.

C. LINEAR INDEPENDENCE OF VECTORS

The first section of this chapter discussed linear independence of a set of algebraic equations. Since these equations can be represented as vector equations there must be corresponding conditions and tests for linear independence of vectors.

A set of vectors, $\mathbf{x}_1, \mathbf{x}_2, \ldots, \mathbf{x}_n$, are linearly independent if the equation

$$c_1\mathbf{x}_1 + c_2\mathbf{x}_2 + \cdots + c_n\mathbf{x}_n = 0 \tag{2-11}$$

can be satisfied only by setting

$$c_1 = c_2 = \cdots = c_n = 0 \tag{2-12}$$

This definition is very similar to that for linear independence of algebraic equations (see Eqs. 2-2 and 2-3). If and only if one of the set of vectors is dependent on others of the set then Eq. 2-11 can be satisfied by nonzero values of some of the constant multipliers. For example, if $\mathbf{x}_1 = a_1\mathbf{e}_1 + a_2\mathbf{e}_2 + a_3\mathbf{e}_3$, $\mathbf{x}_2 = b_1\mathbf{e}_1 + b_2\mathbf{e}_2 + b_3\mathbf{e}_3$ and $\mathbf{x}_3 = 2\mathbf{x}_1 + \mathbf{x}_2 = (2a_1 + b_1)\mathbf{e}_1 + (2a_2 + b_2)\mathbf{e}_2 + (2a_3 + b_3)\mathbf{e}_3$, Eq. 2-11 becomes

$$c_1\mathbf{x}_1 + c_2\mathbf{x}_2 + c_3\mathbf{x}_3 = 0$$

and is satisfied for $c_1 = -2$, $c_2 = -1$, and $c_3 = 1$.

In general, the set of vectors $\mathbf{x}_1, \mathbf{x}_2, \ldots, \mathbf{x}_n$ can be referred to an n-dimensional orthonormal basis $\mathbf{e}_1, \mathbf{e}_2, \ldots, \mathbf{e}_n$ where

$$\mathbf{e}_1 = \begin{bmatrix} 1 \\ 0 \\ 0 \\ \cdot \\ \cdot \\ \cdot \\ 0 \end{bmatrix}, \quad \mathbf{e}_2 = \begin{bmatrix} 0 \\ 1 \\ 0 \\ \cdot \\ \cdot \\ \cdot \\ 0 \end{bmatrix}, \quad \ldots, \quad \mathbf{e}_n = \begin{bmatrix} 0 \\ 0 \\ \cdot \\ \cdot \\ \cdot \\ 0 \\ 1 \end{bmatrix}$$

If we let

$$\begin{aligned}
\mathbf{x}_1 &= a_{11}\mathbf{e}_1 + a_{12}\mathbf{e}_2 + \cdots + a_{1n}\mathbf{e}_n \\
\mathbf{x}_2 &= a_{21}\mathbf{e}_1 + a_{22}\mathbf{e}_2 + \cdots + a_{2n}\mathbf{e}_n \\
&\cdot \qquad\qquad\qquad\qquad \cdot \\
&\cdot \qquad\qquad\qquad\qquad \cdot \\
&\cdot \qquad\qquad\qquad\qquad \cdot \\
\mathbf{x}_n &= a_{n1}\mathbf{e}_1 + a_{n2}\mathbf{e}_2 + \cdots + a_{nn}\mathbf{e}_n
\end{aligned}$$

then Eq. 2-11 can be expanded to be

$$\begin{aligned}
&(c_1 a_{11} + c_2 a_{21} + \cdots + c_n a_{n1})\,\mathbf{e}_1 \\
&+ (c_1 a_{12} + c_2 a_{22} + \cdots + c_n a_{n2})\,\mathbf{e}_2 \\
&\quad\vdots \\
&+ (c_1 a_{1n} + c_2 a_{2n} + \cdots + c_n a_{nn})\,\mathbf{e}_n = 0
\end{aligned}$$

In order that the above vector equation be satisfied, all components must be zero, thus Eq. 2-11 is satisfied only if

$$\begin{aligned}
c_1 a_{11} + c_2 a_{21} + \cdots + c_n a_{n1} &= 0 \\
c_1 a_{12} + c_2 a_{22} + \cdots + c_n a_{n2} &= 0 \\
&\vdots \\
c_1 a_{1n} + c_2 a_{2n} + \cdots + c_n a_{nn} &= 0
\end{aligned}$$

Just as in the case of algebraic equations we must determine what conditions are imposed on the array of vector coefficients to cause $c_1 = c_2 = \cdots = c_n = 0$. Solving for a given c_i results in

$$c_i = \frac{\begin{vmatrix} a_{11} & a_{21} & \cdots & 0 & \cdots & a_{n1} \\ a_{12} & a_{22} & \cdots & 0 & \cdots & a_{n2} \\ \vdots & & & & & \vdots \\ a_{1n} & a_{2n} & \cdots & 0 & \cdots & a_{nn} \end{vmatrix}}{\begin{vmatrix} a_{11} & a_{21} & \cdots & a_{i1} & \cdots & a_{n1} \\ a_{12} & a_{22} & \cdots & a_{i2} & \cdots & a_{n2} \\ \vdots & & & & & \vdots \\ a_{1n} & a_{2n} & \cdots & a_{in} & \cdots & a_{nn} \end{vmatrix}}$$

Since the numerator is zero, a nonzero denominator will insure that all c_i are zero. Again, the condition that ensures linear independence is

$$\det \mathbf{A} \neq 0$$

In this case, however, the matrix $\mathbf{A}$ is made up of the components of the set of vectors to which the test of linear independence is applied. The first column of the matrix $\mathbf{A}$ is made up of the components of $\mathbf{x}_1$, the second column is composed of the components of $\mathbf{x}_2$, and so on.

EXAMPLE 2-4. Test the following sets of vectors for linear independence:

(a)

$$\mathbf{x}_1 = \begin{bmatrix} 1 \\ 2 \\ 0 \end{bmatrix}, \quad \mathbf{x}_2 = \begin{bmatrix} 1 \\ 0 \\ 1 \end{bmatrix}, \quad \mathbf{x}_3 = \begin{bmatrix} 3 \\ 4 \\ 1 \end{bmatrix}$$

(b)

$$\mathbf{x}_1 = \begin{bmatrix} 1 \\ 1 \\ 1 \\ 1 \end{bmatrix}, \quad \mathbf{x}_2 = \begin{bmatrix} 0 \\ 1 \\ 0 \\ 1 \end{bmatrix}, \quad \mathbf{x}_3 = \begin{bmatrix} 0 \\ 0 \\ 1 \\ 1 \end{bmatrix}, \quad \mathbf{x}_4 = \begin{bmatrix} 1 \\ 0 \\ 0 \\ 1 \end{bmatrix}$$

Solution. (a) The matrix $\mathbf{A}$ is formed, giving

$$\mathbf{A} = \begin{bmatrix} 1 & 1 & 3 \\ 2 & 0 & 4 \\ 0 & 1 & 1 \end{bmatrix}$$

The determinant is

$$\det \mathbf{A} = 0$$

Thus one of the vectors is linearly dependent on the other two.

(b) The matrix $\mathbf{A}$ is

$$\mathbf{A} = \begin{bmatrix} 1 & 0 & 0 & 1 \\ 1 & 1 & 0 & 0 \\ 1 & 0 & 1 & 0 \\ 1 & 1 & 1 & 1 \end{bmatrix}$$

which has a determinant value of 2, thus linear independence is insured.

If a set of n vectors are linearly independent, they span the n-dimensional space. These vectors can then be used to form an orthonormal basis by the Gram-Schmidt procedure.

Throughout the remainder of this text a vector will be assumed to be represented in terms of the orthonormal basis

$$\mathbf{e}_1 = \begin{bmatrix} 1 \\ 0 \\ \cdot \\ \cdot \\ \cdot \\ 0 \end{bmatrix}, \quad \mathbf{e}_2 = \begin{bmatrix} 0 \\ 1 \\ \cdot \\ \cdot \\ \cdot \\ 0 \end{bmatrix}, \quad \ldots, \quad \mathbf{e}_n = \begin{bmatrix} 0 \\ 0 \\ \cdot \\ \cdot \\ \cdot \\ 1 \end{bmatrix}$$

unless otherwise noted.

The vector expressed as

$$\mathbf{x}_1 = \begin{bmatrix} 1 \\ 0 \\ 2 \\ 1 \end{bmatrix}$$

means

$$\mathbf{x}_1 = \mathbf{e}_1 + 2\mathbf{e}_3 + \mathbf{e}_4$$

D. CHANGE OF BASIS

We have seen that a given vector basis can be changed to a new basis for the purpose of simplifying succeeding mathematical operations. In Sec. B we noted that a vector expressed in terms of the original basis will have a new representation in terms of the second basis. This section is concerned with the determination of this new representation.

Assume that a vector $\mathbf{z}$ is expressed in terms of the basis vectors $\mathbf{x}_1$, $\mathbf{x}_2$, and $\mathbf{x}_3$ as

$$\mathbf{z}^x = z_1^x \mathbf{x}_1 + z_2^x \mathbf{x}_2 + z_3^x \mathbf{x}_3$$

The vectors $\mathbf{x}_1$, $\mathbf{x}_2$, and $\mathbf{x}_3$ of course are expressible in terms of $\mathbf{e}_1$, $\mathbf{e}_2$, and $\mathbf{e}_3$ as

$$\mathbf{x}_1 = a_{11}\mathbf{e}_1 + a_{12}\mathbf{e}_2 + a_{13}\mathbf{e}_3$$

$$\mathbf{x}_2 = a_{21}\mathbf{e}_1 + a_{22}\mathbf{e}_2 + a_{23}\mathbf{e}_3$$

$$\mathbf{x}_3 = a_{31}\mathbf{e}_1 + a_{32}\mathbf{e}_2 + a_{33}\mathbf{e}_3$$

The vector $\mathbf{z}$ can be represented by

$$\mathbf{z}^x = \begin{bmatrix} z_1^x \\ z_2^x \\ z_3^x \end{bmatrix} \tag{2-13}$$

where the superscript denotes the basis used. If a new basis is constructed, related to the x basis by a linear transformation, the new basis is

$$\mathbf{y}_1 = d_{11}\mathbf{x}_1 + d_{12}\mathbf{x}_2 + d_{13}\mathbf{x}_3$$
$$\mathbf{y}_2 = d_{21}\mathbf{x}_1 + d_{22}\mathbf{x}_2 + d_{23}\mathbf{x}_3$$
$$\mathbf{y}_3 = d_{31}\mathbf{x}_1 + d_{32}\mathbf{x}_2 + d_{33}\mathbf{x}_3$$

In matrix form this equation becomes

$$\mathbf{y} = \mathbf{D}\mathbf{x} \tag{2-14}$$

where

$$\mathbf{D} = \begin{bmatrix} d_{11} & d_{12} & d_{13} \\ d_{21} & d_{22} & d_{23} \\ d_{31} & d_{32} & d_{33} \end{bmatrix}$$

The matrix $\mathbf{D}$ is referred to as the basis transformation matrix. Inverting Eq. 2-14 gives $\mathbf{x}$ in terms of $\mathbf{y}$ as

$$\mathbf{x} = \mathbf{D}^{-1}\mathbf{y} \tag{2-15}$$

where $\mathbf{D}^{-1}$ has elements as follows:

$$\mathbf{D}^{-1} = \begin{bmatrix} d'_{11} & d'_{12} & d'_{13} \\ d'_{21} & d'_{22} & d'_{23} \\ d'_{31} & d'_{32} & d'_{33} \end{bmatrix}$$

If the old basis is orthogonal the matrix $\mathbf{D}$ can be found by noting that

$$\begin{aligned} d_{11} &= \langle \mathbf{y}_1, \mathbf{x}_1 \rangle / \langle \mathbf{x}_1, \mathbf{x}_1 \rangle \\ d_{12} &= \langle \mathbf{y}_1, \mathbf{x}_2 \rangle / \langle \mathbf{x}_2, \mathbf{x}_2 \rangle \\ &\cdot \\ &\cdot \\ &\cdot \\ d_{ij} &= \langle \mathbf{y}_i, \mathbf{x}_j \rangle / \langle \mathbf{x}_j, \mathbf{x}_j \rangle \end{aligned} \tag{2-16}$$

If the vectors $\mathbf{x}_i$ are also normalized so that $\langle \mathbf{x}_j, \mathbf{x}_j \rangle = 1$, Eq. 2-16 becomes

$$d_{ij} = \langle \mathbf{y}_i, \mathbf{x}_j \rangle \tag{2-17}$$

Once $\mathbf{D}$ has been found the inverse matrix can easily be determined. If the new basis is orthogonal, it may be easier to evaluate $\mathbf{D}^{-1}$ directly by the appropriate scalar products, since

$$d'_{ij} = \langle \mathbf{x}_i, \mathbf{y}_j \rangle / \langle \mathbf{y}_j, \mathbf{y}_j \rangle \tag{2-18}$$

Again if $\langle \mathbf{y}_j, \mathbf{y}_j \rangle = 1$ this equation simplifies to

$$d'_{ij} = \langle \mathbf{x}_i, \mathbf{y}_j \rangle \tag{2-19}$$

If neither the **x** basis nor the **y** basis is orthogonal, the equations for **y** in terms of **x** must be deduced by other means to construct **D** or $\mathbf{D}^{-1}$.

Now let us return to consider the expression for **z** in terms of the **y** basis. Since

$$\mathbf{z}^x = z_1^x \mathbf{x}_1 + z_2^x \mathbf{x}_2 + z_3^x \mathbf{x}_3$$

we can substitute the expressions for $\mathbf{x}_1$, $\mathbf{x}_2$, and $\mathbf{x}_3$ in terms of $\mathbf{y}_1$, $\mathbf{y}_2$, and $\mathbf{y}_3$ given by Eq. 2-15 to express **z** in terms of the **y** basis. This gives

$$\mathbf{z}^y = (z_1^x d'_{11} + z_2^x d'_{21} + z_3^x d'_{31})\mathbf{y}_1 + (z_1^x d'_{12} + z_2^x d'_{22} + z_3^x d'_{32})\mathbf{y}_2 + (z_1^x d'_{13} + z_2^x d'_{23} + z_3^x d'_{33})\mathbf{y}_3$$

or

$$\mathbf{z}^y = \begin{bmatrix} z_1^y \\ z_2^y \\ z_3^y \end{bmatrix}$$

where

$$z_1^y = (z_1^x d'_{11} + z_2^x d'_{21} + z_3^x d'_{31})$$

$$z_2^y = (z_1^x d'_{12} + z_2^x d'_{22} + z_3^x d'_{32})$$

$$z_3^y = (z_1^x d'_{13} + z_2^x d'_{23} + z_3^x d'_{33})$$

Rather than carrying out the operations required above to find $\mathbf{z}^y$ we can note that the components of $\mathbf{z}^y$ are related to the components of $\mathbf{z}^x$ by

$$\mathbf{z}^y = (\mathbf{D}^{-1})^t \, \mathbf{z}^x \tag{2-20}$$

In words, if we are given the matrix **D** relating the old basis to the new, any vector expressed in terms of the old basis can be transformed to the new basis by multiplying the original vector by the transpose of $\mathbf{D}^{-1}$. We can denote $\mathbf{D}^t$ by the matrix **F** so that

$$(\mathbf{D}^{-1})^t = \mathbf{F}^{-1} \tag{2-21}$$

EXAMPLE 2-5. If

$$\mathbf{z}^x = \begin{bmatrix} 1 \\ 2 \\ -3 \end{bmatrix}$$

express **z** in terms of the orthonormal basis vectors $\mathbf{y}_1$, $\mathbf{y}_2$, and $\mathbf{y}_3$. Use the vectors $\mathbf{x}_1$, $\mathbf{x}_2$, $\mathbf{x}_3$, $\mathbf{y}_1$, $\mathbf{y}_2$, and $\mathbf{y}_3$ given in Example 2-3.

Solution. Since **y** forms an orthonormal basis, Eq. 2-19 can be used to evaluate the elements of $\mathbf{D}^{-1}$. This gives

$$\mathbf{D}^{-1} = \begin{bmatrix} \sqrt{2} & 0 & 0 \\ 2\sqrt{2} & \sqrt{2} & 0 \\ 3\sqrt{2}/2 & \sqrt{2}/2 & 2 \end{bmatrix}$$

The new vector is

$$\mathbf{z}^y = (\mathbf{D}^{-1})^t \mathbf{z}^x = (\mathbf{F}^{-1})\, \mathbf{z}^x = \begin{bmatrix} \sqrt{2}/2 \\ \sqrt{2}/2 \\ -6 \end{bmatrix}$$

We should note that we can proceed in the opposite direction also. That is, if

$$\mathbf{y} = \mathbf{D}\mathbf{x}$$

and we have the vector $\mathbf{z}^y$ given, we can find $\mathbf{z}^x$ by inverting Eq. 2-20 to give

$$\mathbf{z}^x = \mathbf{D}^t\, \mathbf{z}^y \qquad (2\text{-}22)$$

or

$$\mathbf{z}^x = \mathbf{F}\, \mathbf{z}^y \qquad (2\text{-}23)$$

EXAMPLE 2-6. Express the vector

$$\mathbf{z}^y = \begin{bmatrix} \sqrt{2}/2 \\ \sqrt{2}/2 \\ -6 \end{bmatrix}$$

in terms of the **x** basis if the same **x** and **y** bases are used as in Example 2-5.

Solution. The vector $\mathbf{F}^{-1}$ from Example 2-5 is inverted to give

$$\mathbf{F} = \begin{bmatrix} 1/\sqrt{2} & -\sqrt{2} & -1/4 \\ 0 & 1/\sqrt{2} & -1/4 \\ 0 & 0 & 1/2 \end{bmatrix}$$

Equation 2-23 is applied to give

$$\mathbf{z}^x = \begin{bmatrix} 1 \\ 2 \\ -3 \end{bmatrix}$$

which is the original vector given in Example 2-5.

E. LINEAR TRANSFORMATIONS

The preceding section considered the case where a vector expressed in terms of a given basis is transformed to a new basis representation. This represents an important problem in the field of electrical engineering. The types of linear transformations encountered in practical problems are (1) rotation of the space about a point, and (2) uniform expansion of the space about a point. These operations can always be represented by a matrix equation. Furthermore, a given matrix equation of the form of Eq. 2-20 or Eq. 2-22 can be considered to consist of a mapping or transforming operation with the possibility of applying the previously developed theory.

A second problem of great importance occurs when the basis for two vectors related by a matrix equation is changed. For example, if

$$\omega^x = \mathbf{A}\,\mathbf{z}^x \tag{2-24}$$

and a change of basis from $\mathbf{x}$ to $\mathbf{y}$ is effected, the corresponding equation for ω^y as a function of $\mathbf{z}^y$ is required. We can derive this new relationship by considering Eqs. 2-20 and 2-21.

Assuming that

$$\mathbf{y} = \mathbf{D}\mathbf{x}$$

and

$$\mathbf{F}^{-1} = (\mathbf{D}^{-1})^t$$

we can write

$$\omega^y = \mathbf{F}^{-1}\,\omega^x \quad \text{or} \quad \omega^x = \mathbf{F}\,\omega^y$$

and

$$\mathbf{z}^y = \mathbf{F}^{-1}\,\mathbf{z}^x \quad \text{or} \quad \mathbf{z}^x = \mathbf{F}\,\mathbf{z}^y$$

Substitution of these equations into Eq. 2-24 gives

$$\mathbf{F}\,\omega^y = \mathbf{A}\mathbf{F}\mathbf{z}^y$$

Solving for ω^y results in

$$\omega^y = \mathbf{F}^{-1}\,\mathbf{A}\mathbf{F}\mathbf{z}^y = \mathbf{B}\,\mathbf{z}^y \tag{2-25}$$

The new matrix $\mathbf{B}$ is formed by premultiplying $\mathbf{A}$ by $\mathbf{F}^{-1}$ and multiplying $\mathbf{F}$ by this result, thus

$$\mathbf{B} = \mathbf{F}^{-1}\,\mathbf{A}\mathbf{F} \tag{2-26}$$

The procedure for converting a matrix equation from one basis to another is

1. Calculate the basis transformation matrix $\mathbf{D}$.
2. Construct $\mathbf{F} = \mathbf{D}^t$.
3. Calculate $\mathbf{F}^{-1}$.
4. Convert the matrix $\mathbf{A}$ of the original equation to $\mathbf{B} = \mathbf{F}^{-1}\,\mathbf{A}\mathbf{F}$.

EXAMPLE 2-7. If

$$\omega^x = \begin{bmatrix} \omega_1 \\ \omega_2 \\ \omega_3 \end{bmatrix} = \mathbf{A}\mathbf{z}^x = \begin{bmatrix} 1 & 0 & 1 \\ 0 & 2 & -1 \\ 3 & -1 & 2 \end{bmatrix} \begin{bmatrix} z_1 \\ z_2 \\ z_3 \end{bmatrix}$$

and

$$\mathbf{x}_1 = \begin{bmatrix} 1 \\ 1 \\ 0 \end{bmatrix}, \quad \mathbf{x}_2 = \begin{bmatrix} 3 \\ 1 \\ 0 \end{bmatrix}, \quad \mathbf{x}_3 = \begin{bmatrix} 2 \\ 1 \\ 2 \end{bmatrix}, \quad \mathbf{y}_1 = \frac{1}{\sqrt{2}} \begin{bmatrix} 1 \\ 1 \\ 0 \end{bmatrix}, \quad \mathbf{y}_2 = \frac{1}{\sqrt{2}} \begin{bmatrix} 1 \\ -1 \\ 0 \end{bmatrix}$$

and $\mathbf{y}_3 = \begin{bmatrix} 0 \\ 0 \\ 1 \end{bmatrix}$, find the equation relating ω^y to $\mathbf{z}^y$.

Solution. The $\mathbf{D}^{-1}$ matrix was found in Example 2-5 to be

$$\mathbf{D}^{-1} = \begin{bmatrix} \sqrt{2} & 0 & 0 \\ 2\sqrt{2} & \sqrt{2} & 0 \\ 3\sqrt{2}/2 & \sqrt{2}/2 & 2 \end{bmatrix}$$

The matrix $\mathbf{D}$ can be found to be

$$\mathbf{D} = \begin{bmatrix} 1/\sqrt{2} & 0 & 0 \\ -\sqrt{2} & 1/\sqrt{2} & 0 \\ -1/4 & -1/4 & 1/2 \end{bmatrix}$$

The values of $\mathbf{F}$ and $\mathbf{F}^{-1}$ are

$$\mathbf{F}^{-1} = \begin{bmatrix} \sqrt{2} & 2\sqrt{2} & 3\sqrt{2}/2 \\ 0 & \sqrt{2} & \sqrt{2}/2 \\ 0 & 0 & 2 \end{bmatrix}$$

$$\mathbf{F} = \begin{bmatrix} 1/\sqrt{2} & -\sqrt{2} & -1/4 \\ 0 & 1/\sqrt{2} & -1/4 \\ 0 & 0 & 1/2 \end{bmatrix}$$

The matrix $\mathbf{B}$, called the linear transformation of $\mathbf{A}$, is

$$\mathbf{B} = \begin{bmatrix} \sqrt{2} & 2\sqrt{2} & 3\sqrt{2}/2 \\ 0 & \sqrt{2} & \sqrt{2}/2 \\ 0 & 0 & 2 \end{bmatrix} \begin{bmatrix} 1 & 0 & 1 \\ 0 & 2 & -1 \\ 3 & -1 & 2 \end{bmatrix} \begin{bmatrix} 1/\sqrt{2} & -\sqrt{2} & -1/4 \\ 0 & 1/\sqrt{2} & -1/4 \\ 0 & 0 & 1/2 \end{bmatrix}$$

$$= \begin{bmatrix} 11/2 & -17/2 & -\sqrt{2} \\ 3/2 & -3/2 & -3\sqrt{2}/4 \\ 3\sqrt{2} & -7\sqrt{2} & 1 \end{bmatrix}$$

F. EIGENVECTORS AND EIGENVALUES

We have considered transformations in the general case; now we can discuss the special case that is quite important to computer analysis of networks. As previously mentioned, eigenvector theory is useful as a result of the use of the digital computer to solve network equations. This theory is merely a tool that can minimize computing time for certain types of circuit problems.

In general, the matrix equation

$$\mathbf{y} = \mathbf{A}\mathbf{x}$$

transforms a vector **x** to a second vector **y**. A special case of some interest occurs when the vector **y** has the same direction as **x**. In this case **y** is proportional to **x** and can be expressed as

$$\mathbf{y} = \lambda \mathbf{x} \tag{2-27}$$

where λ is a constant. The well-known eigenvector equation results from equating the two expressions for **y** giving

$$\mathbf{A}\,\mathbf{x} = \lambda \mathbf{x} \tag{2-28}$$

When this equation is satisfied, the vector **x** is called the eigenvector and the constant λ is called the *eigenvalue.* Usually, there will be several eigenvectors and eigenvalues corresponding to a given matrix **A**.

The set of algebraic equations corresponding to Eq. 2-28 is given by

$$\begin{aligned}
a_{11}x_1 + a_{12}x_2 + \cdots + a_{1n}x_n &= \lambda x_1 \\
a_{21}x_1 + a_{22}x_2 + \cdots + a_{2n}x_n &= \lambda x_2 \\
\vdots \qquad\qquad & \qquad \vdots \\
a_{n1}x_1 + a_{n2}x_2 + \cdots + a_{nn}x_n &= \lambda x_n
\end{aligned}$$

The transformation matrix **A** will always be an $n \times n$ matrix, since **y** must have the same number of components as **x**. Transposing to form a set of homogeneous equations gives

$$\begin{aligned}
(a_{11} - \lambda)x_1 + a_{12}x_2 + \cdots + a_{1n}x_n &= 0 \\
a_{21}x_1 + (a_{22} - \lambda)x_2 + \cdots + a_{2n}x_n &= 0 \\
\vdots \qquad\qquad & \qquad \vdots \\
a_{n1}x_1 + a_{n2}x_2 + \cdots + (a_{nn} - \lambda)x_n &= 0
\end{aligned}$$

The matrix expression representing this set of equations is

$$(\mathbf{A} - \lambda \mathbf{U}) \mathbf{x} = 0 \tag{2-29}$$

In Sec. **A** of this chapter it was shown that n homogeneous equations have a nontrivial solution only if the determinant of the coefficients is zero. In this case the set contains q linearly dependent equations and $n - q$ variables are expressible in terms of the remaining q variables.

Most physical problems specify the matrix **A** in terms of network parameters and the calculation of **x** is the desired goal. The procedure used to determine **x** is to first find the eigenvalues of **A**, then by applying the condition of Eq. 2-29, the eigenvectors of **A** can be determined. Evaluation of det $(\mathbf{A} - \lambda\mathbf{U})$ results in a polynomial in λ that is referred to as the characteristic polynomial. The equation in λ formed by equating det $(\mathbf{A} - \lambda\mathbf{U})$ to zero is called the characteristic equation of the matrix **A**. The roots of this equation are the eigenvalues. If **A** is an $n \times n$ matrix, the characteristic equation will contain λ to the nth power and thus there will be n roots or eigenvalues. These eigenvalues can be simple or multiple (repeated) and they can be real or complex depending on the given problem. The remainder of this section will be concerned with solving Eq. 2-28 for the case of simple eigenvalues. Succeeding sections will consider repeated and complex eigenvalues.

EXAMPLE 2-8. Find the eigenvalues and eigenvectors corresponding to

$$\mathbf{A} = \begin{bmatrix} 3 & 2 \\ 1 & 2 \end{bmatrix}$$

Solution. The characteristic equation of **A** is given by

$$\begin{vmatrix} (3 - \lambda) & 2 \\ 1 & (2 - \lambda) \end{vmatrix} = \lambda^2 - 5\lambda + 4 = 0$$

This equation factors to become

$$(\lambda - 4)(\lambda - 1) = 0$$

giving eigenvalues of $\lambda_1 = 1$ and $\lambda_2 = 4$. The fact that there are two roots to the characteristic equation means that there are two and only two values of λ that will lead to a nontrivial solution of the eigenvector equation. For both values of λ, we can find some vector **x** that will be transformed by **A** into a vector in the same direction as **x**, but multiplied by λ.

To find the eigenvectors Eq. 2-28 or Eq. 2-29 is used along with the actual eigenvalue. For $\lambda_1 = 1$ we have

$$\begin{bmatrix} 3 & 2 \\ 1 & 2 \end{bmatrix} \begin{bmatrix} x_1 \\ x_2 \end{bmatrix} = (1) \begin{bmatrix} x_1 \\ x_2 \end{bmatrix}$$

Carrying out the multiplication gives

$$3x_1 + 2x_2 = x_1$$

and

$$x_1 + 2x_2 = x_2$$

Note that the equations are linearly dependent, allowing either component to be written in terms of the other. For this case $x_2 = -x_1$. The eigenvector of **A** corresponding to an eigenvalue of 1 is

$$\mathbf{x} = \begin{bmatrix} x_1 \\ -x_1 \end{bmatrix}$$

which is a vector having components equal in magnitude, but opposite in sign.

The eigenvector corresponding to $\lambda_2 = 4$ is found to be

$$\mathbf{x} = \begin{bmatrix} x_1 \\ (1/2)x_1 \end{bmatrix}$$

In this case any vector whose second component equals 1/2 the value of the first component is an eigenvector.

Often the eigenvectors will be chosen to be normalized vectors. Thus

$$\mathbf{x} = \frac{1}{\sqrt{2}} \begin{bmatrix} 1 \\ -1 \end{bmatrix} \quad \text{and} \quad \mathbf{x} = \frac{2}{\sqrt{5}} \begin{bmatrix} 1 \\ 1/2 \end{bmatrix}$$

are the normalized eigenvectors of **A**.

There are several reasons for using an eigenvector approach in solving a set of simultaneous equations. One important reason is that it is useful in evaluating a function of a matrix. In working with linear networks or systems a power series representation of a matrix is often required. A power series can be expressed as

$$\mathbf{F}(\mathbf{A}) = k_0 \mathbf{U} + k_1 \mathbf{A} + k_2 \mathbf{A}^2 + \cdots + k_n \mathbf{A}^n + \cdots \tag{2-30}$$

where **A** is an $n \times n$ matrix and **U** is the $n \times n$ identity matrix.

The Cayley-Hamilton Theorem

This is a useful theorem that allows the function of a matrix to be expressed in terms of a power series with a finite number of terms rather than an infinite number as in Eq. 2-30. The Cayley-Hamilton theorem is expressed as follows:

Let **A** be an $n \times n$ matrix, and let

$$p(\lambda) = b_0 + b_1\lambda + b_2\lambda^2 + \cdots + (-1)^n \lambda^n$$

be its characteristic polynomial. Then

$$b_0\mathbf{U} + b_1\mathbf{A} + b_2\mathbf{A}^2 + \cdots + (-1)^n \mathbf{A}^n = 0$$

This theorem states that a matrix satisfies its own characteristic equation. The proof of this theorem is contained in Appendix 1.

Because $\mathbf{A}$ satisfies its characteristic equation, $\mathbf{A}^n$ can be expressed in terms of lower powers of $\mathbf{A}$. Solving for $\mathbf{A}^n$ gives

$$\mathbf{A}^n = (-1)^{n-1} \; [b_{n-1} \mathbf{A}^{n-1} + \cdots + b_1 \mathbf{A} + b_0 \mathbf{U}] \tag{2-31}$$

From Eq. 2-31 it is easy to see that all powers of $\mathbf{A}$ higher than n can be written in terms of powers of $\mathbf{A}$ equal to or lower than $n - 1$. For example,

$$\begin{aligned} \mathbf{A}^{n+1} &= \mathbf{A}(-1)^{n-1} \; [b_{n-1} \mathbf{A}^{n-1} + \cdots + b_1 \mathbf{A} + b_0 \mathbf{U}] \\ &= (-1)^{n-1} \; [b_{n-1} \mathbf{A}^{n} + \cdots + b_1 \mathbf{A}^2 + b_0 \mathbf{A}] \\ &= (-1)^{n-1} \; [b_{n-1} (-1)^{n-1} \{b_{n-1} \mathbf{A}^{n-1} + \cdots + b_1 \mathbf{A} + b_0 \mathbf{U}\} \\ &\qquad + b_{n-2} \mathbf{A}^{n-1} + \cdots + b_1 \mathbf{A}^2 + b_0 \mathbf{A}] \end{aligned} \tag{2-32}$$

The importance of this result is that any function of $\mathbf{A}$ can be expressed in terms of powers of $\mathbf{A}$ less than or equal to $n - 1$ where n is the dimension of the matrix. This follows from the fact that any function of a matrix encountered in engineering practice can be represented by a power series expansion in $\mathbf{A}$ as given by Eq. 2-30. Using Eqs. 2-31 and 2-32 allows Eq. 2-30 to be written as

$$\mathbf{F}(\mathbf{A}) = K_0 \mathbf{U} + K_1 \mathbf{A} + K_2 \mathbf{A}^2 + \cdots + K_{n-1} \mathbf{A}^{n-1} = \sum_{i=0}^{n-1} K_i \mathbf{A}^i \tag{2-33}$$

EXAMPLE 2-9. If

$$\mathbf{A} = \begin{bmatrix} 3 & 2 \\ 1 & 2 \end{bmatrix}$$

find $\mathbf{A}^4 + 2\mathbf{A}^3 + \mathbf{A}$ (a) by matrix multiplication, (b) by applying Eq. 2-33.

Solution

(a)

$$\mathbf{A}^2 = \begin{bmatrix} 3 & 2 \\ 1 & 2 \end{bmatrix} \begin{bmatrix} 3 & 2 \\ 1 & 2 \end{bmatrix} = \begin{bmatrix} 11 & 10 \\ 5 & 6 \end{bmatrix}$$

$$\mathbf{A}^3 = \begin{bmatrix} 3 & 2 \\ 1 & 2 \end{bmatrix} \begin{bmatrix} 11 & 10 \\ 5 & 6 \end{bmatrix} = \begin{bmatrix} 43 & 42 \\ 21 & 22 \end{bmatrix}$$

$$\mathbf{A}^4 = \begin{bmatrix} 3 & 2 \\ 1 & 2 \end{bmatrix} \begin{bmatrix} 43 & 42 \\ 21 & 22 \end{bmatrix} = \begin{bmatrix} 171 & 170 \\ 85 & 86 \end{bmatrix}$$

$$\begin{aligned} \mathbf{A}^4 + 2\mathbf{A}^3 + \mathbf{A} &= \begin{bmatrix} 171 & 170 \\ 85 & 86 \end{bmatrix} + 2 \begin{bmatrix} 43 & 42 \\ 21 & 22 \end{bmatrix} + \begin{bmatrix} 3 & 2 \\ 1 & 2 \end{bmatrix} \\ &= \begin{bmatrix} 260 & 256 \\ 128 & 132 \end{bmatrix} \end{aligned}$$

(b) The characteristic equation from Example 2-8 is $p(\lambda) = \lambda^2 - 5\lambda + 4 = 0$. Substituting $\mathbf{A}$ for λ and solving for $\mathbf{A}^2$ gives

$$\mathbf{A}^2 = 5\mathbf{A} - 4\mathbf{U}$$

Multiplying both sides by $\mathbf{A}$ yields

$$\mathbf{A}^3 = 5\mathbf{A}^2 - 4\mathbf{A} = 5(5\mathbf{A} - 4\mathbf{U}) - 4\mathbf{A} = 21\mathbf{A} - 20\mathbf{U}$$

Multiplying both sides again by $\mathbf{A}$ results in

$$\mathbf{A}^4 = 21\mathbf{A}^2 - 20\mathbf{A} = 21(5\mathbf{A} - 4\mathbf{U}) - 20\mathbf{A} = 85\mathbf{A} - 84\mathbf{U}$$

Using these values in the original equation gives

$$\mathbf{A}^4 + 2\mathbf{A}^3 + \mathbf{A} = 128\mathbf{A} - 124\mathbf{U}$$

$$= \begin{bmatrix} 260 & 256 \\ 128 & 132 \end{bmatrix}$$

The second method reduces the problem to nothing more than algebra with a minimum of matrix multiplication.

There is another method by which the function of a matrix can be simply evaluated. We observe that any function $f(\lambda)$ can be expressed in terms of the characteristic polynomial as

$$f(\lambda) = k(\lambda) \cdot p(\lambda) + r(\lambda) \tag{2-34}$$

where $r(\lambda)$ is a remainder polynomial of lower order than the characteristic equation. Given $f(\lambda)$ and $p(\lambda)$, both $r(\lambda)$ and $k(\lambda)$ can be found by noting that

$$\frac{f(\lambda)}{p(\lambda)} = k(\lambda) + \frac{r(\lambda)}{p(\lambda)}$$

Once $k(\lambda)$ and $r(\lambda)$ are found we can express the function of $\mathbf{A}$ as

$$\mathbf{F}(\mathbf{A}) = \mathbf{k}(\mathbf{A})\,\mathbf{p}(\mathbf{A}) + \mathbf{r}(\mathbf{A}) = \mathbf{r}(\mathbf{A}) \tag{2-35}$$

since $\mathbf{p}(\mathbf{A}) = 0$.

EXAMPLE 2-10. Evaluate the function

$$\mathbf{F}(\mathbf{A}) = \mathbf{A}^4 + 2\mathbf{A}^3 + \mathbf{A}$$

where $\mathbf{A} = \begin{bmatrix} 3 & 2 \\ 1 & 2 \end{bmatrix}$, using the remainder method.

Solution. From Example 2-8 $p(\lambda) = \lambda^2 - 5\lambda + 4$. Then

$$\frac{f(\lambda)}{p(\lambda)} = \frac{\lambda^4 + 2\lambda^3 + \lambda}{\lambda^2 - 5\lambda + 4} = \lambda^2 + 7\lambda + 31 + \frac{128\lambda - 124}{\lambda^2 - 5\lambda + 4}$$

The remainder term is $r(\lambda) = 128\lambda - 124$. From Eq. 2-35,

$$\mathbf{F}(\mathbf{A}) = \mathbf{r}(\mathbf{A}) = 128\mathbf{A} - 124\mathbf{U} = \begin{bmatrix} 260 & 256 \\ 128 & 132 \end{bmatrix}$$

The remainder method is particularly useful in evaluating $\mathbf{F}(\mathbf{A})$ when a closed-form series expansion is not given. The function $\mathbf{e}^{\mathbf{A}t}$ is a good example of this case.

We must consider the case of complex eigenvectors since *RLC* networks can often lead to eigenvectors which are complex. The matrix

$$\mathbf{A} = \begin{bmatrix} -1 & 3 \\ -1 & -2 \end{bmatrix}$$

has eigenvalues of $\lambda_1 = -3/2 + j\sqrt{11}/2$ and $\lambda_2 = -3/2 - j\sqrt{11}/2$. The eigenvector corresponding to λ_1 can be found by applying Eq. 2-29 to give

$$\begin{bmatrix} 1/2 - j\sqrt{11}/2 & 3 \\ -1 & -1/2 - j\sqrt{11}/2 \end{bmatrix} \begin{bmatrix} x_1 \\ x_2 \end{bmatrix} = \begin{bmatrix} 0 \\ 0 \end{bmatrix}$$

Again the two equations involving x_1 and x_2 are linearly dependent. If x_1 is selected to equal unity then x_2 is found to be

$$x_2 = -1/6 + j\sqrt{11}/6$$

The eigenvector is then

$$\mathbf{x}(\lambda_1) = \begin{bmatrix} 1 \\ -1/6 + j\sqrt{11}/6 \end{bmatrix}$$

A similar solution for the eigenvector corresponding to the second eigenvalue results in

$$\mathbf{x}(\lambda_2) = \begin{bmatrix} 1 \\ -1/6 - j\sqrt{11}/6 \end{bmatrix}$$

There are other possible choices for these eigenvectors, but since the eigenvalues are complex conjugates the eigenvectors are also chosen in such a way that each component of one eigenvector is complex conjugate of the corresponding component of the other. This choice is always possible for a pair of complex eigenvalues.

G. THE EXPONENTIAL FUNCTION OF A MATRIX

When a network or a system is specified in terms of state variables, the equation

$$\mathbf{x} = \mathbf{A}\mathbf{x} \tag{2-36}$$

commonly occurs. We might suspect that the solution to this equation is similar to that of the first-order, scalar differential equation

$$\dot{x} = ax$$

namely

$$x = e^{at} x_0$$

If this is the case, Eq. 2-36 has the solution

$$\mathbf{x} = \mathbf{e}^{\mathbf{A}t} \mathbf{x}_0 \tag{2-37}$$

Differentiating Eq. 2-37 with respect to time gives

$$\dot{\mathbf{x}} = \frac{d\mathbf{e}^{\mathbf{A}t}}{dt} \mathbf{x}_0 = \frac{d(\mathbf{e}^{\mathbf{A}t})}{d(\mathbf{A}t)} \frac{d(\mathbf{A}t)}{dt} \mathbf{x}_0 = \mathbf{e}^{\mathbf{A}t} \mathbf{A} \mathbf{x}_0$$

From Eq. 2-36 the derivative of **x** also equals **Ax**, thus

$$\mathbf{e}^{\mathbf{A}t} \mathbf{A}\mathbf{x}_0 = \mathbf{A}\mathbf{e}^{\mathbf{A}t} \mathbf{x}_0 \tag{2-38}$$

Before one can conclude that Eq. 2-38 validates the proposed solution for **x** at least two points must be considered: it must be shown (1) that $\mathbf{e}^{\mathbf{A}t} \mathbf{A} = \mathbf{A}\mathbf{e}^{\mathbf{A}t}$, and (2) that $\frac{d}{dt} [\mathbf{e}^{\mathbf{A}t}] = \mathbf{e}^{\mathbf{A}t} \mathbf{A}$. That $\mathbf{e}^{\mathbf{A}t} \mathbf{A} = \mathbf{A}\mathbf{e}^{\mathbf{A}t}$ can be proven by considering the power-series expansion for $\mathbf{e}^{\mathbf{A}t}$. In general terms $\mathbf{e}^{\mathbf{A}t}$ can be expressed by Eq. 2-33 as

$$\mathbf{e}^{\mathbf{A}t} = \sum_{i=0}^{n-1} K_i \mathbf{A}^i t^i$$

The product $\mathbf{A}\mathbf{e}^{\mathbf{A}t}$ is then

$$\mathbf{A}\mathbf{e}^{\mathbf{A}t} = \mathbf{A}\left[\sum_{i=0}^{n-1} K_i \mathbf{A}^i t^i\right] = \sum_{i=0}^{n-1} K_i \mathbf{A} \mathbf{A}^i t^i = \sum_{i=0}^{n-1} K_i \mathbf{A}^{i+1} t^i$$

Reversing the position of the terms gives

$$\mathbf{e}^{\mathbf{A}t} \mathbf{A} = \left[\sum_{i=0}^{n-1} K_i \mathbf{A}^i t^i\right] \mathbf{A} = \sum_{i=0}^{n-1} K_i \mathbf{A}^i t^i \mathbf{A} = \sum_{i=0}^{n-1} K_i \mathbf{A}^{i+1} t^i$$

Thus $\mathbf{A}\mathbf{e}^{\mathbf{A}t} = \mathbf{e}^{\mathbf{A}t} \mathbf{A}$, and in general

$$\mathbf{A}\mathbf{F}(\mathbf{A}) = \mathbf{F}(\mathbf{A}) \mathbf{A} \tag{2-39}$$

Several approaches can be taken in discussing the derivative of the exponential function of a matrix. We could propose a series expression for $\mathbf{e}^{\mathbf{A}t}$ and check to see if the derivative with respect to time equals $\mathbf{e}^{\mathbf{A}t} \mathbf{A}$. We will use a second approach however, solving for the required series expansion for $\mathbf{e}^{\mathbf{A}t}$ based on the assumptions that

$$\frac{d\mathbf{e}^{\mathbf{A}t}}{dt} = \mathbf{e}^{\mathbf{A}t}\,\mathbf{A} \tag{2-40}$$

and

$$\mathbf{e}^{\mathbf{A}t}\big|_{t=0} = \mathbf{U}$$

If $\mathbf{e}^{\mathbf{A}t}$ is represented by an infinite series expansion we have

$$\mathbf{e}^{\mathbf{A}t} = k_0\mathbf{U} + k_1\mathbf{A}t + k_2\,\mathbf{A}^2t^2 + \cdots + k_n\,\mathbf{A}^nt^n + \cdots$$

Differentiating this series gives

$$\begin{aligned}\frac{d\mathbf{e}^{\mathbf{A}t}}{dt} &= k_1\mathbf{A} + 2k_2\mathbf{A}^2t + \cdots + nk_n\,\mathbf{A}^nt^{n-1} + \cdots \\ &= \mathbf{A}\,[k_1\,\mathbf{U} + 2k_2\,\mathbf{A}t + \cdots + nk_n\,\mathbf{A}^{n-1}\,t^{n-1} + \cdots]\end{aligned} \tag{2-41}$$

The derivative given by Eq. 2-41 will equal $\mathbf{Ae}^{\mathbf{A}t}$ only if the coefficients of the series expansion are related by

$$\begin{aligned} k_1 &= k_0 \\ k_2 &= k_1/2 \\ &\vdots \\ k_n &= k_{n-1}/n \\ &\vdots \end{aligned}$$

Noting that $\mathbf{e}^{\mathbf{0}} = \mathbf{U}$ gives a value of unity for k_0. Thus

$$\mathbf{e}^{\mathbf{A}t} = \sum_{n=0}^{\infty} \frac{1}{n!}\,\mathbf{A}^nt^n \tag{2-42}$$

The state equation can then be solved as we initially expected with the solution given by Eq. 2-37,

$$\mathbf{x} = \mathbf{e}^{\mathbf{A}t}\,\mathbf{x}_0$$

EXAMPLE 2-11. If $\dot{\mathbf{x}} = \mathbf{Ax}$ where

$$\mathbf{A} = \begin{bmatrix} -2 & 2 \\ 1 & -3 \end{bmatrix} \quad \text{and} \quad \mathbf{x}_0 = \begin{bmatrix} v(0) \\ i(0) \end{bmatrix}$$

find $\mathbf{x}$ as a function of time.

Solution. Equation 2-37 indicates that

$$\mathbf{x} = \mathbf{e}^{\mathbf{A}t}\,\mathbf{x}_0$$

is a solution of the original equation, however, we must now evaluate $\mathbf{e}^{\mathbf{A}t}$. Applying the finite power-series expansion theory to $\mathbf{e}^{\mathbf{A}t}$ (Eq. 2-33) shows that $\mathbf{e}^{\mathbf{A}t}$ can be represented by 2 terms of the expansion. We know from the previous section that

$$f(\lambda) = k(\lambda)\,p(\lambda) + r(\lambda)$$

and

$$\mathbf{F(A)} = \mathbf{k(A)}\,\mathbf{p(A)} + \mathbf{r(A)} = \mathbf{r(A)}$$

At the eigenvalues, where $p(\lambda) = 0$,

$$f(\lambda) = r(\lambda)$$

Thus we can evaluate the eigenvalues and then find $r(\lambda)$ and $\mathbf{r(A)}$. The eigenvalues are found by evaluating

$$\det(\mathbf{A} - \lambda\mathbf{U}) = \begin{vmatrix} -2-\lambda & 2 \\ 1 & -3-\lambda \end{vmatrix}$$

$$= \lambda^2 + 5\lambda + 4 = (\lambda + 1)(\lambda + 4) = 0$$

The eigenvalues are $\lambda_1 = -1$ and $\lambda_2 = -4$. Since $r(\lambda)$ only contains 2 terms we can write

$$f(\lambda) = e^{\lambda t} = K_0 + K_1\lambda$$

where λ is either eigenvalue. Substituting the eigenvalues into this equation gives

$$e^{-t} = K_0 - K_1$$

$$e^{-4t} = K_0 - 4K_1$$

The coefficients must then be $K_0 = (4/3)e^{-t} - (1/3)e^{-4t}$ and $K_1 = (1/3)e^{-t} - (1/3)e^{-4t}$.

The function $\mathbf{F(A)} = \mathbf{e}^{\mathbf{A}t}$ is then

$$\begin{aligned} \mathbf{e}^{\mathbf{A}t} &= K_0\,\mathbf{U} + K_1\,\mathbf{A} \\ &= [(4/3)e^{-t} - (1/3)e^{-4t}]\,\mathbf{U} + [(1/3)e^{-t} - (1/3)e^{-4t}]\,\mathbf{A} \\ &= \begin{bmatrix} [(2/3)e^{-t} + (1/3)e^{-4t}] & [(2/3)e^{-t} - (2/3)e^{-4t}] \\ [(1/3)e^{-t} - (1/3)e^{-4t}] & [(1/3)e^{-t} + (2/3)e^{-4t}] \end{bmatrix} \end{aligned}$$

The complete solution is then

$$\begin{aligned} \mathbf{x} &= \mathbf{e}^{\mathbf{A}t}\,\mathbf{x}_0 \\ &= \begin{bmatrix} e^{-t}\,[(2/3)\,v(0) + (2/3)\,i(0)] + e^{-4t}\,[(1/3)\,v(0) - (2/3)\,i(0)] \\ e^{-t}\,[(1/3)\,v(0) + (1/3)\,i(0)] + e^{-4t}\,[-(1/3)\,v(0) + (2/3)\,i(0)] \end{bmatrix} \end{aligned}$$

We have restricted our discussion thus far to matrices having real, distinct eigenvalues. In practice, both complex and multiple eigenvalues can occur. Before treating multiple eigenvalues an example will demonstrate that the complex case can be treated in a manner similar to that of the real-value case.

EXAMPLE 2-12. If $\dot{\mathbf{x}} = \mathbf{Ax}$ where

$$\mathbf{A} = \begin{bmatrix} -1 & 3 \\ -1 & -2 \end{bmatrix} \quad \text{and} \quad x_0 = \begin{bmatrix} v(0) \\ i(0) \end{bmatrix}$$

find **x** as a function of time.

Solution. The characteristic equation is

$$p(\lambda) = \lambda^2 + 3\lambda + 5 = 0$$

giving eigenvalues of $\lambda_1 = -3/2 + j\sqrt{11}/2$ and $\lambda_2 = -3/2 - j\sqrt{11}/2$. If

$$\mathbf{F(A)} = K_0\,\mathbf{U} + K_1\mathbf{A}$$

then

$$f(\lambda) = K_0 + K_1\lambda$$

giving

$$e^{-(3/2)t}\,e^{j(\sqrt{11}/2)t} = K_0 - (3/2)K_1 + j(\sqrt{11}/2)\,K_1$$
$$e^{-(3/2)t}\,e^{-j(\sqrt{11}/2)t} = K_0 - (3/2)K_1 - j(\sqrt{11}/2)K_1$$

Solving for K_0 and K_1 yields

$$K_0 = e^{-(3/2)t}\,[\cos(\sqrt{11}/2)t + (3/\sqrt{11})\sin(\sqrt{11}/2)t]$$

and

$$K_1 = (2/\sqrt{11})e^{-(3/2)t}\sin(\sqrt{11}/2)t$$

The function **F(A)** is then

$$\begin{aligned} \mathbf{e}^{\mathbf{A}t} &= K_0\,\mathbf{U} + K_1\mathbf{A} \\ &= e^{-(3/2)t} \\ &\times \begin{bmatrix} [\cos(\sqrt{11}/2)t + (1/\sqrt{11})\sin(\sqrt{11}/2)t] & [(6/\sqrt{11})\sin(\sqrt{11}/2)t] \\ [-(2/\sqrt{11})\sin(\sqrt{11}/2)t] & [\cos(\sqrt{11}/2)t - (1/\sqrt{11})\sin(\sqrt{11}/2)t] \end{bmatrix} \end{aligned}$$

Complex roots simply lead to a function which is oscillatory rather than one that is exponential.

H. SOLUTION OF STATE EQUATION BY LAPLACE TRANSFORMS

Starting with the equation

$$\dot{\mathbf{x}} = \mathbf{Ax}$$

we can take the Laplace transform of both sides, giving

$$s\mathbf{X}(s) - \mathbf{x}_0 = \mathbf{A}\mathbf{X}(s)$$

or

$$\mathbf{X}(s) = [s\mathbf{U} - \mathbf{A}]^{-1}\, \mathbf{x}_0 \tag{2-43}$$

In this case, the matrix $[s\mathbf{U} - \mathbf{A}]$ must be inverted to arrive at a solution for $\mathbf{X}(s)$. The result can then be transformed back to the time domain to find $\mathbf{x}$. An example will demonstrate this procedure.

EXAMPLE 2-13. Rework Example 2-11, using Laplace transforms.

Solution. The matrix $[s\mathbf{U} - \mathbf{A}]$ is

$$\begin{bmatrix} s+2 & -2 \\ -1 & s+3 \end{bmatrix}$$

The inverse of this matrix is

$$[s\mathbf{U} - \mathbf{A}]^{-1} = \frac{1}{s^2 + 5s + 4} \begin{bmatrix} s+3 & 2 \\ 1 & s+2 \end{bmatrix}$$

The variable $\mathbf{X}(s)$ has a value of

$$\mathbf{X}(s) = \frac{1}{s^2 + 5s + 4} \begin{bmatrix} s+3 & 2 \\ 1 & s+2 \end{bmatrix} \begin{bmatrix} v(0) \\ i(0) \end{bmatrix}$$

Transforming $\mathbf{X}(s)$ to the time domain gives

$$\mathbf{x} = \begin{bmatrix} [(2/3)e^{-t} + (1/3)e^{-4t}] & [(2/3)e^{-t} - (2/3)e^{-4t}] \\ [(1/3)e^{-t} - (1/3)e^{-4t}] & [(1/3)e^{-t} + (2/3)e^{-4t}] \end{bmatrix} \begin{bmatrix} v(0) \\ i(0) \end{bmatrix}$$

which agrees with Example 2-11.

An interesting point should be emphasized here. When the matrix $[s\mathbf{U} - \mathbf{A}]$ is inverted we can write*

$$[s\mathbf{U} - \mathbf{A}]^{-1} = \frac{\text{adjoint } [s\mathbf{U} - \mathbf{A}]}{\det\, [s\mathbf{U} - \mathbf{A}]} \tag{2-44}$$

Note that the denominator is similar to that formed to calculate the eigenvalues. The characteristic equation is

$$p(\lambda) = \det\, [\mathbf{A} - \lambda\mathbf{U}] = 0$$

Replacing λ by s and taking the negative of the matrix difference causes $[\mathbf{A} - \lambda\mathbf{U}]$ to become $[s\mathbf{U} - \mathbf{A}]$. If the elements of adjoint $[s\mathbf{U} - \mathbf{A}]$ have no factors common to det $[s\mathbf{U} - \mathbf{A}]$ then the values of s causing det $[s\mathbf{U} - \mathbf{A}] = 0$

*The adjoint of a matrix is defined in Appendix 2-1.

are the poles of the function. These same values correspond to the eigenvalues and thus we can ascribe some physical significance to the eigenvalues. The eigenvalues of a system or a network correspond to the poles of the system or network.

In Chapter 5 the question of which method of solution to apply to the state equation will be considered.

I. MULTIPLE EIGENVALUES

Certain modifications of the preceding theory are necessary when **A** gives rise to multiple eigenvalues. Some of the difficulties encountered can be demonstrated by considering the matrix

$$\mathbf{A} = \begin{bmatrix} -2 & -2 \\ 2 & -6 \end{bmatrix}$$

The characteristic equation is

$$p(\lambda) = \lambda^2 + 8\lambda + 16 = (\lambda + 4)(\lambda + 4) = 0$$

giving $\lambda_1 = \lambda_2 = -4$. An eigenvector corresponding to $\lambda = -4$ can be found by substituting into the equation

$$(\mathbf{A} - \lambda\mathbf{U})\mathbf{x} = 0$$

yielding

$$\begin{bmatrix} 2 & -2 \\ 2 & -2 \end{bmatrix} \begin{bmatrix} x_1 \\ x_2 \end{bmatrix} = 0$$

The eigenvector is

$$\mathbf{x} = \begin{bmatrix} x_1 \\ x_1 \end{bmatrix}$$

Other eigenvectors, referred to as *generalized eigenvectors* [4], can be found, but are not appropriate to the present purpose. We are usually more interested in using the eigenvalues to evaluate the function of a matrix, therefore our attention will be directed to this end. Equations 2-34 and 2-35 show the relationship between a function and the characteristic equation for scalars and matrices respectively. That is

$$f(\lambda) = k(\lambda)\,p(\lambda) + r(\lambda) \tag{2-34}$$

and

$$\mathbf{F(A) = k(A)\,p(A) + r(A) = r(A)} \tag{2-35}$$

We note that for an $n \times n$ matrix $r(\lambda)$ can be expressed as a polynomial in λ with highest power equal to $n - 1$. Hence, there are n coefficients to be evaluated to find $r(\lambda)$ or $\mathbf{r}(\mathbf{A})$. If the eigenvalues of $\mathbf{A}$ are distinct, there will be exactly n eigenvalues. Substituting these n values into Eq. 2-34 yields n equations with n unknowns [coefficients of $r(\lambda)$] allowing the determination of $r(\lambda)$. If $\mathbf{A}$ has one multiple eigenvalue of multiplicity q and $n - q$ distinct eigenvalues, only $n - q + 1$ relations result from application of Eq. 2-34. The additional required equations can be found by using a method based on the differentiation of Eq. 2-34.

Let

$$p(\lambda) = (\lambda - 1)(\lambda - \lambda_2) \ldots (\lambda - \lambda_{n-q})(\lambda - \lambda_n)^q$$

Equation 2-34 becomes

$$f(\lambda) = k(\lambda)(\lambda - \lambda_1)(\lambda - \lambda_2) \cdots (\lambda - \lambda_{n-q})(\lambda - \lambda_n)^q + r(\lambda)$$

We can obtain $n - q + 1$ relations by evaluating $f(\lambda)$ at each different eigenvalue, giving

$$\begin{aligned} f(\lambda_1) &= r(\lambda_1) \\ f(\lambda_2) &= r(\lambda_2) \\ &\;\;\vdots \\ f(\lambda_{n-q}) &= r(\lambda_{n-q}) \\ f(\lambda_n) &= r(\lambda_n) \end{aligned}$$

Letting $k(\lambda)(\lambda - \lambda_1)(\lambda - \lambda_2) \cdots (\lambda - \lambda_{n-q}) = m(\lambda)$, Eq. 2-34 can be written

$$f(\lambda) = m(\lambda)(\lambda - \lambda_n)^q + r(\lambda)$$

Differentiating $f(\lambda)$ with respect to λ results in

$$\frac{df}{d\lambda} = \frac{dm}{d\lambda}(\lambda - \lambda_n)^q + qm(\lambda)(\lambda - \lambda_n)^{q-1} + \frac{dr}{d\lambda}$$

Evaluating this expression at $\lambda = \lambda_n$ gives

$$\left.\frac{df}{d\lambda}\right|_{\lambda=\lambda_n} = \left.\frac{dr}{d\lambda}\right|_{\lambda=\lambda_n}$$

which adds one more relationship to the $n - q + 1$ previously found. It is obvious that all derivatives of $f(\lambda)$ with respect to λ up to the $(q - 1)$st order can be expressed as

$$\left.\frac{d^l f}{d\lambda^l}\right|_{\lambda=\lambda_n} = \left.\frac{d^l r}{d\lambda^l}\right|_{\lambda=\lambda_n} \tag{2-45}$$

since all other terms in the equation will contain the factor $(\lambda - \lambda_n)$. This equation will supply the necessary remaining $q - 1$ relationships to find all the coefficients of $r(\lambda)$. An example will demonstrate the use of this method.

EXAMPLE 2-14. Evaluate $\mathbf{e}^{\mathbf{A}t}$, where

$$\mathbf{A} = \begin{bmatrix} 5 & -1 & 1 & 1 & 0 & 0 \\ 1 & 3 & -1 & -1 & 0 & 0 \\ 0 & 0 & 4 & 0 & 1 & 1 \\ 0 & 0 & 0 & 4 & -1 & -1 \\ 0 & 0 & 0 & 0 & 3 & 1 \\ 0 & 0 & 0 & 0 & 1 & 3 \end{bmatrix}$$

Solution. The characteristic equation of this matrix is

$$p(\lambda) = (\lambda - 4)^5 (\lambda - 2) = 0$$

The function $e^{\lambda t}$ is expressible as

$$f(\lambda) = e^{\lambda t} = K_0 + K_1\lambda + K_2\lambda^2 + K_3\lambda^3 + K_4\lambda^4 + K_5\lambda^5$$

For $\lambda = 2$ and $\lambda = 4$ we obtain

$$e^{2t} = K_0 + 2K_1 + 4K_2 + 8K_3 + 16K_4 + 32K_5$$

and

$$e^{4t} = K_0 + 4K_1 + 16K_2 + 64K_3 + 256K_4 + 1024K_5$$

Applying Eq. 2-45 gives

$$\frac{df}{d\lambda} = te^{\lambda t} = K_1 + 2\lambda K_2 + 3\lambda^2 K_3 + 4\lambda^3 K_4 + 5\lambda^4 K_5$$

$$\frac{d^2 f}{d\lambda^2} = t^2 e^{\lambda t} = 2K_2 + 6\lambda K_3 + 12\lambda^2 K_4 + 20\lambda^3 K_5$$

$$\frac{d^3 f}{d\lambda^3} = t^3 e^{\lambda t} = 6K_3 + 24\lambda K_4 + 60\lambda^2 K_5$$

$$\frac{d^4 f}{d\lambda^4} = t^4 e^{\lambda t} = 24K_4 + 120\lambda K_5$$

The six resulting equations are

$$\begin{aligned}
e^{2t} &= K_0 + 2K_1 + 4K_2 + 8K_3 + 16K_4 + 32K_5 \\
e^{4t} &= K_0 + 4K_1 + 16K_2 + 64K_3 + 256K_4 + 1024K_5 \\
te^{4t} &= K_1 + 8K_2 + 48K_3 + 256K_4 + 1280K_5 \\
t^2 e^{4t} &= 2K_2 + 24K_3 \quad 192K_4 + 1280K_5 \\
t^3 e^{4t} &= 6K_3 + 96K_4 + 960K_5 \\
t^4 e^{4t} &= 24K_4 + 480K_5
\end{aligned}$$

Rather than solve these equations we will simply comment on the results. The general form of the coefficients will be similar to that of K_5, which is

$$K_5 = (1/48)t^4 e^{4t} - (1/24)t^3 e^{4t} + (1/16)t^2 e^{4t} - (1/16)te^{4t} + (1/32)e^{4t} - (1/32)e^{2t}$$

The matrix function $\mathbf{e}^{\mathbf{A}t}$ is then

$$\mathbf{F}(\mathbf{A}) = K_0 \mathbf{U} + K_1 \mathbf{A} + K_2 \mathbf{A}^2 + K_3 \mathbf{A}^3 + K_4 \mathbf{A}^4 + K_5 \mathbf{A}^5$$

We previously noted that the eigenvalues of the matrix **A** which appears in the state equation correspond to the poles of the response. The **A** matrix of Example 2-14 would not correspond to a stable system, since the poles in this case would be in the right half of the Laplace plane. However, an interesting conclusion can be drawn from this example: namely that repeated eigenvalues lead to a function containing terms of the form $t^n e^{at}$. We recall that multiple poles in a response function lead to a time response containing similar terms, and again the connection between eigenvalues of **A** and poles of the response function is apparent.

J. MATRIX DIAGONALIZATION

In computer-aided design it is often more efficient to diagonalize the matrix **A** before carrying out additional operations to obtain the response [1]. It will be shown that the diagonalization of the matrix **A** in the state equation

$$\dot{\mathbf{x}} = \mathbf{A}\mathbf{x}$$

corresponds to a simple change of basis vectors.

Returning to the eigenvalue equation

$$\mathbf{A}\mathbf{x} = \lambda \mathbf{x} = \lambda \mathbf{U}\mathbf{x}$$

we can write for distinct eigenvalues

$$\begin{aligned} \mathbf{A}\mathbf{x}(\lambda_1) &= \lambda_1 \, \mathbf{x}(\lambda_1) \\ \mathbf{A}\mathbf{x}(\lambda_2) &= \lambda_2 \, \mathbf{x}(\lambda_2) \\ &\vdots \\ \mathbf{A}\mathbf{x}(\lambda_n) &= \lambda_n \, \mathbf{x}(\lambda_n) \end{aligned}$$

where $\mathbf{x}(\lambda_n)$ is the eigenvector corresponding to the eigenvalue λ_n. This set of n matrix equations can be written in terms of a single matrix equation after observing that the left-hand side of the equations can be represented by

$$\mathbf{A}[\mathbf{x}(\lambda_1)\, \mathbf{x}(\lambda_2) \cdots \mathbf{x}(\lambda_n)]$$

or

$$\begin{bmatrix} a_{11} & a_{12} & \cdot & \cdot & \cdot & a_{1n} \\ a_{21} & a_{22} & \cdot & \cdot & \cdot & a_{2n} \\ \cdot & & & & & \cdot \\ \cdot & & & & & \cdot \\ \cdot & & & & & \cdot \\ a_{n1} & a_{n2} & \cdot & \cdot & \cdot & a_{nn} \end{bmatrix} \begin{bmatrix} x_1(\lambda_1) & x_1(\lambda_2) & \cdot & \cdot & \cdot & x_1(\lambda_n) \\ x_2(\lambda_1) & x_2(\lambda_2) & \cdot & \cdot & \cdot & x_2(\lambda_n) \\ \cdot & \cdot & & & & \cdot \\ \cdot & \cdot & & & & \cdot \\ \cdot & \cdot & & & & \cdot \\ x_n(\lambda_1) & x_n(\lambda_2) & \cdot & \cdot & \cdot & x_n(\lambda_n) \end{bmatrix}$$

provided that the right-hand side is expressed as

$$\begin{bmatrix} \lambda_1 x_1(\lambda_1) & \lambda_2 x_1(\lambda_2) & \cdot & \cdot & \cdot & \lambda_n x_1(\lambda_n) \\ \lambda_1 x_2(\lambda_1) & \lambda_2 x_2(\lambda_2) & \cdot & \cdot & \cdot & \lambda_n x_2(\lambda_n) \\ \cdot & \cdot & & & & \cdot \\ \cdot & \cdot & & & & \cdot \\ \cdot & \cdot & & & & \cdot \\ \lambda_1 x_n(\lambda_1) & \lambda_2 x_n(\lambda_2) & \cdot & \cdot & \cdot & \lambda_n x_n(\lambda_n) \end{bmatrix}$$

This equation represents n sets of n simultaneous equations, each set having the same coefficients, but different values of λ and hence different eigenvectors. If the matrix, whose columns correspond to the components of individual eigenvectors, is denoted as $\mathbf{P}$, then the matrix expression becomes

$$\mathbf{AP} = \mathbf{P}\Lambda \tag{2-46}$$

where the matrix Λ is diagonal and given by

$$\Lambda = \begin{bmatrix} \lambda_1 & 0 & \cdot & \cdot & \cdot & 0 \\ 0 & \lambda_2 & & & & 0 \\ \cdot & \cdot & & & & \cdot \\ \cdot & \cdot & & & & \cdot \\ \cdot & \cdot & & & & \cdot \\ 0 & 0 & \cdot & \cdot & \cdot & \lambda_n \end{bmatrix} \tag{2-47}$$

Multiplying both sides of Eq. 2-46 by $\mathbf{P}^{-1}$ gives

$$\mathbf{P}^{-1}\,\mathbf{AP} = \Lambda \tag{2-48}$$

The matrix $\mathbf{A}$ is changed to the diagonal matrix Λ by a linear transformation. Comparing Eq. 2-48 to Eq. 2-26 shows that this particular linear transformation corresponds to that required when changing the basis vectors. Therefore, we conclude that the equations

$$\omega^u = \mathbf{A}\mathbf{z}^u$$

and

$$\omega^v = \Lambda \mathbf{z}^v$$

are equivalent, but are expressed in terms of different basis vectors.

EXAMPLE 2-15. Diagonalize the matrix from Example 2-8,

$$\mathbf{A} = \begin{bmatrix} 3 & 2 \\ 1 & 2 \end{bmatrix}$$

Solution. From Example 2-8, the eigenvalues and normalized eigenvectors are

$$\lambda_1 = 1 \qquad \mathbf{x}(1) = \frac{1}{\sqrt{2}} \begin{bmatrix} 1 \\ -1 \end{bmatrix}$$

$$\lambda_2 = 4 \qquad \mathbf{x}(4) = \frac{2}{\sqrt{5}} \begin{bmatrix} 1 \\ 1/2 \end{bmatrix}$$

The matrix **P** is then

$$\mathbf{P} = \begin{bmatrix} 1/\sqrt{2} & 2/\sqrt{5} \\ -1/\sqrt{2} & 1/\sqrt{5} \end{bmatrix}$$

The inverse is

$$\mathbf{P}^{-1} = \begin{bmatrix} \sqrt{2}/3 & -2\sqrt{2}/3 \\ \sqrt{5}/3 & \sqrt{5}/3 \end{bmatrix}$$

The diagonal matrix is then $\Lambda = \mathbf{P}^{-1}\mathbf{AP}$, so

$$\Lambda = \begin{bmatrix} \sqrt{2}/3 & -2\sqrt{2}/3 \\ \sqrt{5}/3 & \sqrt{5}/3 \end{bmatrix} \begin{bmatrix} 3 & 2 \\ 1 & 2 \end{bmatrix} \begin{bmatrix} 1/\sqrt{2} & 2/\sqrt{5} \\ -1/\sqrt{2} & 1/\sqrt{5} \end{bmatrix}$$

$$= \begin{bmatrix} 1 & 0 \\ 0 & 4 \end{bmatrix}$$

If the diagonalization of **A** corresponds to a change of basis, we should interest ourselves in determining what the new basis is related to. The eigenvector matrix **P** is equivalent to the transpose of the basis transformation matrix **D** (see Sec. **D**).

If the vector **x** is expressible as

$$\mathbf{x}^u = \begin{bmatrix} x_1 \\ x_2 \\ x_3 \end{bmatrix}$$

in terms of the original basis (u basis), then this vector becomes

$$\mathbf{x}^v = (\mathbf{D}^{-1})^t \mathbf{x}^u = \mathbf{P}^{-1} \mathbf{x}^u$$

in the new basis (ν basis). Thus, using the transpose of **P** as the basis transformation matrix causes the operator **A** to transform to a diagonal matrix Λ. We will now establish the importance of this transformation in terms of the resulting simplification in the solution of network problems.

In the case of a-c problems where the input is a steady-state sinusoid, the equation describing the problem is [1]

$$\dot{\mathbf{x}} = \mathbf{A}\mathbf{x} + \mathbf{g} \tag{2-49}$$

(We will derive this relationship in a succeeding chapter.) The vector **g** is related to the input vector. In the steady-state case **g** may be written as

$$\mathbf{g} = 2\mathbf{g}_0 \cos \omega t = \mathbf{g}_0 \left(e^{j\omega t} + e^{-j\omega t}\right)$$

where ω is the radian frequency of the input. We note that the vector **g** is composed of two complex conjugate components. Rather than carry both components through the entire solution, we can calculate the response due to only one component. The response due to the other component will be the complex conjugate of the calculated response. Chapter 5 demonstrates the application of this method to practical problems. For our present purpose we will only consider the response due to one component so that **g** becomes

$$\mathbf{g} = \mathbf{g}_0\, e^{j\omega t} \tag{2-50}$$

The steady-state response will be

$$\mathbf{x} = \mathbf{x}(\omega)\, e^{j\omega t} \tag{2-51}$$

where $\mathbf{x}(\omega)$ is a complex vector to be determined. Substituting Eqs. 2-50 and 2-51 into Eq. 2-49 gives

$$[j\omega\mathbf{U} - \mathbf{A}]\ \mathbf{x}(\omega) = \mathbf{g}_0 \tag{2-52}$$

Since $\mathbf{g}_0$ is a known quantity we can solve for $\mathbf{x}(\omega)$ giving

$$\mathbf{x}(\omega) = [j\omega\mathbf{U} - \mathbf{A}]^{-1}\ \mathbf{g}_0 \tag{2-53}$$

This problem can be solved directly by inverting the matrix $[j\omega\mathbf{U} - \mathbf{A}]$ and multiplying $\mathbf{g}_0$ by the result. Let us assume, however, that we require a frequency response of the circuit or system. This can be obtained by solving Eq. 2-52 or Eq. 2-53 for several values of ω. If 100 frequency points are needed, then 100 such solutions are necessary. Using the Gaussian elimination method Eq. 2-52 can be solved directly with $n^3/3$ operations.

If diagonalization is used, the number of operations can be significantly reduced. We first solve for the eigenvalues and eigenvectors of **A**. This gives us the matrix **P** and the diagonal matrix Λ. Since **A**, **P**, and Λ are constant matrices, they need be calculated only once rather than at each frequency. Now the transformation

$$\mathbf{P}^{-1}\ [j\omega\mathbf{U} - \mathbf{A}]\,\mathbf{P}$$

is equivalent to

$$j\omega\mathbf{U} - \mathbf{P}^{-1}\mathbf{A}\mathbf{P} = j\omega\mathbf{U} - \Lambda$$

therefore we can write

$$\begin{aligned}[j\omega\mathbf{U} - \mathbf{A}]^{-1} &= [j\omega\mathbf{U} - \mathbf{P}\Lambda\mathbf{P}^{-1}]^{-1} \\ &= (\mathbf{P}[j\omega\mathbf{U} - \Lambda]\mathbf{P}^{-1})^{-1}\end{aligned}$$

The inverse of the last term of the above equation is

$$\mathbf{P}[j\omega\mathbf{U} - \Lambda]^{-1}\mathbf{P}^{-1}$$

as can be proven by taking the product of this term and the original term. That is,

$$\begin{aligned}&(\mathbf{P}[j\omega\mathbf{U} - \Lambda]\mathbf{P}^{-1})(\mathbf{P}[j\omega\mathbf{U} - \Lambda]^{-1}\mathbf{P}^{-1} \\ &\qquad = \mathbf{P}[j\omega\mathbf{U} - \Lambda]\ [j\omega\mathbf{U} - \Lambda]^{-1}\mathbf{P}^{-1} \\ &\qquad = \mathbf{P}\mathbf{P}^{-1} = \mathbf{U}\end{aligned}$$

We can now express $\mathbf{x}(\omega)$ in terms of Λ rather than $\mathbf{A}$ as

$$\mathbf{x}(\omega) = \mathbf{P}[j\omega\mathbf{U} - \Lambda]^{-1}\mathbf{P}^{-1}\mathbf{g}_0 \tag{2-54}$$

This expression for $\mathbf{x}(\omega)$ requires that the diagonal matrix $[j\omega\mathbf{U} - \Lambda]$ be inverted at each frequency of interest, but the inversion of an $n \times n$ diagonal matrix requires only n multiplications rather than $n^3/3$. The other quantities, $\mathbf{P}$, $\mathbf{P}^{-1}$, and $\mathbf{g}_0$ are constant with frequency. If we let

$$\mathbf{D} = \mathbf{P}^{-1}\mathbf{g}_0 = \begin{bmatrix} d_1 \\ d_2 \\ \cdot \\ \cdot \\ \cdot \\ d_n \end{bmatrix}$$

then the equation for the components of $\mathbf{x}(\omega)$ can be written as

$$x_i(\omega) = \sum_{j=1}^{n} \frac{p_{ij}\,d_j}{j\omega - \lambda_j} \tag{2-55}$$

where p_{ij} represents the elements of $\mathbf{P}$.

If the variable $\mathbf{x}(\omega)$ is to be calculated at many frequencies, then diagonalization of $\mathbf{A}$ can result in a substantial reduction of computation time. If the calculation is to be done at only a few frequencies, it can be more efficient to solve Eq. 2-52 at each frequency rather than find the eigenvalues and eigenvectors of $\mathbf{A}$ [1].

We digress at this point to prove a general relationship which was used in a specific way in the preceding derivation for $\mathbf{x}_0$. We can show that

$$\mathbf{F}(\mathbf{A}) = \mathbf{P}\mathbf{F}(\mathbf{\Lambda})\,\mathbf{P}^{-1} \tag{2-56}$$

whereas we noted previously that

$$[j\omega\mathbf{U} - \mathbf{A}]^{-1} = \mathbf{P}[j\omega\mathbf{U} - \mathbf{\Lambda}]^{-1}\,\mathbf{P}^{-1}$$

If we express the function $\mathbf{F}(\mathbf{\Lambda})$ in terms of a power series we have

$$\mathbf{F}(\mathbf{\Lambda}) = \sum_{i=0}^{n-1} K_i\,\mathbf{\Lambda}^i$$

Therefore,

$$\mathbf{P}\mathbf{F}(\mathbf{\Lambda})\,\mathbf{P}^{-1} = \mathbf{P}\left[\sum_{i=0}^{n-1} K_i\,\mathbf{\Lambda}^i\right]\mathbf{P}^{-1} = \sum_{i=0}^{n-1} K_i\,\mathbf{P}\mathbf{\Lambda}^i\,\mathbf{P}^{-1}$$

Note since

$$\mathbf{\Lambda} = \mathbf{P}^{-1}\,\mathbf{A}\mathbf{P}$$

that

$$\begin{aligned}\mathbf{\Lambda}^k &= (\mathbf{P}^{-1}\,\mathbf{A}\mathbf{P})\,(\mathbf{P}^{-1}\,\mathbf{A}\mathbf{P})\cdots(\mathbf{P}^{-1}\,\mathbf{A}\mathbf{P})\\ &= \mathbf{P}^{-1}\,\mathbf{A}^k\,\mathbf{P}\end{aligned}$$

Substituting this result into the power series gives

$$\mathbf{P}\mathbf{F}(\mathbf{\Lambda})\,\mathbf{P}^{-1} = \sum_{i=0}^{n-1} K_i\,\mathbf{A}^k = \mathbf{F}(\mathbf{A}) \tag{2-57}$$

We will later show that diagonalization of $\mathbf{A}$ is an effective tool in computing $\mathbf{e}^{\mathbf{A}t}$, a term that often arises in electrical systems or networks. It should now be apparent that matrix diagonalization by using eigenvectors is a topic of some importance. We still have not considered the problem in terms of multiple eigenvalues and thus this topic will be considered next.

K. GENERALIZED EIGENVECTORS

An $n \times n$ matrix can always be diagonalized if n eigenvectors can be found. When this is the case, the diagonal matrix is expressed in terms of a basis related to the eigenvectors of the original matrix. Since the eigenvectors form a transformed basis for any vector in n-dimensional space, they must span the space thus implying that the n eigenvectors are linearly independent. We have seen that each distinct eigenvalue leads to a distinct eigenvector. If multiple eigenvalues occur we have no guarantee that n eigenvectors can be found. For example, the matrix

$$\mathbf{A} = \begin{bmatrix} 3 & 1 \\ -1 & 1 \end{bmatrix}$$

has the eigenvalue $\lambda = 2$ of multiplicity 2. The matrix $\mathbf{A} - \lambda\mathbf{U}$ gives

$$\mathbf{A} - \lambda\mathbf{U} = \begin{bmatrix} 1 & 1 \\ -1 & -1 \end{bmatrix}$$

which yields only one eigenvector, namely

$$\mathbf{x} = \begin{bmatrix} x_1 \\ -x_1 \end{bmatrix}$$

Since there is only one eigenvector associated with the 2×2 matrix $\mathbf{A}$, this matrix cannot be diagonalized with an eigenvector basis. In some cases, n eigenvectors can be found even though a multiple eigenvalue occurs. A diagonal form of the matrix can then be found. If

$$\mathbf{A} = \begin{bmatrix} 1 & 0 & -2 \\ 0 & 0 & 0 \\ -2 & 0 & 4 \end{bmatrix}$$

the eigenvalues are $\lambda_1 = 0$, $\lambda_2 = 0$, and $\lambda_3 = 5$. Using $\lambda_1 = 0$ gives

$$\mathbf{A} - \lambda_1\mathbf{U} = \begin{bmatrix} 1 & 0 & -2 \\ 0 & 0 & 0 \\ -2 & 0 & 4 \end{bmatrix}$$

Allowing this matrix to multiply $\mathbf{x}$ and equating the result to zero gives the three equations

$$x_1 = 2x_3$$
$$0 = 0$$
$$2x_3 = 4x_3$$

The first and third equations impose the requirement on the eigenvector that $x_1 = 2x_3$. The second equation imposes no requirements on the eigenvector. We can choose two eigenvectors satisfying the single requirement, both of which are linearly independent. The vectors

$$\mathbf{x} = \begin{bmatrix} 2 \\ 0 \\ 1 \end{bmatrix}$$

and

$$\mathbf{x} = \begin{bmatrix} 0 \\ 1 \\ 0 \end{bmatrix}$$

are two possible choices of eigenvectors. Thus, two eigenvectors are found corresponding to the multiple eigenvalue $\lambda = 0$. A third eigenvector can be found for $\lambda_3 = 5$ and $\mathbf{A}$ can then be diagonalized.

If the multiplicity of a given eigenvalue λ_0 is m, there can be m linearly independent eigenvectors. Often there will be less than m independent eigenvectors which means the matrix cannot be diagonalized. However, there is a canonical form to which the matrix can be transformed. This form is referred to as the *Jordan canonical form* and can be useful in subsequent calculations involving the original matrix. The matrix that transforms $\mathbf{A}$ to the Jordan canonical form has columns made up of eigenvectors and generalized eigenvectors. Generalized eigenvectors are defined as vectors that satisfy the equation

$$(\mathbf{A} - \lambda \mathbf{U})^k \mathbf{x} = 0 \tag{2-58}$$

but such that

$$(\mathbf{A} - \lambda \mathbf{U})^{k-1} \mathbf{x} \neq 0 \tag{2-59}$$

The vector $\mathbf{x}$ is a generalized eigenvector of rank k.

The procedure for finding the eigenvectors and generalized eigenvectors is as follows:

1. Determine the eigenvalues.
2. Find all eigenvectors associated with distinct eigenvalues.
3. Find all possible eigenvectors associated with a multiple eigenvalue. If the multiplicity of this eigenvalue is m and r eigenvectors are found, then $m - r$ generalized eigenvectors can be found.
4. Find the generalized eigenvectors by forming $(\mathbf{A} - \lambda \mathbf{U})^2 \mathbf{x} = 0$ and solving for the vectors $\mathbf{x}$ that are linearly independent of the eigenvectors. Then form $(\mathbf{A} - \lambda \mathbf{U})^3 \mathbf{x} = 0$ and find the vectors $\mathbf{x}$ that are linearly independent of the generalized eigenvectors of lower rank. Continue this procedure until no new vectors satisfy the equation $(\mathbf{A} - \lambda \mathbf{U})^k \mathbf{x} = 0$.

EXAMPLE 2-16. Find the eigenvectors and generalized eigenvectors of the matrix

$$\mathbf{A} = \begin{bmatrix} 2 & 0 & 0 & 0 & 1 & 0 \\ 0 & 2 & 1 & 1 & 0 & -1 \\ 0 & 0 & 2 & 1 & 0 & 0 \\ 0 & 0 & 0 & 2 & 0 & 0 \\ 0 & 0 & 1 & 0 & 2 & -1 \\ 0 & 0 & 0 & 0 & 0 & 2 \end{bmatrix}$$

Solution. The characteristic equation is found by evaluating det $(\mathbf{A} - \lambda\mathbf{U})$ to be $(2 - \lambda)^6 = 0$. The matrix **A** has only one eigenvalue ($\lambda = 2$) of multiplicity 6. The eigenvectors are found by equating $(\mathbf{A} - 2\mathbf{U})x$ to zero. This gives

$$\begin{bmatrix} 0 & 0 & 0 & 0 & 1 & 0 \\ 0 & 0 & 1 & 1 & 0 & -1 \\ 0 & 0 & 0 & 1 & 0 & 0 \\ 0 & 0 & 0 & 0 & 0 & 0 \\ 0 & 0 & 1 & 0 & 0 & -1 \\ 0 & 0 & 0 & 0 & 0 & 0 \end{bmatrix} \begin{bmatrix} x_1 \\ x_2 \\ x_3 \\ x_4 \\ x_5 \\ x_6 \end{bmatrix} = 0$$

which imposes the following conditions on the components of **x**:

$$x_5 = 0$$

$$x_3 + x_4 - x_6 = 0$$

$$x_4 = 0$$

$$x_3 - x_6 = 0$$

No conditions are imposed on x_1 and x_2, but x_3 must equal x_6 and $x_4 = x_5 = 0$. Three linearly independent eigenvectors can be chosen from the vector with components

$$\begin{bmatrix} x_1 \\ x_2 \\ x_3 \\ 0 \\ 0 \\ x_3 \end{bmatrix}$$

Since only three eigenvectors have been found, there must be three generalized eigenvectors corresponding to $\lambda = 2$. We form

$$(\mathbf{A} - \lambda\mathbf{U})^2\,\mathbf{x} = \begin{bmatrix} 0 & 0 & 1 & 0 & 0 & -1 \\ 0 & 0 & 0 & 1 & 0 & 0 \\ 0 & 0 & 0 & 0 & 0 & 0 \\ 0 & 0 & 0 & 0 & 0 & 0 \\ 0 & 0 & 0 & 1 & 0 & 0 \\ 0 & 0 & 0 & 0 & 0 & 0 \end{bmatrix} \begin{bmatrix} x_1 \\ x_2 \\ x_3 \\ x_4 \\ x_5 \\ x_6 \end{bmatrix}$$

Equating this matrix to zero gives the conditions

$$x_3 - x_6 = 0$$
$$x_4 = 0$$

Only one generalized eigenvector of rank 2 can be chosen, that is

$$\begin{bmatrix} x_1 \\ x_2 \\ x_3 \\ 0 \\ x_5 \\ x_3 \end{bmatrix}$$

Other vectors could be formed satisfying the required conditions on the components, but these vectors would not be linearly independent of the eigenvectors. Since the generalized eigenvector has $x_5 \neq 0$ we know that this vector will not satisfy the equation $(\mathbf{A} - \lambda\mathbf{U})\mathbf{x}$.

We now equate $(\mathbf{A} - \lambda\mathbf{U})^3\mathbf{x}$ to zero, giving

$$\begin{bmatrix} 0 & 0 & 0 & 1 & 0 & 0 \\ 0 & 0 & 0 & 0 & 0 & 0 \\ 0 & 0 & 0 & 0 & 0 & 0 \\ 0 & 0 & 0 & 0 & 0 & 0 \\ 0 & 0 & 0 & 0 & 0 & 0 \\ 0 & 0 & 0 & 0 & 0 & 0 \end{bmatrix} \begin{bmatrix} x_1 \\ x_2 \\ x_3 \\ x_4 \\ x_5 \\ x_6 \end{bmatrix} = 0$$

The only condition resulting from this equation is

$$x_4 = 0$$

A generalized eigenvector of rank 3 can be chosen as

$$\begin{bmatrix} x_1 \\ x_2 \\ x_3 \\ 0 \\ x_5 \\ x_6 \end{bmatrix}$$

The matrix equation $(\mathbf{A} - \lambda\mathbf{U})^4\ \mathbf{x} = 0$ imposes no conditions on the components and the generalized eigenvector of rank 4 can be chosen as

$$\begin{bmatrix} x_1 \\ x_2 \\ x_3 \\ x_4 \\ x_5 \\ x_6 \end{bmatrix}$$

The preceding example demonstrates that linearly independent eigenvectors and generalized eigenvectors of a matrix $\mathbf{A}$ can be found. The utility of the generalized eigenvector remains to be examined. Let us define $\mathbf{x}(k)$ as a generalized eigenvector of rank k satisfying the equations

$$(\mathbf{A} - \lambda\mathbf{U})^k\ \mathbf{x}(k) = 0$$

and

$$(\mathbf{A} - \lambda\mathbf{U})^{k-1}\ \mathbf{x}(k) \neq 0$$

We note that

$$(\mathbf{A} - \lambda\mathbf{U})^{k-1}\ (\mathbf{A} - \lambda\mathbf{U})\ \mathbf{x}(k) = 0$$

thus if we take $\mathbf{x}(k-1)$ equal to $(\mathbf{A} - \lambda\mathbf{U})\ \mathbf{x}(k)$ then $\mathbf{x}(k-1)$ is obviously a generalized eigenvector of rank $k-1$ by definition.

We can form other generalized eigenvectors in this manner to obtain

$$\begin{aligned}
\mathbf{x}(k-1) &= (\mathbf{A} - \lambda\mathbf{U})\ \mathbf{x}(k) \\
\mathbf{x}(k-2) &= (\mathbf{A} - \lambda\mathbf{U})\ \mathbf{x}(k-1) \\
\mathbf{x}(k-3) &= (\mathbf{A} - \lambda\mathbf{U})\ \mathbf{x}(k-2) \\
&\ \ \vdots \\
\mathbf{x}(2) &= (\mathbf{A} - \lambda\mathbf{U})\ \mathbf{x}(3) \\
\mathbf{x}(1) &= (\mathbf{A} - \lambda\mathbf{U})\ \mathbf{x}(2) \\
0 &= (\mathbf{A} - \lambda\mathbf{U})\ \mathbf{x}(1)
\end{aligned} \tag{2-60}$$

Manipulating the above equations gives the relations

$$\begin{aligned}
\mathbf{Ax}(k) &= \lambda\, \mathbf{x}(k) + \mathbf{x}(k-1) \\
\mathbf{Ax}(k-1) &= \lambda\, \mathbf{x}(k-1) + \mathbf{x}(k-2) \\
&\ \ \vdots \\
\mathbf{Ax}(2) &= \lambda\, \mathbf{x}(2) + \mathbf{x}(1) \\
\mathbf{Ax}(1) &= \lambda\, \mathbf{x}(1)
\end{aligned} \tag{2-61}$$

If the matrix **A** happens to be of dimension $k \times k$, λ has multiplicity k, and there is one eigenvector of each rank up to rank k, then we can write the above set of equations as

$$\mathbf{AM} = \mathbf{M}\begin{bmatrix} \lambda & 1 & 0 & 0 & 0 & \cdot & \cdot & \cdot & 0 \\ 0 & \lambda & 1 & 0 & 0 & \cdot & \cdot & \cdot & 0 \\ 0 & 0 & \lambda & 1 & 0 & \cdot & \cdot & \cdot & 0 \\ \cdot & & & & & & & & \cdot \\ \cdot & & & & & & & & \cdot \\ \cdot & & & & & & & & \cdot \\ 0 & \cdot & \cdot & \cdot & & 0 & \lambda & 1 & 0 \\ 0 & \cdot & \cdot & \cdot & & 0 & 0 & \lambda & 1 \\ 0 & & & & & 0 & 0 & 0 & \lambda \end{bmatrix} = \mathbf{MA}^J \tag{2-62}$$

The matrix **M** is made up of the eigenvector and generalized eigenvectors corresponding to the eigenvalue λ, arranged such that

$$\mathbf{M} = [\mathbf{x}(1)\ \mathbf{x}(2)\ \mathbf{x}(3) \cdots \mathbf{x}(k)] \tag{2-63}$$

The near-diagonal matrix $\mathbf{A}^J$ is the Jordan canonical form of **A**. This matrix can be expressed as

$$\mathbf{A}^J = \mathbf{M}^{-1}\, \mathbf{AM} \tag{2-64}$$

If **A** is an $n \times n$ matrix, where n is larger than k, there will be other "chains" of generalized eigenvectors, each chain resulting from the successive application of Eq. 2-60. The matrix **M** will then be made up of all such chains and the matrix A^J may contain some zero elements in the superdiagonal rather than all ones.

The important point here is that the matrix **A** can be transformed to a form that can be easily manipulated. A time saving is often effected in certain problems by using $\mathbf{A}^J$ rather than **A**, as will be shown later. A few examples should clarify the preceding theory.

EXAMPLE 2-17. Find the Jordan canonical form of the matrix of Example 2-16.

Solution. The matrix **A** has the eigenvalue 2 with a multiplicity of 6. There are three eigenvectors and three generalized eigenvectors of rank 2, 3, and 4. The general forms of these vectors have been found in Example 2-16. We shall construct a chain of eigenvectors by starting with

$$\mathbf{x}(4) = \begin{bmatrix} x_1(4) \\ x_2(4) \\ x_3(4) \\ x_4(4) \\ x_5(4) \\ x_6(4) \end{bmatrix}$$

Applying Eq. 2-60 to $\mathbf{x}(4)$ gives

$$\mathbf{x}(3) = (\mathbf{A} - 2\mathbf{U})\,\mathbf{x}(4) = \begin{bmatrix} x_5(4) \\ x_3(4) + x_4(4) - x_6(4) \\ x_4(4) \\ 0 \\ x_3(4) - x_6(4) \\ 0 \end{bmatrix}$$

Continuing this process gives

$$\mathbf{x}(2) = \begin{bmatrix} x_3(4) - x_6(4) \\ x_4(4) \\ 0 \\ 0 \\ x_4(4) \\ 0 \end{bmatrix} \quad \text{and} \quad \mathbf{x}(1) = \begin{bmatrix} x_4(4) \\ 0 \\ 0 \\ 0 \\ 0 \\ 0 \end{bmatrix}$$

In addition to this chain consisting of three generalized eigenvectors and an eigenvector, there are two other eigenvectors which can be chosen as

$$\begin{bmatrix} 0 \\ x_2 \\ 0 \\ 0 \\ 0 \\ 0 \end{bmatrix} \quad \text{and} \quad \begin{bmatrix} 0 \\ 0 \\ x_3 \\ 0 \\ 0 \\ x_3 \end{bmatrix}$$

The matrix **M** can then be chosen to be (in this instance $x_1 = x_2 = x_3 = x_4 = x_5 = x_6 = 1$)

$$\begin{bmatrix} 1 & 0 & 1 & 1 & 0 & 0 \\ 0 & 1 & 1 & 1 & 1 & 0 \\ 0 & 0 & 1 & 1 & 0 & 1 \\ 0 & 0 & 0 & 1 & 0 & 0 \\ 0 & 1 & 0 & 1 & 0 & 0 \\ 0 & 0 & 0 & 1 & 0 & 1 \end{bmatrix}$$

In choosing the components, care must be taken to insure that none of the eigenvectors or generalized eigenvectors result in the zero vector and that all vectors are linearly independent. Taking the inverse of this matrix gives

$$\mathbf{M}^{-1} = \begin{bmatrix} 1 & 0 & -1 & -1 & 0 & 1 \\ 0 & 0 & 0 & -1 & 1 & 0 \\ 0 & 0 & 1 & 0 & 0 & -1 \\ 0 & 0 & 0 & 1 & 0 & 0 \\ 0 & 1 & -1 & 0 & -1 & 1 \\ 0 & 0 & 0 & -1 & 0 & 1 \end{bmatrix}$$

The matrix $\mathbf{A}^J$ is found to be

$$\mathbf{A}^J = \mathbf{M}^{-1}\,\mathbf{A}\mathbf{M} = \left[\begin{array}{cccc:cc} 2 & 1 & 0 & 0 & 0 & 0 \\ 0 & 2 & 1 & 0 & 0 & 0 \\ 0 & 0 & 2 & 1 & 0 & 0 \\ 0 & 0 & 0 & 2 & 0 & 0 \\ \hdashline 0 & 0 & 0 & 0 & 2 & 0 \\ 0 & 0 & 0 & 0 & 0 & 2 \end{array}\right]$$

The submatrix enclosed by dashed lines in the previous example is called a *Jordan block.* The chain of eigenvectors included in the transformation matrix **M** determines the arrangement of the Jordan block. The remaining first-order eigenvectors result in the diagonal part of the matrix $\mathbf{A}^J$. It is possible to have more than one Jordan block in $\mathbf{A}^J$. The next example demonstrates this point.

EXAMPLE 2-18. A 10×10 matrix is found to have three eigenvalues. The multiplicity of λ_0 is 6, the multiplicity of λ_1 is three, and the multiplicity of λ_2 is unity. Furthermore, it is known that three eigenvectors correspond to the eigenvalue λ_0 along with generalized eigenvectors of rank 2, 3, and 4. Two eigenvectors plus a generalized eigenvector of rank 2 correspond to λ_1. Write the Jordan canonical form of this matrix.

Solution. One chain can be formed from those eigenvectors corresponding to λ_0. This chain includes the three generalized eigenvectors and one eigenvector. This chain will result in a 4×4 Jordan block in $\mathbf{A}^J$. Another chain including the generalized eigenvector and one eigenvector corresponding to λ_1 will result in a 2×2 Jordan block. The overall form can be chosen as shown.

$$\mathbf{A}^J = \begin{bmatrix} \lambda_0 & 1 & 0 & 0 & 0 & 0 & 0 & 0 & 0 & 0 \\ 0 & \lambda_0 & 1 & 0 & 0 & 0 & 0 & 0 & 0 & 0 \\ 0 & 0 & \lambda_0 & 1 & 0 & 0 & 0 & 0 & 0 & 0 \\ 0 & 0 & 0 & \lambda_0 & 0 & 0 & 0 & 0 & 0 & 0 \\ 0 & 0 & 0 & 0 & \lambda_0 & 0 & 0 & 0 & 0 & 0 \\ 0 & 0 & 0 & 0 & 0 & \lambda_0 & 0 & 0 & 0 & 0 \\ 0 & 0 & 0 & 0 & 0 & 0 & \lambda_1 & 1 & 0 & 0 \\ 0 & 0 & 0 & 0 & 0 & 0 & 0 & \lambda_1 & 0 & 0 \\ 0 & 0 & 0 & 0 & 0 & 0 & 0 & 0 & \lambda_1 & 0 \\ 0 & 0 & 0 & 0 & 0 & 0 & 0 & 0 & 0 & \lambda_2 \end{bmatrix}$$

Except for the arrangement of the blocks, the form of the Jordan canonical matrix is unique [4]. There are methods of finding the matrix $\mathbf{A}^J$ without solving for the eigenvectors [2, 5] however these methods will not be considered here.

We should now consider the use of the Jordan canonical matrix. When a matrix has distinct eigenvalues we can write the set of equations

$$\begin{aligned} \mathbf{A}\mathbf{x}(\lambda_1) &= \lambda_1 \mathbf{x}(\lambda_1) \\ \mathbf{A}\mathbf{x}(\lambda_2) &= \lambda_2 \mathbf{x}(\lambda_2) \\ &\vdots \\ \mathbf{A}\mathbf{x}(\lambda_n) &= \lambda_n \mathbf{x}(\lambda_n) \end{aligned}$$

which can be expressed as the matrix equation

$$\mathbf{AP} = \mathbf{P}\Lambda \tag{2-46}$$

The diagonal matrix Λ can then be obtained from Eq. 2-46. Hopefully it has been demonstrated that this matrix is quite useful in decreasing the effort required to evaluate certain functions of the matrix $\mathbf{A}$. Likewise, the Jordan canonical matrix $\mathbf{A}^J$ can also lead to simplified mathematical operations in evaluating specific functions of $\mathbf{A}$ when multiple eigenvalues occur. Introducing the concept of generalized eigenvalues does nothing more for us than allow the equations

$$\mathbf{Ax}(k) = \lambda\, \mathbf{x}(k) + \mathbf{x}(k-1)$$
$$\mathbf{Ax}(k-1) = \lambda\, \mathbf{x}(k-1) + \mathbf{x}(k-2)$$
$$\vdots$$
$$\mathbf{Ax}(2) = \lambda\, \mathbf{x}(2) + \mathbf{x}(1)$$
$$\mathbf{Ax}(1) = \lambda\, \mathbf{x}(1)$$

to be written. While these equations are not quite as simple as those corresponding to the distinct eigenvalue case, they allow us to write Eq. 2-62:

$$\mathbf{AM} = \mathbf{MA}^J$$

The matrix $\mathbf{A}^J$ is not diagonal, but is near-diagonal and will lead to ease of manipulation.

We have shown that

$$\mathbf{F(A)} = \mathbf{P}\, \mathbf{F}(\Lambda)\mathbf{P}^{-1} \qquad [2\text{-}56]$$

Let us assume that we want to evaluate $\mathbf{e}^{\mathbf{A}t}$ for an $n \times n$ matrix $\mathbf{A}$ with distinct eigenvalues. We can express this function as

$$\mathbf{e}^{\mathbf{A}t} = \sum_{i=0}^{n-1} K_i\, \mathbf{A}^i$$

Using the remainder method we can evaluate the necessary coefficients K_0, $K_1, \cdots K_{n-1}$. We must now raise $\mathbf{A}$ to all powers up to $(n - 1)$ to apply the preceding formula. To avoid this necessity we could solve for the eigenvalues and eigenvectors of $\mathbf{A}$. If $\mathbf{A}$ has distinct eigenvalues then we can work with $\mathbf{F}(\Lambda)$ rather than $\mathbf{F(A)}$. We write

$$\mathbf{e}^{\Lambda t} = \sum_{i=0}^{n-1} K_i\, \Lambda^i$$

and evaluate the coefficients as previously. Evaluating all powers of Λ is now simplified considerably since we can now perform scalar exponentiation of the diagonal elements (eigenvalues) to obtain powers of Λ. Once $\mathbf{e}^{\Lambda t}$ is found we can apply Eq. 2-56 giving

$$\mathbf{e}^{\mathbf{A}t} = \mathbf{P}\, \mathbf{e}^{\Lambda t}\, \mathbf{P}^{-1}$$

This method of approach may lead to less computation, but the eigenvector problem must be first solved thus diminishing the savings in computational effort. However, since $\mathbf{e}^{\mathbf{A}t}$ is a commonly occurring function in electrical engineering it is worthwhile to pursue alternate methods of evaluation. If we express $\mathbf{e}^{\Lambda t}$ in terms of an infinite series with the aid of Eq. 2-42 we have

$$\mathbf{e}^{\Lambda t} = \sum_{n=0}^{\infty} \frac{1}{n!} \Lambda^n t^n \tag{2-65}$$

Collecting terms, we see that Eq. 2.65 reduces to

$$\mathbf{e}^{\Lambda t} = \begin{bmatrix} \sum_{n=0}^{\infty} \frac{1}{n!} \lambda_1^n t^n & 0 & \cdots & 0 \\ 0 & \sum_{n=0}^{\infty} \frac{1}{n!} \lambda_2^n t^n & \cdots & 0 \\ \vdots & & & \vdots \\ 0 & \cdots & 0 & \sum_{n=0}^{\infty} \frac{1}{n!} \lambda_n^n t^n \end{bmatrix}$$

$$= \begin{bmatrix} \mathbf{e}^{\lambda_1 t} & 0 & \cdots & 0 \\ 0 & e^{\lambda_2 t} & \cdots & 0 \\ \vdots & & & \vdots \\ 0 & \cdots & 0 & e^{\lambda_n t} \end{bmatrix} \tag{2-66}$$

The function $\mathbf{e}^{\Lambda t}$ can be expressed in terms of a diagonal matrix with scalar exponentials as the diagonal elements. Equation 2-66 can be written immediately once the eigenvalues have been found. Equation 2-56 can again be applied to give

$$\mathbf{e}^{\mathbf{A}t} = \mathbf{P}\mathbf{e}^{\Lambda t}\mathbf{P}^{-1}$$

avoiding the necessity of evaluating the coefficients in the finite series expansion and powers of Λ. For larger matrices this method can be quite effective.

We might investigate the existence of a similar scheme applicable to the case where multiple eigenvalues of $\mathbf{A}$ occur. It can be shown [4] when evaluating $\mathbf{e}^{\mathbf{A}^J t}$ (where $\mathbf{A}^J$ is assumed to contain ones along the superdiagonal) that

$$\mathbf{e}^{\mathbf{A}^J t} = \begin{bmatrix} e^{\lambda t} & te^{\lambda t} & t^2 e^{\lambda t}/2 & \cdot & \cdot & \cdot \\ 0 & e^{\lambda t} & te^{\lambda t} & \cdot & \cdot & \cdot \\ 0 & 0 & e^{\lambda t} & \cdot & \cdot & \cdot \\ \cdot & \cdot & \cdot & \cdot & \cdot & \cdot \\ \cdot & \cdot & \cdot & \cdot & \cdot & \cdot \end{bmatrix}$$

The exponential function of the Jordan canonical matrix can be written by inspection and upon applying the correct transformation we can find $\mathbf{e}^{\mathbf{A}t}$ to be

$$\mathbf{e}^{\mathbf{A}t} = \mathbf{M}\mathbf{e}^{\mathbf{A}^J t}\,\mathbf{M}^{-1}$$

While we have assumed that the matrix consists of one Jordan block, the same theory can be applied to a matrix consisting of several blocks [4].

Knowing Λ or $\mathbf{A}^J$ allows us to calculate $\mathbf{e}^{\mathbf{A}t}$ without resorting to series expansions provided the eigenvector problem is solved. While there may be additional uses of eigenvectors, we can conclude for our purposes that the definition of eigenvectors and generalized eigenvectors lead to matrix forms that are useful in calculating functions of the matrix **A**.

APPENDIX 2-1

The Cayley-Hamilton Theorem*

Let **A** be an $n \times n$ matrix and let $p(\lambda) = b_0 + b_1\lambda + b_2\lambda^2 + \cdots + (-1)^n\lambda^n$ be its characteristic polynomial. Then

$$b_0\,\mathbf{U} + b_1\,\mathbf{A} + \mathbf{b}_2\,\mathbf{A}^2 + \cdots + (-1)^n\,\mathbf{A}^n = 0$$

To prove this theorem it is necessary to develop some preliminary theory regarding polynomials with matrix coefficients. A matrix polynomial in λ is expressible as

$$f(\lambda) = \mathbf{A}_0 + \mathbf{A}_1\lambda + \mathbf{A}_2\lambda^2 + \cdots + \mathbf{A}_n\lambda^n \tag{A1-1}$$

where all $\mathbf{A}_n$ are $n \times n$ matrices. Given values for all $\mathbf{A}_n$ the polynomial can be written for example as

$$f(\lambda) = \begin{bmatrix} 1 & 1 \\ 0 & -2 \end{bmatrix} + \begin{bmatrix} 2 & 0 \\ -2 & 1 \end{bmatrix}\lambda + \begin{bmatrix} -1 & 1 \\ 3 & 2 \end{bmatrix}\lambda^2$$

$$= \begin{bmatrix} 1 + 2\lambda - \lambda^2 & 1 + \lambda^2 \\ -2\lambda + 3\lambda^2 & -2 + \lambda + 2\lambda^2 \end{bmatrix}$$

A polynomial with matrix coefficients is equivalent to a matrix with polynomial elements in λ. We can consider the result of substituting a matrix **C** for the

*This proof follows that outlined by M. J. Weiss in *Higher Algebra for the Undergraduate* (New York: John Wiley & Sons, Inc., 1949).

unknown scalar λ. This yields

$$f(\mathbf{C}) = \mathbf{A}_0 + \mathbf{A}_1\mathbf{C} + \mathbf{A}_2\mathbf{C}^2 + \cdots + \mathbf{A}_n\mathbf{C}^n \tag{A1-2}$$

Since matrices do not commute in general we can note that if the position of the coefficient matrix is interchanged with that of $\mathbf{C}$, $f(\mathbf{C})$ will have a different value. The function given by Eq. A1-2 is called the *right functional value*, while the second function is called a *left functional value.* The following discussion will treat right functional values, noting that the same results can be shown valid for the left functional case. The next theorem is used in proving the Cayley-Hamilton theorem.

Theorem: Let $f(\lambda)\cdot g(\lambda) = h(\lambda)$, where $f(\lambda)$, $g(\lambda)$, and $h(\lambda)$ are polynomials with matrix coefficients. Let $\mathbf{C}$ be a matrix such that $g(\mathbf{C}) = 0$. Then $h(\mathbf{C}) = 0$.

Proof: If

$$f(\lambda) = \mathbf{A}_0 + \mathbf{A}_1\lambda + \mathbf{A}_2\lambda^2 + \cdots + \mathbf{A}_n\lambda^n$$

and

$$g(\lambda) = \mathbf{B}_0 + \mathbf{B}_1\lambda + \mathbf{B}_2\lambda^2 + \cdots + \mathbf{B}_m\lambda^m$$

then

$$h(\mathbf{C}) = \mathbf{A}_0\mathbf{B}_0 + (\mathbf{A}_0\mathbf{B}_1 + \mathbf{A}_1\mathbf{B}_0)\mathbf{C} + (\mathbf{A}_0\mathbf{B}_2 + \mathbf{A}_1\mathbf{B}_1 + \mathbf{A}_2\mathbf{B}_0)\mathbf{C}^2 + \cdots + \mathbf{A}_n\mathbf{B}_m\mathbf{C}^{n+m}$$

Rearranging terms gives

$$\begin{aligned} h(\mathbf{C}) &= \mathbf{A}_0\,(\mathbf{B}_0 + \mathbf{B}_1\,\mathbf{C} + \mathbf{B}_2\mathbf{C}^2 + \cdots + \mathbf{B}_m\mathbf{C}^m) \\ &\quad + \mathbf{A}_1\,(\mathbf{B}_0 + \mathbf{B}_1\,\mathbf{C} + \mathbf{B}_2\mathbf{C}^2 + \cdots + \mathbf{B}_m\mathbf{C}^m)\,\mathbf{C} + \cdots \\ &\quad + \mathbf{A}_n\,(\mathbf{B}_0 + \mathbf{B}_1\,\mathbf{C} + \mathbf{B}_2\mathbf{C}^2 + \cdots + \mathbf{B}_m\mathbf{C}^m)\,\mathbf{C}^n \\ &= \mathbf{A}_0 g(\mathbf{C}) + \mathbf{A}_1 g(\mathbf{C})\,\mathbf{C} + \cdots + \mathbf{A}_n g(\mathbf{C})\,\mathbf{C}^n \end{aligned}$$

It is obvious that if $g(\mathbf{C}) = 0$, then $h(\mathbf{C}) = 0$.

A second theorem will simplify the proof of the Cayley-Hamilton theorem.

Theorem: $\mathbf{A}$(adjoint $\mathbf{A}$) = (adjoint $\mathbf{A}$)$\mathbf{A}$ = $|\mathbf{A}|\,\mathbf{U}$, where $|\mathbf{A}|$ is the determinant of $\mathbf{A}$.

Proof: The matrix (adjoint $\mathbf{A}$) is defined such that

$$\mathbf{A}^{-1} = \frac{(\text{adjoint } \mathbf{A})}{|\mathbf{A}|}$$

Multiplying by $\mathbf{A}$ gives

$$\mathbf{A}\,\mathbf{A}^{-1} = \mathbf{U} = \frac{\mathbf{A}\,(\text{adjoint } \mathbf{A})}{|\mathbf{A}|}$$

Thus $|\mathbf{A}|\mathbf{U} = \mathbf{A}$ (adjoint $\mathbf{A}$). The second relation given by the theorem is trivial to prove.

We are now ready to prove the Cayley-Hamilton theorem.

Proof: The determinant of $(\mathbf{A} - \lambda\mathbf{U})$ is

$$|\mathbf{A} - \lambda\mathbf{U}| = p(\lambda)$$

which is the characteristic polynomial. From the previous theorem we can write

$$[\text{adjoint } (\mathbf{A} - \lambda\mathbf{U})]\ (\mathbf{A} - \lambda\mathbf{U}) = p(\lambda)\,\mathbf{U}$$

The matrices $(\mathbf{A} - \lambda\mathbf{U})$ and [adjoint $(\mathbf{A} - \lambda\mathbf{U})$] are both matrices with polynomial elements which can also be written as a polynomial in λ with matrix coefficients. That is, the above equation could be written as

$$f(\lambda)g(\lambda) = h(\lambda)$$

Now if we substitute $\mathbf{A}$ for λ, the matrix $(\mathbf{A} - \lambda\mathbf{U})$ becomes $(\mathbf{A} - \mathbf{AU})$ which obviously equals zero. Since $g(\mathbf{A}) = (\mathbf{A} - \mathbf{AU}) = 0$, then from the first of the two preceding theorems $h(\mathbf{A}) = p(\mathbf{A})\,\mathbf{U} = p(\mathbf{A}) = 0$. Thus it is seen that the matrix $\mathbf{A}$ satisfies its own characteristic equation or

$$p(\mathbf{A}) = b_0\mathbf{U} + b_1\mathbf{A} + b_2\mathbf{A}^2 + \cdots + (-1)^n\,\mathbf{A}^n = 0$$

BIBLIOGRAPHY

1. Branin, F. H., Jr., "Computer Methods of Network Analysis," *Proceedings of the Institute of Electrical and Electronics Engineers*, Vol. 55, No. 11 (November 1967).
2. De Russo, P. M., R. J. Roy, and C. M. Close, *State Variables for Engineers*, John Wiley & Sons, Inc., New York, 1965.
3. Finkbeiner, D. T., II, *Introduction to Matrices and Linear Transformations*, 2d ed., W. H. Freeman and Company, San Francisco, Calif., 1966.
4. Friedman, B., *Principles and Techniques of Applied Mathematics*, John Wiley & Sons, Inc., New York, 1956.
5. Gupta, S. C., *Transform and State Variable Methods in Linear Systems*, John Wiley & Sons, Inc., New York, 1966.
6. Weiss, M. J., *Higher Algebra for the Undergraduate*, John Wiley & Sons, Inc., New York, 1949.
7. Zadeh, L. A., and C. A. Desoer, *Linear System Theory*, McGraw-Hill Book Company, New York, 1963.

PROBLEMS

2-1 Determine if the set of equations are linearly independent.

$$x_1 + 2x_2 - 4x_3 = 0$$
$$3x_1 - 4x_2 + 10x_3 = 0$$
$$2x_1 - x_2 + 3x_3 = 0$$

2-2 Determine if the set of equations are linearly independent.

$$2x_1 - x_2 + 2x_3 = 0$$
$$3x_1 + x_2 - x_3 = 0$$
$$x_1 + x_2 + x_3 = 0$$

2-3 Solve for x_1, x_2, x_3, and x_4.

$$x_1 + x_2 + x_3 + x_4 = 0$$
$$x_1 + 2x_2 + x_3 - x_4 = 0$$
$$2x_1 + x_2 - x_3 + 2x_4 = 0$$
$$3x_1 - 2x_2 + 4x_3 - x_4 = 0$$

2-4 Solve for as many variables as possible in terms of the remaining variables.

$$2x_1 + x_2 + x_3 - x_4 = 0$$
$$x_1 + 2x_2 - x_3 - 2x_4 = 0$$
$$-x_1 - x_2 - 6x_3 + x_4 = 0$$
$$2x_1 + 2x_2 - 6x_3 - 2x_4 = 0$$

2-5 If $x_4 = 3$ in Prob. 2-4, find x_1, x_2, and x_3.

2-6 Solve for x_1, x_2, x_3, and x_4 if possible. If not, solve for as many lower components as possible in terms of remaining components.

$$x_1 + x_2 + x_3 + x_4 = 5$$
$$2x_1 - x_2 - 2x_3 + x_4 = 11$$
$$5x_1 - x_2 - 3x_3 + 3x_4 = 27$$
$$x_1 - 2x_2 - 3x_3 = 6$$

2-7 If $\mathbf{x} = \mathbf{e}_1 + 2\mathbf{e}_2 + 3\mathbf{e}_3$ and $\mathbf{y} = -2\mathbf{e}_1 + 3\mathbf{e}_2 + 3\mathbf{e}_3$ calculate
(a) the unit vector in the $\mathbf{x}$ direction
(b) the unit vector in the $\mathbf{y}$ direction
(c) the scalar product $\langle \mathbf{x}, \mathbf{y} \rangle$

2-8 Using the vectors in Prob. 2-7, calculate
(a) the projection of $\mathbf{x}$ on $\mathbf{y}$
(b) the projection of $\mathbf{y}$ on $\mathbf{x}$

2-9 If $\mathbf{x} = 3\mathbf{e}_1 - \mathbf{e}_2 + 2\mathbf{e}_3$, $\mathbf{y} = +\mathbf{e}_1 + 2\mathbf{e}_2 + \mathbf{e}_3$, and $\mathbf{z} = 2\mathbf{e}_1 + \mathbf{e}_2 + 3\mathbf{e}_3$ calculate
(a) the projection of $\mathbf{x}$ on $\mathbf{z}$
(b) the projection of $\mathbf{x}$ on $\mathbf{y}$

2-10 Using the vectors in Prob. 2-9, calculate
(a) the projection of $\mathbf{y}$ on $\mathbf{x}$
(b) the projection of $\mathbf{y}$ on $\mathbf{z}$

2-11 Using the vectors in Prob. 2-9, calculate
(a) the projection of $\mathbf{z}$ on $\mathbf{x}$
(b) the projection of $\mathbf{z}$ on $\mathbf{y}$

2-12 Using the vectors in Prob. 2-9, calculate an orthonormal basis using the direction of $\mathbf{x}$ as the direction of one of the new basis vectors.

2-13 Repeat Prob. 2-12 using the direction of $\mathbf{y}$ as the direction of one of the new basis vectors.

2-14 Repeat Prob. 2-12 using the direction of $\mathbf{z}$ as the direction of one of the new basis vectors.

2-15 If $\omega = \mathbf{e}_1 + \mathbf{e}_2 + \mathbf{e}_3 + \mathbf{e}_4$, $\mathbf{x} = 2\mathbf{e}_1 + \mathbf{e}_2 - \mathbf{e}_3 + 2\mathbf{e}_4$, $\mathbf{y} = 3\mathbf{e}_1 - \mathbf{e}_2 + 2\mathbf{e}_2 + 5\mathbf{e}_4$, and $\mathbf{z} = -2\mathbf{e}_1 - 2\mathbf{e}_2 + 3\mathbf{e}_2 - \mathbf{e}_4$ calculate an orthonormal basis using the direction of ω as the direction of one of the basis vectors.

2-16 Repeat Prob. 2-15 using the direction of $\mathbf{x}$ as the direction of one of the basis vectors.

2-17 Repeat Prob. 2-15 using the direction of $\mathbf{y}$ as the direction of one of the basis vectors.

2-18 Repeat Prob. 2-15 using the direction of $\mathbf{z}$ as the direction of one of the basis vectors.

2-19 If

$$\mathbf{x}_1 = \begin{bmatrix} 1 \\ 0 \\ 2 \end{bmatrix}, \quad \mathbf{x}_2 = \begin{bmatrix} 1 \\ 1 \\ 0 \end{bmatrix}, \quad \mathbf{x}_3 = \begin{bmatrix} 0 \\ 0 \\ 3 \end{bmatrix},$$

$$\mathbf{y}_1 = \begin{bmatrix} 1 \\ 1 \\ 1 \end{bmatrix}, \quad \mathbf{y}_2 = \begin{bmatrix} 0 \\ 4 \\ 2 \end{bmatrix}, \quad \mathbf{y}_3 = \begin{bmatrix} 2 \\ 0 \\ 2 \end{bmatrix}$$

(expressed in terms of the orthonormal vector basis $\mathbf{e}_1$, $\mathbf{e}_2$, and $\mathbf{e}_3$) and the vector $\mathbf{z}$ is

$$\mathbf{z} = \begin{bmatrix} 1 \\ 2 \\ -1 \end{bmatrix}$$

expressed in terms of the $\mathbf{x}$ basis, find the expression for $\mathbf{z}$ in terms of the $\mathbf{y}$ basis.

2-20 If the vector $\mathbf{z}$ is equal to $\begin{bmatrix} 2 \\ 1 \\ 3 \end{bmatrix}$ expressed in terms of the $\mathbf{y}$ basis, express $\mathbf{z}$ in terms of the $\mathbf{x}$ basis using the $\mathbf{x}$ and $\mathbf{y}$ vectors given in Prob. 2-19.

2-21 Given that

$$\mathbf{x}_1 = \begin{bmatrix} 2 \\ 2 \\ -1 \end{bmatrix}, \quad \mathbf{x}_2 = \begin{bmatrix} 1 \\ 0 \\ 4 \end{bmatrix}, \quad \mathbf{x}_3 = \begin{bmatrix} 1 \\ -1 \\ 0 \end{bmatrix}$$

construct an orthonormal basis with the direction of $\mathbf{x}_1$ used as the direction of one of the basis vectors. If

$$\mathbf{z} = \begin{bmatrix} 2 \\ -1 \\ -2 \end{bmatrix}$$

in terms of the original three **x** basis vectors, express **z** in terms of the new orthonormalized basis.

2-22 Using the **x** basis vectors of Prob. 2-21, find a suitable **y** basis such that if

$$\mathbf{z}^x = \begin{bmatrix} 2 \\ -1 \\ 2 \end{bmatrix}$$

then

$$\mathbf{z}^y = \begin{bmatrix} 1 \\ 1 \\ 1 \end{bmatrix}$$

2-23 Assume that a matrix **A** is expressed in terms of its operation on a vector having the **x** basis given in Prob. 2-19 as

$$\mathbf{A} = \begin{bmatrix} 3 & 2 & 1 \\ 0 & 4 & 2 \\ 1 & 0 & -3 \end{bmatrix}$$

Find the matrix **B** that expresses the same information when premultiplying a vector having the **y** basis of Prob. 2-19.

2-24 Repeat Prob. 2-23 if

$$\mathbf{A} = \begin{bmatrix} -2 & -1 & 4 \\ 3 & 6 & -2 \\ 1 & 1 & 7 \end{bmatrix}$$

2-25 Find the eigenvalues and eigenvectors corresponding to

$$\mathbf{A} = \begin{bmatrix} -4 & 2 \\ 3 & -5 \end{bmatrix}$$

2-26 (a) Develop a subroutine to find the real, distinct eigenvalues of a matrix. Assume that the maximum size of the matrix is 6×6.

(b) Develop a subroutine to evaluate a set of normalized eigenvectors given the eigenvalues as input. Normalize the eigenvectors to unit length.

2-27 Using the programs developed in Prob. 2-26 find the real eigenvalues and the eigenvectors of

$$\mathbf{A} = \begin{bmatrix} -4 & 3 & 2 \\ 0 & -1 & 2 \\ 1 & 0 & -6 \end{bmatrix}$$

2-28 With the programs developed in Prob. 2-26, find the eigenvalues and eigenvectors corresponding to

$$\mathbf{A} = \begin{bmatrix} 2 & 1 & 0 \\ -3 & -1 & 2 \\ 1 & 0 & -3 \end{bmatrix}$$

2-29 If

$$\mathbf{A} = \begin{bmatrix} 3 & 2 \\ 1 & 2 \end{bmatrix}$$

find

$$\mathbf{A}^6 + 4\mathbf{A}^5 + 2\mathbf{A}^4 + \mathbf{A}^3 + \mathbf{A}^2 + 6\mathbf{A}$$

2-30 Using the matrix **A** given in Prob. 2-29, evaluate

$$\mathbf{B} = 10 \sin \mathbf{A}t$$

(a) by a series expansion in **A**
(b) by evaluating $e^{j\Lambda t}$ and applying Euler's equation

2-31 Using the matrix **A** given in Prob. 2-29, evaluate

$$\mathbf{B} = 5 \cos \mathbf{A}t$$

(a) by a series expansion in **A**
(b) by evaluating $e^{j\Lambda t}$ and applying Euler's equation

2-32 Using the matrix **A** given in Prob. 2-29, evaluate

$$\mathbf{B} = e^{\mathbf{A}t}$$

2-33 Evaluate $\sin \mathbf{A}t$ where

$$\mathbf{A} = \begin{bmatrix} 5 & -1 & 1 & 1 & 0 & 0 \\ 1 & 3 & -1 & -1 & 0 & 0 \\ 0 & 0 & 4 & 0 & 1 & 1 \\ 0 & 0 & 0 & 4 & -1 & -1 \\ 0 & 0 & 0 & 0 & 3 & 1 \\ 0 & 0 & 0 & 0 & 1 & 3 \end{bmatrix}$$

2-34 Evaluate $\cos \mathbf{A}t$ using the **A** matrix of Prob. 2-33.

2-35 Find the eigenvalues and a matrix **P** made up of 2 eigenvectors corresponding to

$$\mathbf{A} = \begin{bmatrix} -5 & 2 \\ 2 & -8 \end{bmatrix}$$

Express **A** in diagonal form. Evaluate $e^{\Lambda t}$ and then transform using **P** to find $e^{\mathbf{A}t}$.

2-36 Repeat Prob. 2-35 if

$$\mathbf{A} = \begin{bmatrix} 4.5 & -4.0 \\ 11.0 & -9.0 \end{bmatrix}$$

2-37 Find the eigenvalues and generalized eigenvectors corresponding to

$$\mathbf{A} = \begin{bmatrix} 1 & 1 & 2 \\ 0 & 1 & 3 \\ 0 & 0 & 2 \end{bmatrix}$$

Express the matrix **A** in Jordan canonical form.

2-38 Find the eigenvalues and generalized eigenvectors corresponding to

$$\mathbf{A} = \begin{bmatrix} 0 & 0 & 0 & 0 & 1 & 0 \\ 0 & 0 & 1 & 1 & 0 & -1 \\ 0 & 0 & 0 & 0 & 0 & 0 \\ 0 & 0 & 0 & 0 & 0 & 0 \\ 0 & 0 & 1 & 0 & 0 & -1 \\ 0 & 0 & 0 & 0 & 0 & 0 \end{bmatrix}$$

Express the matrix **A** in Jordan canonical form.

2-39 Write a Fortran subroutine to evaluate the real or complex eigenvalues (distinct) of a matrix. Assume that the maximum matrix dimension is 6×6. Use the program to calculate the eigenvalues of

$$\mathbf{A} = \begin{bmatrix} -0.5 & -1.0 & -1.0 & -0.5 & -1.0 \\ 1.0 & 0 & 0 & 0 & 0 \\ 0 & 1.0 & 0 & 0 & 0 \\ 0 & 0 & 1.0 & 0 & 0 \\ 0 & 0 & 0 & 1.0 & 0 \end{bmatrix}$$

2-40 Use a Fortran subroutine to calculate the eigenvalues of

$$\mathbf{A} = \begin{bmatrix} -2 & 1 & 1 & 1 \\ -7 & -5 & -2 & -4 \\ 0 & -1 & -3 & -2 \\ -1 & 0 & -1 & 0 \end{bmatrix}$$

chapter **3**

NETWORK AND GRAPH THEORY

An engineering student once remarked that he had an instructor who could introduce the simplest concept and then add so much theory relating to this concept that no one could understand the original point. Unnecessary theory can often obscure a major concept. A useful approach in engineering courses is to start with the specific and proceed to the general. Specific examples or specific methods are discussed before developing the general methods based on complex theory. General applications resulting from a given theory are often less important than a specific application. For example, we apply Kirchhoff's voltage and current laws to individual loops or nodes to write equations which can be solved simultaneously. We could write a matrix equation, following rules prescribed by theory, that expresses all of the Kirchhoff voltage or current conditions in a single equation. For simple problems there is usually more work involved in arriving at the matrix equation than in writing equations for individual loops and nodes. There are cases however, when it is advantageous to use the matrix-equation approach. In the solution of complex networks this approach is useful, since the "bookkeeping" chores associated with circuit analysis are handled by matrices automatically. In computer-aided circuit design the computer will accept data in matrix form and can easily manipulate matrix equations. This chapter will develop the elementary graph theory necessary in many computer-aided design applications. The application of graph theory allows the computer to formulate equations for analysis based on a minimal amount of information regarding the network.

The computer can be programmed to utilize information related to circuit topology such as the nodes to which each element is connected. Key topological matrices are then generated by the computer, allowing efficient formulation of the analysis equations of interest. This chapter not only discusses topology but considers methods of formulating and solving loop or node equations by computer.

A. ELEMENTARY TOPOLOGY

Before developing some important topological concepts, it is necessary to precisely define a number of new terms.

Graph. The structure or configuration of a network is given by the graph of the network. A graph is constructed by replacing each element of the network by a line segment. There are cases, to be discussed later, where a line segment might replace a voltage source and/or current source and/or one or more passive elements.

Node. The junction of two or more line segments is a node.

Branch. Any line segment of a graph connecting two nodes is a branch. A branch may replace a single element or the combination of several elements of the network.

Loop. A set of branches that form a closed path.

Tree. A tree is a subgraph of the original graph. To qualify as a tree, the subgraph must contain all nodes of the graph, form a connected subgraph, and yet contain no loops.

Tree branch. Any branch of the particular tree under consideration is a tree branch.

Link branch. Any branch of the graph not belonging to the particular tree under consideration is a link branch.

Figure 3-1 shows examples of the preceding definitions. In order that any

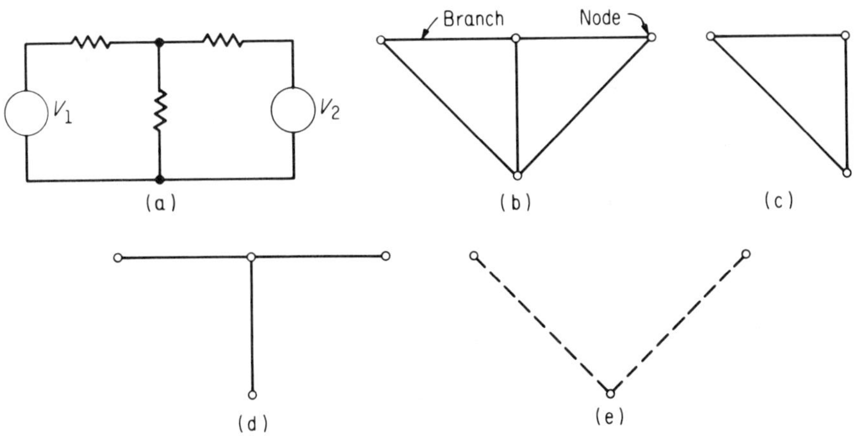

FIG. 3-1. (a) Electrical network. (b) Graph of network. (c) Loop. (d) Tree. (e) Link branches corresponding to tree shown in (d).

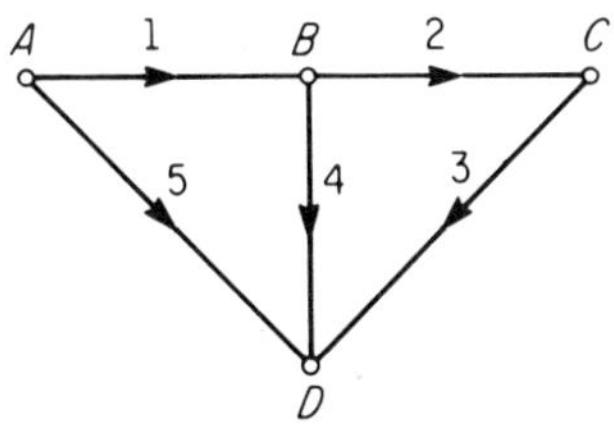

FIG. 3-2. Oriented graph.

particular node or branch can be identified, each node and branch is numbered. The graph can also have arbitrary directions assigned to the branches, and thus becomes an oriented graph like that of Fig. 3-2.

It should be noted that many different trees exist for a given graph. Figure 3-3 demonstrates several trees corresponding to the graph of Fig. 3-2. Each

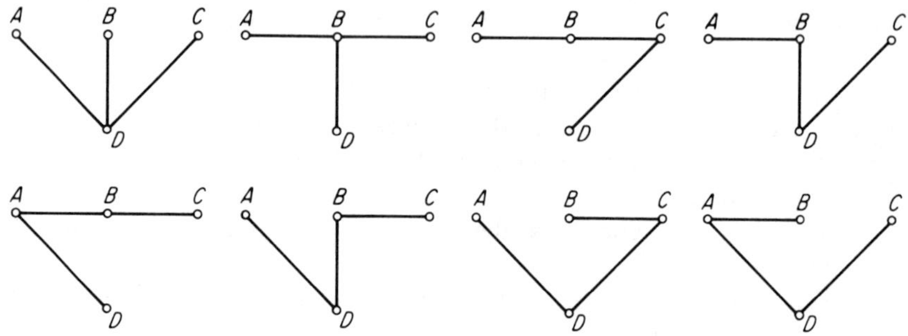

FIG. 3-3. Several trees of the same graph.

tree contains $N - 1$ branches where N is the number of nodes. In certain problems it is desirable to find all the trees of a network. Obviously this is a difficult problem for a network of appreciable size. There are computer programs available that will generate several possible trees given the graph of the network.

Cutset. A set of a minimum number of branches that can be cut by an imaginary surface to divide the graph into two separate parts is called a *cutset*. The addition to the subgraphs of any branch of this cutset will create a connected graph. Some cutsets of the graph of Fig. 3-2 are shown in Fig. 3-4.

Branches 1 and 5 make up cutset A, and branches 2, 4 and 5 make up cutset E.

Note that cutset E is not an independent cutset. All branches included in cutset E have already been included in other cutsets. The branch voltage is defined such that the voltage at the node from which the current is assumed to

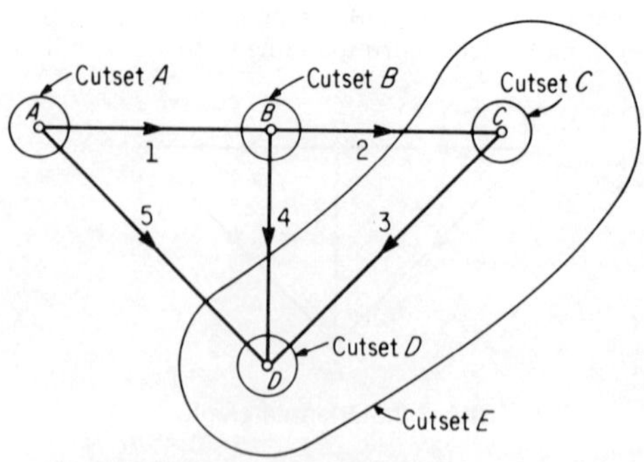

FIG. 3-4. Examples of cutsets.

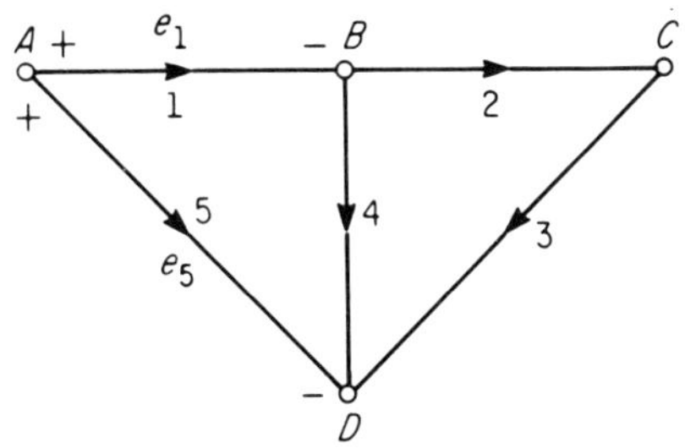

FIG. 3-5. Graph demonstrating current and voltage convention.

flow is more positive than the node voltage of the other end of the branch. Figure 3-5 demonstrates this convention.

There are three topological matrices of special interest in network problems [1]. The first of these is the incidence matrix **A**. This matrix can be obtained in one of two ways. The matrix **A** tells which branch is connected to each node and the direction of the branch relative to that node. The information relating to whichever node is selected as the reference node is deleted. As in most topological matrices, it is convenient to order the information such that the matrix can be partitioned into two submatrices, one dealing with tree branches and the other with link branches. Consider the graph of Fig. 3-2 along with a tree of this graph shown in Fig. 3-6.

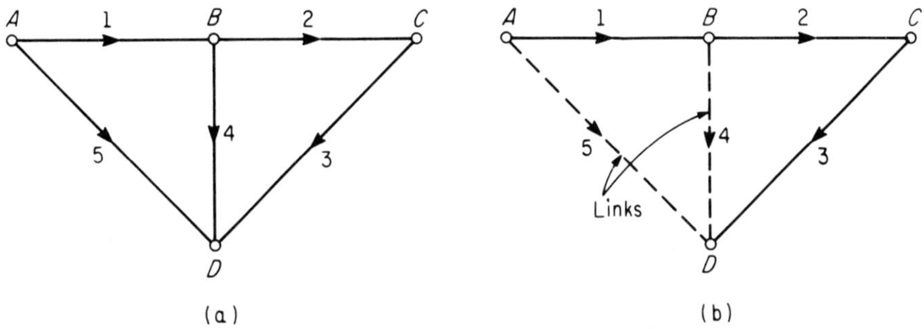

FIG. 3-6. (a) Graph of Fig. 3-2. (b) Tree of this graph.

If we select D as the reference node, the matrix **A** can be formed by identifying the nodes with rows of the matrix and branches with columns as follows. The signs of the nonzero elements of this matrix are determined by corresponding branch directions. A branch directed away from a node results in a positive sign.

Nodes	1	2	3	4	5	Branch
A	+1	0	0	0	+1	
B	−1	+1	0	+1	0	
C	0	−1	+1	0	0	

Note that the tree branch information appears as a distinct submatrix as a result of numbering the tree branches consecutively. Thus

$$\mathbf{A} = \begin{array}{c} \quad \mathbf{A}_T \qquad\quad \mathbf{A}_L \\ \left[\begin{array}{ccc:cc} 1 & 0 & 0 & 0 & 1 \\ -1 & 1 & 0 & 1 & 0 \\ 0 & -1 & 1 & 0 & 0 \end{array}\right] \end{array}$$

where $\mathbf{A}_T$ is the tree submatrix and $\mathbf{A}_L$ is the link submatrix. The **A** matrix can also be formed by including rows corresponding to all nodes and then deleting the row corresponding to the reference node. Since there are $N - 1$ tree branches giving $N - 1$ columns in $\mathbf{A}_T$ this submatrix will always be square. Kirchhoff's current law can be expressed in terms of **A** as

$$\mathbf{Ai} = 0 \tag{3-1}$$

where **i** is a column vector of branch currents.

A second matrix of interest is the branch-mesh, tieset, or circuit matrix **B**. A tieset consists of all branches in the loop formed by adding one link to a tree. Therefore, any tieset will include only one link. The number of the tieset by convention is the same as the number of the link that creates the loop and the loop direction corresponds to the link direction. The rows of **B** are identified with the tiesets and the columns indicate whether a given branch is included in the tieset. The two tiesets of the previous graphs are shown in Fig. 3-7. If a

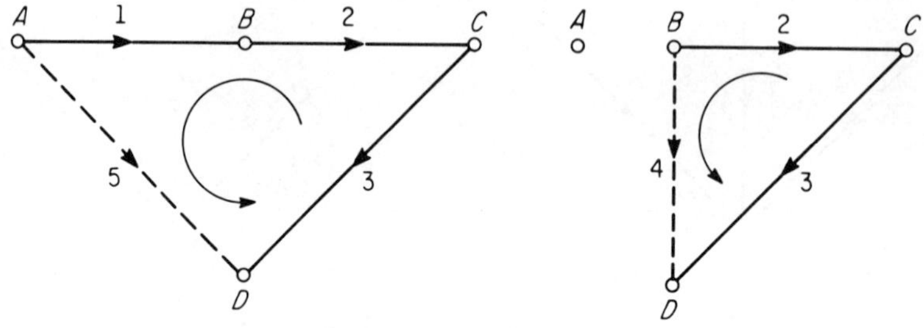

FIG. 3-7. Tiesets of a graph.

given branch direction corresponds to that of the loop current, a positive element results. A negative element appears in **B** when a branch direction opposes the loop-current direction.

The matrix **B** is formed as follows:

Tieset	1	2	3	4	5	Branch
4	0	-1	-1	1	0	
5	-1	-1	-1	0	1	

Thus

$$\mathbf{B} = \begin{array}{c} \qquad \mathbf{B}_T \qquad\qquad\quad \mathbf{B}_L \\ \left[\begin{array}{ccc:cc} 0 & -1 & -1 & 1 & 0 \\ -1 & -1 & -1 & 0 & 1 \end{array}\right] \end{array}$$

where $\mathbf{B}_T$ is the submatrix associated with the tree branches and $\mathbf{B}_L$ is associated with the links. It is obvious from the definition that $\mathbf{B}_L$ will be the identity matrix of dimension equal to the number of links in the graph.

Kirchhoff's voltage law can be expressed as

$$\mathbf{Be} = 0 \tag{3-2}$$

where **e** is the column vector of branch voltages.

The cutset matrix **C** is the third useful topological matrix. Each row of **C** corresponds to a cutset while the columns indicate which branches are included in that cutset. In order to insure that we have the correct number of independent equations, each cutset should include a single distinct tree branch. By convention, the cutsets are numbered to correspond to the number of the tree branch they include and such that this branch has a positive direction referenced

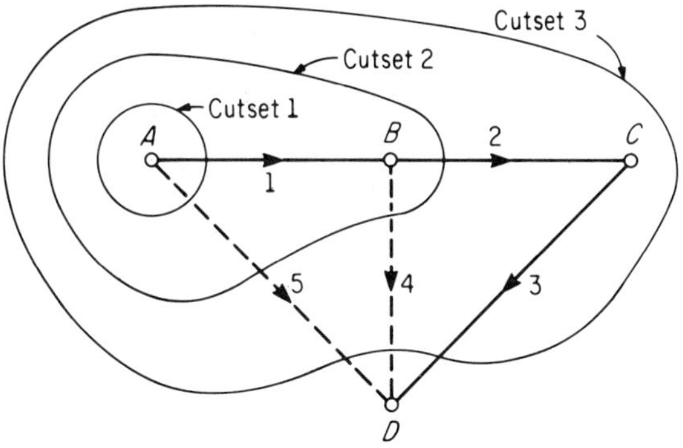

FIG. 3-8. Three cutsets of a graph.

to the cutset (away from the cutset). Figure 3-8 shows 3 cutsets used to obtain the matrix **C**.

The matrix **C** is formed as follows:

Cutset	1	2	3	4	5	Branch
1	1	0	0	0	1	
2	0	1	0	1	1	
3	0	0	1	1	1	

Again the matrix is partitioned into one submatrix related to tree branches and one related to links. The **C** matrix is

$$\mathbf{C} = \begin{array}{c} \begin{matrix} \mathbf{C}_T & & \mathbf{C}_L \end{matrix} \\ \left[\begin{array}{ccc:cc} 1 & 0 & 0 & 0 & 1 \\ 0 & 1 & 0 & 1 & 1 \\ 0 & 0 & 1 & 1 & 1 \end{array}\right] \end{array}$$

This matrix is related to the **A** matrix which was used to express the fact that the sum of the branch currents at any node must equal zero. The **C** matrix can be used to equate the total current entering a given portion of the network (defined by the cutset) to zero. Thus

$$\mathbf{Ci} = 0 \tag{3-3}$$

where **i** is the vector of branch currents. Note that since a cutset divides the graph into separate parts, the total current flow into either part must equal zero. The submatrix $\mathbf{C}_T$ will always be the unit matrix of dimension $N - 1$.

Each of the three matrices developed in the preceding paragraphs can be obtained by following specified rules. This means that a computer could generate these matrices based on a minimum amount of information regarding the graph structure.

Some basic relations between the matrices can be developed quite easily. For example, the **A** matrix is so constructed that the branches connected to each node are displayed. On the other hand, we can transpose this matrix so that each row corresponds to a branch and the entries in this row show which nodes the branch spans. Since polarity information is also indicated, the matrix $\mathbf{A}^t$ relates the branch voltage to the node voltage. Thus

$$\mathbf{A}^t = \begin{bmatrix} 1 & -1 & 0 \\ 0 & 1 & -1 \\ 0 & 0 & 1 \\ 0 & 1 & 0 \\ 1 & 0 & 0 \end{bmatrix}$$

and the equation

$$\mathbf{e} = \mathbf{A}^t\,\mathbf{v} \tag{3-4}$$

represents the set of equations

$$e_1 = v_A - v_B$$
$$e_2 = v_B - v_C$$

$$e_3 = v_C$$
$$e_4 = v_B$$
$$e_5 = v_A$$

The vector **v** is the column vector of node voltages.

If we combine Eqs. 3-2 and 3-4, we find that

$$\mathbf{Be} = \mathbf{B}\,\mathbf{A}^t\,\mathbf{v} = 0$$

This result shows that

$$\mathbf{B}\,\mathbf{A}^t = 0 \tag{3-5}$$

since the components of **v** are linearly independent.

We can proceed in a similar manner to derive a relationship between the matrices **B** and **C**. The matrix **B** indicates which branches occur in a given loop defined by a tieset. If **B** is transposed the rows correspond to branches and the columns indicate which loops contain a given branch. In the original graph the current through a given branch can be written as the algebraic sum of all loop currents through this branch. The transpose of **B** then relates the loop current to the branch current as

$$\mathbf{i} = \mathbf{B}^t\,\mathbf{i}_L \tag{3-6}$$

where $\mathbf{i}_L$ is the column vector of loop or tieset currents.

Combining Eqs. 3-3 and 3-6 results in

$$\mathbf{Ci} = \mathbf{C}\,\mathbf{B}^t\,\mathbf{i}_L = 0$$

Thus

$$\mathbf{C}\,\mathbf{B}^t = 0 \tag{3-7}$$

Equations 3-5 and 3-7 allow us to establish some useful results regarding the calculation of the matrices **B** and **C** from a knowledge of **A**. If we express Eq. 3-5 in terms of submatrices, we have

$$[\mathbf{B}_T\ \ \mathbf{B}_L]\begin{bmatrix}\mathbf{A}_T^t\\ \mathbf{A}_L^t\end{bmatrix} = \mathbf{B}_T\,\mathbf{A}_T^t + \mathbf{B}_L\,\mathbf{A}_L^t = 0 \tag{3-8}$$

Since $\mathbf{B}_L$ is the unit matrix we can write

$$\mathbf{B}_T = -\,\mathbf{A}_L^t\,[\mathbf{A}_T^t]^{-1} \tag{3-9}$$

The matrix **B** can then be determined from **A**.

Expressing Eq. 3-7 in terms of submatrices gives

$$[\mathbf{C}_T\ \ \mathbf{C}_L]\begin{bmatrix}\mathbf{B}_T^t\\ \mathbf{B}_L^t\end{bmatrix} = \mathbf{C}_T\,\mathbf{B}_T^t + \mathbf{C}_L\,\mathbf{B}_L^t = 0$$

Using the knowledge that $\mathbf{C}_T$ and $\mathbf{B}_L^t$ are unit matrices we can find $\mathbf{C}_L$ to be

$$\mathbf{C}_L = -\mathbf{B}_T^t \tag{3-10}$$

Equation 3-9 allows $\mathbf{C}_L$ to be written as

$$\mathbf{C}_L = \mathbf{A}_T^{-1}\mathbf{A}_L \tag{3-11}$$

We now note that

$$\mathbf{A}_T^{-1}\mathbf{A} = \mathbf{A}_T^{-1}[\mathbf{A}_T \ \mathbf{A}_L] = [\mathbf{U} \ \mathbf{A}_T^{-1}\mathbf{A}_L]$$

where **U** is the unit matrix. The unit matrix is designated **U** to avoid confusion with the source-current matrix **I** appearing later in this chapter. Since

$$\mathbf{C} = [\mathbf{C}_T \ \mathbf{C}_L] = [\mathbf{U} \ \mathbf{C}_L]$$

the previous two equations can be combined to give the result

$$\mathbf{C} = \mathbf{A}_T^{-1}\mathbf{A} \tag{3-12}$$

An example will demonstrate the application of several of the preceding equations.

EXAMPLE 3-1. Determine the **A**, **B**, and **C** matrices for the graph of Fig. 3-9 using the tree identified by solid lines.

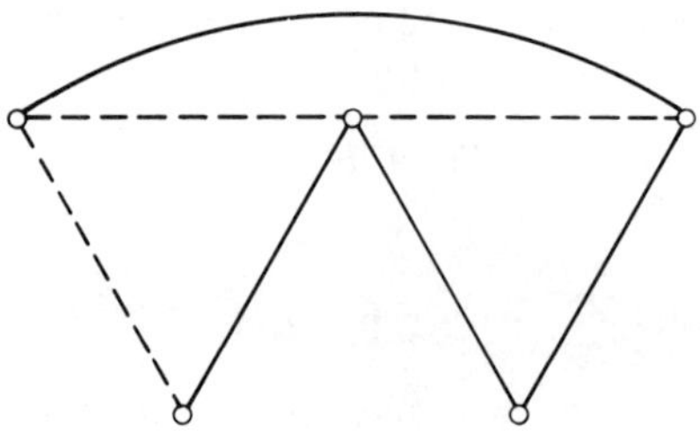

FIG. 3-9. Network graph for Example 3-1.

Solution. The graph is first oriented and nodes and branches are labeled. It is convenient to number the tree branches consecutively as shown in Fig. 3-10.

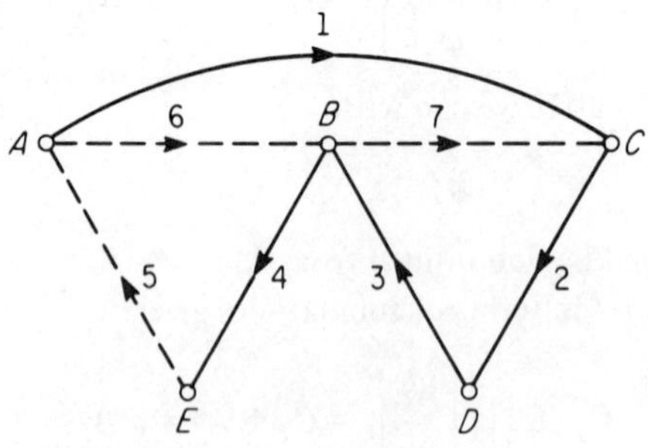

FIG. 3-10. Oriented graph of Fig. 3-9.

The **A** matrix is formed with node E chosen as the reference node. This gives

Node	1	2	3	4	5	6	7	Branch
A	1	0	0	0	−1	1	0	
B	0	0	−1	1	0	−1	1	
C	−1	1	0	0	0	0	−1	
D	0	−1	1	0	0	0	0	

or

$$\mathbf{A} = \left[\begin{array}{cccc:ccc} 1 & 0 & 0 & 0 & -1 & 1 & 0 \\ 0 & 0 & -1 & 1 & 0 & -1 & 1 \\ -1 & 1 & 0 & 0 & 0 & 0 & -1 \\ 0 & -1 & 1 & 0 & 0 & 0 & 0 \end{array}\right] = [\mathbf{A}_T \ \ \mathbf{A}_L]$$

Using Eq. 3-9, the submatrix $\mathbf{B}_T$ can be found. Since

$$\mathbf{A}_L^t = \begin{bmatrix} -1 & 0 & 0 & 0 \\ 1 & -1 & 0 & 0 \\ 0 & 1 & -1 & 0 \end{bmatrix}$$

and

$$[\mathbf{A}_T^t]^{-1} = \begin{bmatrix} 1 & 1 & 1 & 1 \\ 0 & 0 & 0 & 1 \\ 0 & 1 & 1 & 1 \\ 0 & 0 & 1 & 1 \end{bmatrix}$$

then

$$\mathbf{B}_T = \begin{bmatrix} 1 & 0 & 0 & 0 \\ -1 & 1 & 0 & 0 \\ 0 & -1 & 1 & 0 \end{bmatrix} \begin{bmatrix} 1 & 1 & 1 & 1 \\ 0 & 0 & 0 & 1 \\ 0 & 1 & 1 & 1 \\ 0 & 0 & 1 & 1 \end{bmatrix}$$

$$= \begin{bmatrix} 1 & 1 & 1 & 1 \\ -1 & -1 & -1 & 0 \\ 0 & 1 & 1 & 0 \end{bmatrix}$$

Now $\mathbf{B}_L$ will be an identity matrix with dimension equal to the number of links of the graph, which is three in this case. Thus

$$\mathbf{B} = [\mathbf{B}_T \ \ \mathbf{B}_L] = \left[\begin{array}{cccc:ccc} 1 & 1 & 1 & 1 & 1 & 0 & 0 \\ -1 & -1 & -1 & 0 & 0 & 1 & 0 \\ 0 & 1 & 1 & 0 & 0 & 0 & 1 \end{array}\right]$$

Equation 3-12 will allow **C** to be calculated as

$$\mathbf{C} = \mathbf{A}_T^{-1}\,\mathbf{A} = \begin{bmatrix} 1 & 0 & 0 & 0 \\ 1 & 0 & 1 & 0 \\ 1 & 0 & 1 & 1 \\ 1 & 1 & 1 & 1 \end{bmatrix} \begin{bmatrix} 1 & 0 & 0 & 0 & -1 & 1 & 0 \\ 0 & 0 & -1 & 1 & 0 & -1 & 1 \\ -1 & 1 & 0 & 0 & 0 & 0 & -1 \\ 0 & -1 & 1 & 0 & 0 & 0 & 0 \end{bmatrix}$$

$$= \begin{bmatrix} 1 & 0 & 0 & 0 & -1 & 1 & 0 \\ 0 & 1 & 0 & 0 & -1 & 1 & -1 \\ 0 & 0 & 1 & 0 & -1 & 1 & -1 \\ 0 & 0 & 0 & 1 & -1 & 0 & 0 \end{bmatrix}$$

The matrices **B** and **C** can be checked by constructing them from the methods previously outlined.

Thus far we have discussed only the structure of networks and various methods of describing this structure. The elements of each branch are important in determining the response of the network of course, and the method of relating the elements to the structure must be considered. One useful method is to represent the elements of each branch in a general way, allowing for almost all possible element configurations. Consider the ith branch of a network [1] shown in Fig. 3-11.

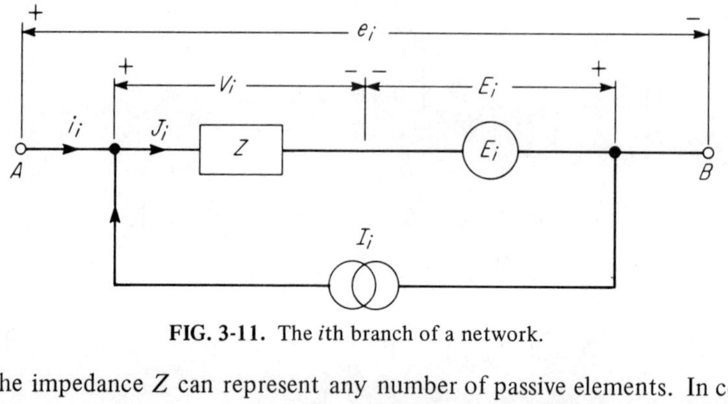

FIG. 3-11. The ith branch of a network.

The impedance Z can represent any number of passive elements. In certain cases it may be desirable to express the passive elements in terms of an admittance Y. The source E_i allows the possibility of a distinct voltage source in all branches. Similarly, I_i can represent a current source in any or all branches. Both of these sources are independent. We will shortly see that dependent sources are treated as transimpedances or transadmittances.

The current through Z is equal to the sum of i_i and I_i, thus

$$J_i = I_i + i_i \tag{3-13}$$

The element voltage V_i is given by

$$V_i = E_i + e_i \tag{3-14}$$

We then have a vector of branch current **[i]**, a vector of source current **[I]**, and a vector of element current **[J]**. Similarly, we have a vector of branch voltage **[e]**, a vector of source voltage **[E]**, and a vector of element voltage **[V]**. For a passive network, Ohm's law for each element can be expressed as

$$V_i = ZJ_i$$

or as

$$J_i = YV_i$$

Consider the passive circuit excited by three independent voltage sources and an independent current source of Fig. 3-12. For the analysis methods to be

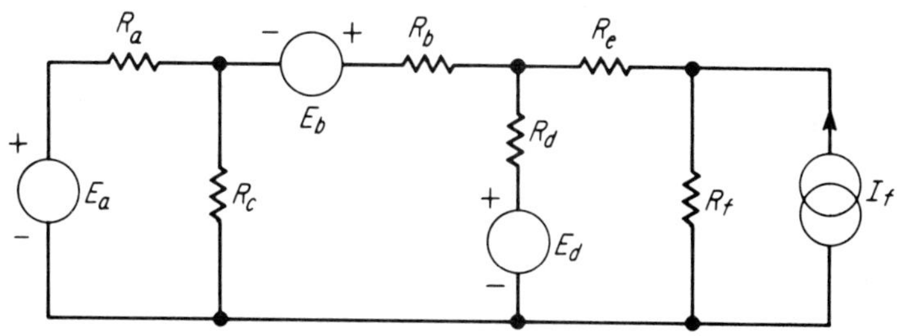

FIG. 3-12. Passive circuit excited by independent sources.

considered in this chapter, the standard branch must contain a finite impedance or admittance, even though either or both sources can be zero. In cases where it is necessary to include an open circuit in a branch, a small admittance (large resistance) can often be added to the network with little effect on the circuit performance. The branches for the network of Fig. 3-12 can be chosen such that finite impedances are included in each branch as shown in Fig. 3-13.

Using the standard branches as defined in Fig. 3-13, the impedance matrix is

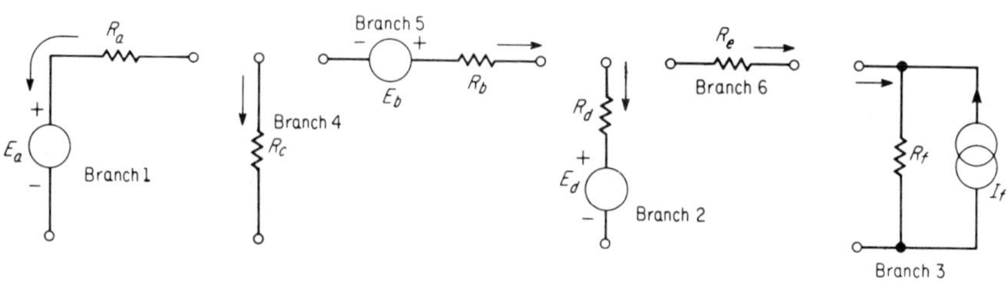

FIG. 3-13. Standard branches.

$$\mathbf{Z} = \begin{bmatrix} R_a & 0 & 0 & 0 & 0 & 0 \\ 0 & R_d & 0 & 0 & 0 & 0 \\ 0 & 0 & R_f & 0 & 0 & 0 \\ 0 & 0 & 0 & R_c & 0 & 0 \\ 0 & 0 & 0 & 0 & R_b & 0 \\ 0 & 0 & 0 & 0 & 0 & R_e \end{bmatrix}$$

The source voltage vector for this case (branch directions indicated by the arrows) is

$$\mathbf{E} = \begin{bmatrix} -E_a \\ -E_d \\ 0 \\ 0 \\ E_b \\ 0 \end{bmatrix}$$

while the current source vector is

$$\mathbf{I} = \begin{bmatrix} 0 \\ 0 \\ I_f \\ 0 \\ 0 \\ 0 \end{bmatrix}$$

In order to handle dependent sources as will be encountered in active circuits or mutual impedances, we must make Ohm's law more general. The circuit of Fig. 3-14 demonstrates the problem arising in active circuits when the above equations are applied. The dependent current source is proportional to the element voltage V_2 which in this case equals the branch voltage e_2. This

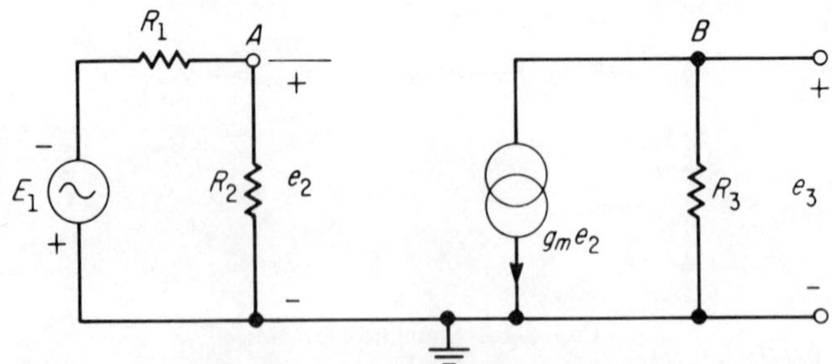

FIG. 3-14. Active network.

circuit is redrawn in Fig. 3-15 to show the correspondence to the standard branches as previously defined.

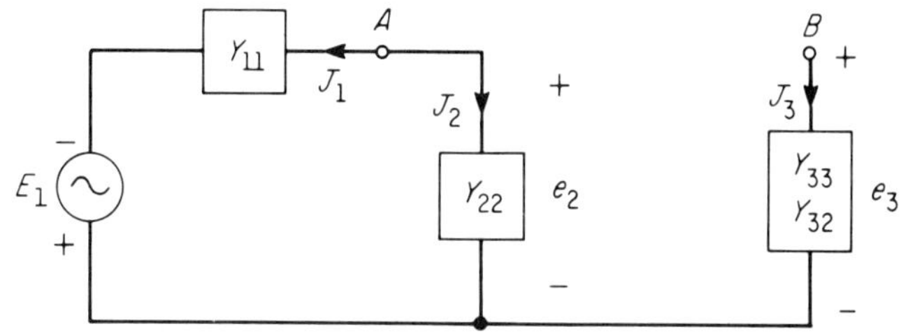

FIG. 3-15. Network of Fig. 3-14 showing three branches.

Note that the dependent source is included as an admittance rather than a source. The effects of dependent sources are accounted for by lumping the transadmittance or transimpedance with the self-admittance or self-impedance and treating the combination as a single element. The three equations for the element currents are

$$J_1 = Y_{11}(E_1 + e_1) = Y_{11} V_1$$
$$J_2 = Y_{22} e_2 = Y_{22} V_2$$
$$J_3 = g_m e_2 + Y_{33} e_3 = g_m V_2 + Y_{33} V_3$$

The current J_3 is not equal to $Y_{33} V_3$, since the transadmittance must be included. In matrix form, the **J** vector can be written as

$$\mathbf{J} = \begin{bmatrix} Y_{11} & 0 & 0 \\ 0 & Y_{22} & 0 \\ 0 & Y_{32} & Y_{33} \end{bmatrix} \mathbf{V}$$

where $Y_{32} = g_m$. The transadmittance has given rise to the off-diagonal term Y_{32}.

Ohm's law for any network is then expressed as

$$\mathbf{J} = \mathbf{Y}\mathbf{V} \tag{3-15}$$

or

$$\mathbf{V} = \mathbf{Z}\mathbf{J} \tag{3-16}$$

Diagonal terms in **Y** or **Z** represent self-admittances and self-impedances while an off-diagonal term Y_{ij} or Z_{ij} represents transadmittance or transimpedance. A mutual admittance gives rise to two off-diagonal terms such that

$$Y_{ij} = Y_{ji}$$

The use of dependent sources to represent behavior of active devices can lead to slight problems on occasion. For example, if the current or voltage generated by the source is proportional to a branch variable rather than an element variable, the equation for **J** or **V** must be modified. Consider the case depicted by Fig. 3-16. The current generated by the dependent source is pro-

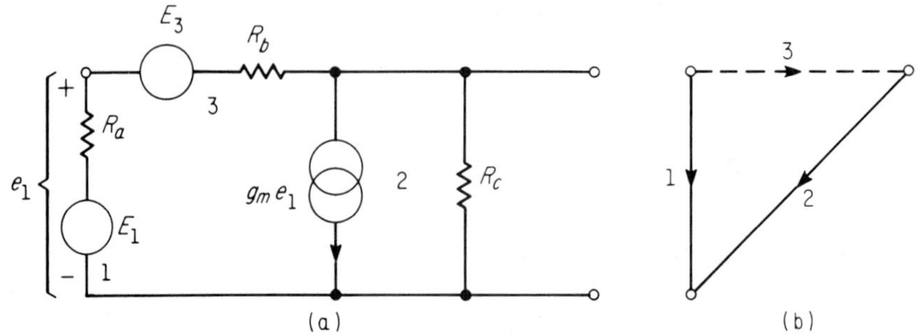

FIG. 3-16. Current source dependent on branch voltage. (a) Actual circuit. (b) Oriented graph.

portional to the branch voltage e_1, which cannot be expressed in terms of element voltages. The element current equations are

$$J_1 = V_1/R_a$$
$$J_2 = g_m e_1 + V_2/R_c$$
$$J_3 = V_3/R_b$$

Since J_2 depends on e_1, we cannot express this effect simply by means of the element admittance matrix. However, we can reconsider the equation that relates element voltages and currents from a different viewpoint to account for this branch variable dependence. The original equation for element current is

$$\mathbf{J} = \mathbf{Y}\,\mathbf{V} \qquad [3\text{-}15]$$

Applying Eq. 3-14 gives

$$\mathbf{J} = \mathbf{Y}\,\mathbf{E} + \mathbf{Y}\,\mathbf{e} \qquad (3\text{-}17)$$

Equation 3-15 is valid when dependent sources are controlled by element voltages, but for the network of Fig. 3-16 this equation does not apply. For this type of network we can write

$$\mathbf{J} = \mathbf{Y}\,\mathbf{V} + \mathbf{Y}'\,\mathbf{e} = \mathbf{Y}\mathbf{E} + (\mathbf{Y} + \mathbf{Y}')\,\mathbf{e}$$
$$= \mathbf{Y}_S\,\mathbf{E} + \mathbf{Y}_B\,\mathbf{e} \qquad (3\text{-}18)$$

The matrix $\mathbf{Y}_S$ accounts for all currents proportional to source voltages while $\mathbf{Y}_B$ accounts for currents proportional to branch voltages. Under normal circumstances $\mathbf{Y}_S$ and $\mathbf{Y}_B$ will both equal **Y**, but if a current source is controlled by a branch voltage then $\mathbf{Y}_S = \mathbf{Y}$ and $\mathbf{Y}_B$ is equal to the sum of **Y** and $\mathbf{Y}'$. It is obvious that a current source controlled by a source voltage would lead to a matrix $\mathbf{Y}_S$ that differs from $\mathbf{Y}$ also. The matrix $\mathbf{Y}'$ accounts for all current sources that are dependent upon branch voltages.

For the network of Fig. 3-16 the admittance matrices are

$$\mathbf{Y}_S = \mathbf{Y} = \begin{bmatrix} 1/R_a & 0 & 0 \\ 0 & 1/R_c & 0 \\ 0 & 0 & 1/R_b \end{bmatrix}$$

and

$$\mathbf{Y}_B = \mathbf{Y} + \mathbf{Y}' = \begin{bmatrix} 1/R_a & 0 & 0 \\ 0 & 1/R_c & 0 \\ 0 & 0 & 1/R_b \end{bmatrix} + \begin{bmatrix} 0 & 0 & 0 \\ g_m & 0 & 0 \\ 0 & 0 & 0 \end{bmatrix}$$

$$= \begin{bmatrix} 1/R_a & 0 & 0 \\ g_m & 1/R_c & 0 \\ 0 & 0 & 1/R_b \end{bmatrix}$$

If the current source in Fig. 3-16 were controlled by the source voltage E_1—that is, $i = g_m E_1$—the element current equations would be

$$J_1 = V_1/R_a$$
$$J_2 = g_m E_1 + V_2/R_c$$
$$J_3 = V_3/R_b$$

The admittance matrices would then be

$$\mathbf{Y}_S = \mathbf{Y} + \mathbf{Y}' = \begin{bmatrix} 1/R_a & 0 & 0 \\ 0 & 1/R_c & 0 \\ 0 & 0 & 1/R_b \end{bmatrix} + \begin{bmatrix} 0 & 0 & 0 \\ g_m & 0 & 0 \\ 0 & 0 & 0 \end{bmatrix}$$

$$= \begin{bmatrix} 1/R_a & 0 & 0 \\ g_m & 1/R_c & 0 \\ 0 & 0 & 1/R_b \end{bmatrix}$$

and $\mathbf{Y}_B = \mathbf{Y}$. Corresponding modifications can be made for the case of transimpedances.

B. SOLUTION OF THE NETWORK PROBLEM [1]

Network-analysis problems are largely of a single type. We are given the information to determine the matrices **A**, **B**, and **C**, and we can determine **Y** or **Z** from given element values. We either know or assume the excitation and this fixes the source vectors **E** and **I**. The problem is to determine branch or loop currents or node or branch voltages. Regardless of the unknown vector, the approach proceeds by applying Ohm's and Kirchhoff's law in a specific way to arrive at a solution for the desired vector.

Tieset Method

This approach is used to find the loop-current vector or the branch current vector. Equation 3-16

$$\mathbf{V} = \mathbf{ZJ} \tag{3-16}$$

is the starting point for this method. Using the vector expressions for Eqs. 3-13 and 3-14 in Eq. 3-16 gives

$$\mathbf{E} + \mathbf{e} = \mathbf{ZI} + \mathbf{Zi} \tag{3-19}$$

This equation can be written as

$$[\mathbf{E} - \mathbf{ZI}] + \mathbf{e} = \mathbf{Zi}$$

We now multiply both sides of the equation by the matrix $\mathbf{B}$ and note from Eq. 3-2 that $\mathbf{Be} = 0$ to get

$$\mathbf{BE} - \mathbf{BZI} = \mathbf{BZi} \tag{3-20}$$

One might be tempted to multiply both sides of Eq. 3-20 by $\mathbf{B}^{-1}$, but since $\mathbf{B}$ is a nonsquare matrix it has no inverse. If we apply Eq. 3-6 we obtain

$$\mathbf{B}[\mathbf{E} - \mathbf{ZI}] = \mathbf{BZB}^t\, \mathbf{i}_L \tag{3-21}$$

The product of $\mathbf{B}\,\mathbf{Z}\,\mathbf{B}^t$ is a square matrix of dimension equal to the number of link branches and possesses an inverse. The vector $\mathbf{i}_L$ is then expressible in terms of known quantities as

$$\mathbf{i}_L = [\mathbf{BZB}^t]^{-1}\, \mathbf{B}[\mathbf{E} - \mathbf{ZI}] \tag{3-22}$$

The branch current is then

$$\mathbf{i} = \mathbf{B}^t\, \mathbf{i}_L = \mathbf{B}^t [\mathbf{BZB}^t]^{-1}\, \mathbf{B}[\mathbf{E} - \mathbf{ZI}] \tag{3-23}$$

The voltage $\mathbf{e}$ is found from Eq. 3-19 to be

$$\mathbf{e} = [\mathbf{ZB}^t\, [\mathbf{BZB}^t]^{-1}\, \mathbf{B} - \mathbf{U}]\; [\mathbf{E} - \mathbf{ZI}] \tag{3-24}$$

where $\mathbf{U}$ is the unit or identity matrix of proper dimension. An example will demonstrate the application of these equations.

EXAMPLE 3-2. Find the midband gain of the circuit in Fig. 3-17a assuming that the biasing resistor can be neglected and that the equivalent circuit of Fig. 3-17b applies. Calculate the impedance seen by the input source.

Solution. We will first find the branch current through R_c and then note that e_{out} equals the product of this current and R_c. The graph of this network is shown in Fig. 3-18.

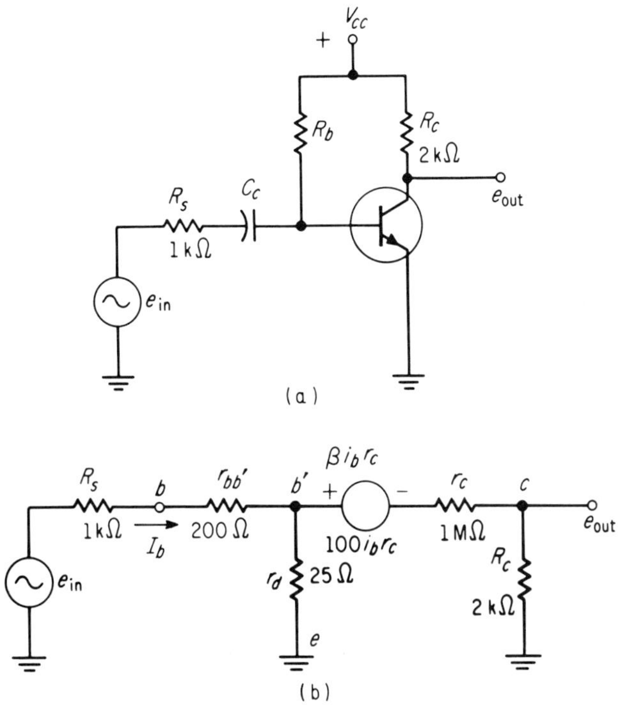

FIG. 3-17. (a) Actual amplifier circuit. (b) Equivalent circuit.

Branch 1 consists of the input source and R_s, branch 2 consists of r_d, branch 3 consists of R_c, branch 4 consists of r_{bb}' and branch 5 contains the dependent source and r_c. The tieset matrix **B** is found as

$$\mathbf{B} = \begin{bmatrix} 1 & 1 & 0 & 1 & 0 \\ 0 & -1 & 1 & 0 & 1 \end{bmatrix}$$

The source vector $\bar{\mathbf{E}}$ is

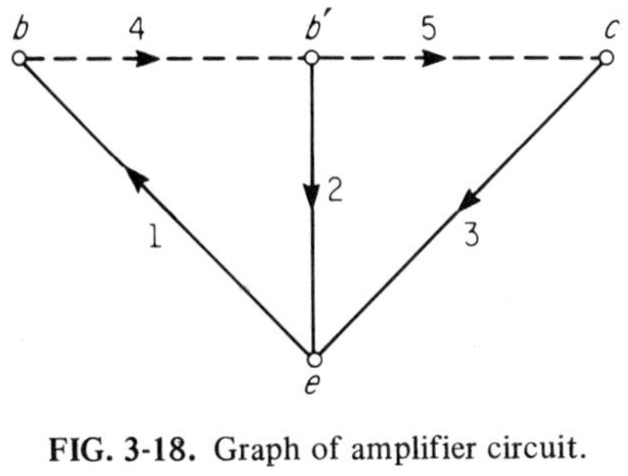

FIG. 3-18. Graph of amplifier circuit.

$$\mathbf{E} = \begin{bmatrix} e_{\text{in}} \\ 0 \\ 0 \\ 0 \\ 0 \end{bmatrix}$$

The impedance matrix is

$$\mathbf{Z} = \begin{bmatrix} R_s & 0 & 0 & 0 & 0 \\ 0 & r_d & 0 & 0 & 0 \\ 0 & 0 & R_c & 0 & 0 \\ 0 & 0 & 0 & r_{bb'} & 0 \\ 0 & 0 & 0 & \beta r_c & r_c \end{bmatrix}$$

where the off-diagonal term βr_c accounts for the dependent source. Note that this source must represent a transimpedance to allow $\mathbf{Z}$ to include the effect. If a dependent current source appears, conversion to a voltage source is required to use an impedance representation.

The matrix $\mathbf{BZB}^t$ is found to be

$$\begin{bmatrix} R_s + r_d + r_{bb'} & -r_d \\ \beta r_c - r_d & r_d + R_c + r_c \end{bmatrix}$$

The inverse of this matrix is

$$[\mathbf{BZB}^t]^{-1} = \frac{1}{D} \begin{bmatrix} r_d + R_c + r_c & r_d \\ r_d - \beta r_c & R_s + r_d + r_{bb'} \end{bmatrix}$$

where

$$D = (R_s + r_{bb'})(r_c + R_c + r_d) + r_d([\beta + 1]\, r_c + R_c)$$

Applying Eq. 3-22 gives

$$\mathbf{i}_L = \frac{1}{D} \begin{bmatrix} e_{\text{in}}(r_d + R_c + r_c) \\ e_{\text{in}}(r_d - \beta r_c) \end{bmatrix}$$

and the branch currents are

$$\mathbf{i} = \mathbf{B}^t\, \mathbf{i}_L = \frac{e_{\text{in}}}{D} \begin{bmatrix} r_d + R_c + r_c \\ R_c + (\beta + 1) r_c \\ r_d - \beta r_c \\ r_d + R_c + r_c \\ r_d - \beta r_c \end{bmatrix}$$

The voltage gain equals $i_3 R_c/e_{in}$ and is

$$A_v = \frac{(r_d - \beta r_c) R_c}{(R_s + r_{bb'})(r_c + R_c + r_d) + r_d([\beta + 1] r_c + R_c)}$$

The impedance seen by the source equals e_{in}/i_1 and is

$$R_{in} = \frac{(R_s + r_{bb'})(r_c + R_c + r_d) + r_d([\beta + 1] r_c + R_c)}{(r_c + R_c + r_d)}$$

Substituting the given values into these equations gives $A_v = -53.7$ and $R_{in} =$ 3.72 kΩ.

It should be noted that this problem could be simplified by combining R_s and $r_{bb'}$ as a single resistance and likewise R_c and r_c. In this case only three branches are present. While the resulting method is no longer a general one, some savings of computing time can result.

Node Method

The starting point for this method is the expression of Ohm's law in the admittance form,

$$\mathbf{J} = \mathbf{YV} \qquad [3\text{-}15]$$

Substituting the vector expressions for **J** and **V** gives

$$\mathbf{I} + \mathbf{i} = \mathbf{YE} + \mathbf{Ye} \qquad (3\text{-}25)$$

This equation can be written

$$\mathbf{I} - \mathbf{YE} + \mathbf{i} = \mathbf{Ye}$$

We now multiply both sides of the equation by the matrix **A**, noting that $\mathbf{Ai} = 0$, to get

$$\mathbf{A}[\mathbf{I} - \mathbf{YE}] = \mathbf{AYe} \qquad (3\text{-}26)$$

Using the fact that $\mathbf{e} = \mathbf{A}^t \mathbf{v}$ allows us to write Eq. 3-26 as

$$\mathbf{A}[\mathbf{I} - \mathbf{YE}] = \mathbf{AYA}^t \mathbf{v} \qquad (3\text{-}27)$$

and solve for the node-voltage vector **v** as

$$\mathbf{v} = [\mathbf{AYA}^t]^{-1} \mathbf{A}[\mathbf{I} - \mathbf{YE}] \qquad (3\text{-}28)$$

We note that Eq. 3-26 cannot be solved for **e** directly, since the product **AY** will not be square in general, and thus possesses no inverse. The product $\mathbf{AYA}^t$ will always be square and will possess an inverse allowing the solution of Eq. 3-28. This method of solution is useful when transadmittances are present in the circuit to be analyzed. Example 3-3 will demonstrate the preceding method.

EXAMPLE 3-3. Evaluate the output voltage for the circuit of Fig. 3-19 if the input signal is a sinusoid. Assume proper bias.

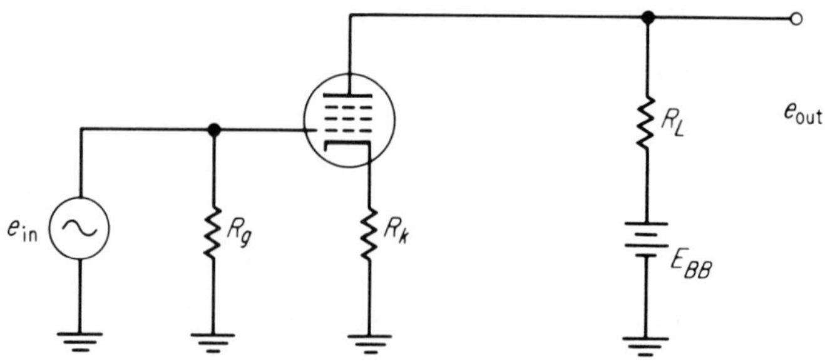

FIG. 3-19. Pentode amplifier.

Solution. The equivalent circuit to be used for this analysis is shown in Fig. 3-20.

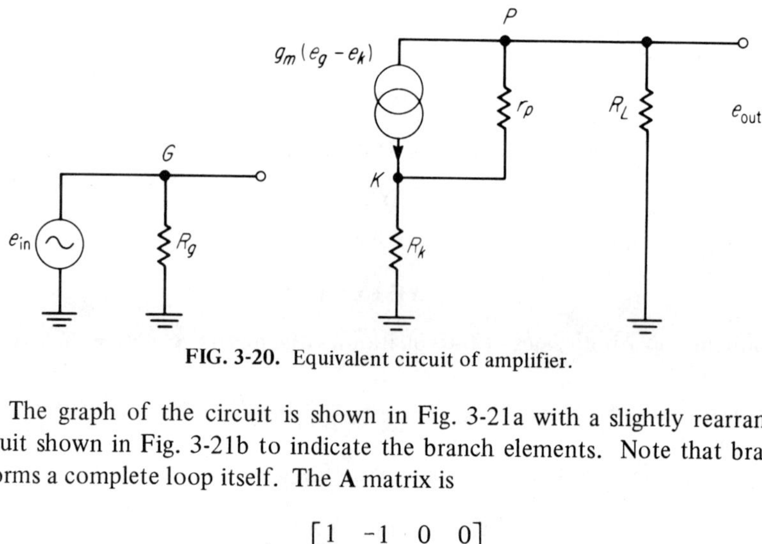

FIG. 3-20. Equivalent circuit of amplifier.

The graph of the circuit is shown in Fig. 3-21a with a slightly rearranged circuit shown in Fig. 3-21b to indicate the branch elements. Note that branch 4 forms a complete loop itself. The **A** matrix is

$$\mathbf{A} = \begin{bmatrix} 1 & -1 & 0 & 0 \\ 0 & 1 & 1 & 0 \end{bmatrix}$$

The admittance matrix is

$$\mathbf{Y} = \begin{bmatrix} 1/R_k & 0 & 0 & 0 \\ -g_m & 1/r_p & 0 & g_m \\ 0 & 0 & 1/R_L & 0 \\ 0 & 0 & 0 & 1/R_g \end{bmatrix}$$

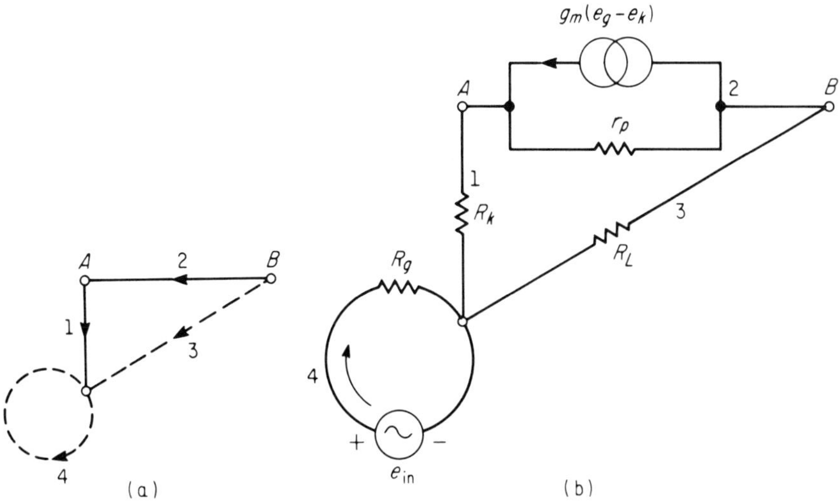

FIG. 3-21. (a) Graph of circuit. (b) Original circuit showing four branches.

The inverse of the product $\mathbf{AYA}^t$ is

$$N\begin{bmatrix}(1/r_p + 1/R_L) & 1/r_p \\ (g_m + 1/r_p) & (1/R_k + g_m + 1/r_p)\end{bmatrix}$$

where $N = R_k r_p R_L/(R_L + r_p + R_k + g_m r_p R_k)$. For this example the current source vector is zero and the voltage source vector is

$$\mathbf{E} = \begin{bmatrix}0 \\ 0 \\ 0 \\ e_{\text{in}}\end{bmatrix}$$

Equation 3-28 can now be applied to give

$$\mathbf{v} = N\begin{bmatrix}g_m\, e_{\text{in}}/R_L \\ -g_m\, e_{\text{in}}/R_k\end{bmatrix}$$

The output voltage of the amplifier corresponds to $\mathbf{v}_B$ and is

$$e_{\text{out}} = \frac{-g_m r_p R_L e_{\text{in}}}{R_L + r_p + R_k + g_m r_p R_k}$$

Often a grid resistance will not appear in the circuit, since the grid is tied to d-c ground through the voltage source. When this occurs a very large value for R_g can be assumed (e.g., $R_g = 10\ \text{M}\Omega$) in order to force the grid circuit to correspond to the standard branch.

The node method is applied in ECAP to calculate d-c and a-c quantities. It will be useful to consider a possible program to analyze simple circuits with this method. From Eq. 3-28 we see that, given the input data, the computer must form the matrices **A**, **Y**, **I**, and **E**. Manipulation of these matrices as indicated by Eq. 3-28 will result in the node-voltage vector **v**. A flow chart of the necessary steps is shown in Fig. 3-22.

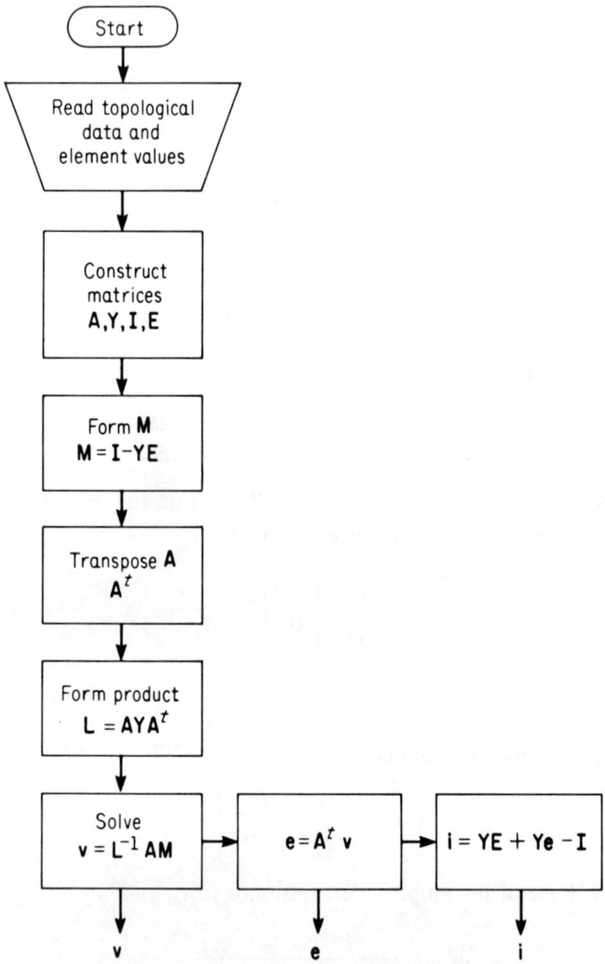

FIG. 3-22. Flow chart for node method.

A simple program to implement this flow chart with FORTRAN II is shown in Table 3-1 along with the input data required to analyze the circuit of Fig. 3-19. The parameter values are $R_g = 1\ \text{M}\Omega$, $R_k = 100\ \Omega$, $R_L = 10\ k\Omega$, $r_p = 500\ k\Omega$, $g_m = 8000\ \mu$mhos, and $e_{in} = 0.2$ V (peak to peak). The output voltages calculated

TABLE 3-1. Program Utilizing Node Method.

```
      DIMENSION A(10,10),B(10,1),C(10),D(10),E(10),F(10),R(10,10)
      DIMENSION T(10,10),Y(10,10),G(10,10)
      COMMON A,B,MN
C     READ NUMBER OF BRANCHES-NUMBER OF NODES
C     (EXCEPT REFERENCE)
C     NUMBER OF VOLTAGES UPON WHICH CURRENT SOURCE
C     DEPENDS
      READ 10,MN,MB,MI
10    FORMAT(3I2)
C     ZERO THE INCIDENCE MATRIX G
      DO 20 I= 1,MN
      DO 20 J= 1,MB
20    G(I,J) = 0.
C     ZERO THE ADMITTANCE MATRIX Y
      DO 30 I= 1,MB
      DO 30 J= 1,MB
30    Y(I,J)= 0.
C     FORM INCIDENCE MATRIX G, VOLTAGE SOURCE VECTOR E,
C     CURRENT SOURCE VECTOR C, DIAGONAL ELEMENTS OF Y
      DO 6 I = 1,MB
    1 READ 40,L,M,Y(I,I),E(I),C(I)
40    FORMAT (6X,I1,1X,I1,10X,3F10.0)
    2 IF(L)4,4,3
3     G(L,I) = 1.
    4 IF(M)6,6,5
5     G(M,I) = -1.
6     CONTINUE
C     ADD NONDIAGONAL TERMS TO Y
      DO 8 I= 1,MI
7     READ 50,L,M,GM
   50 FORMAT(6X,I1,1X,I1,1X,F10.0)
8     Y(M,L)= GM
      PRINT 60,((Y(I,J),I= 1,MB),J= 1,MB)
60    FORMAT(1H ,4F10.8/)
      PRINT 70,(E(I),C(I),I=1,MB)
70    FORMAT(1H ,2F10.4/)
C     FORM PRODUCT YE     (D)
      DO 25 I=1,MB
      D(I)= 0.0
      DO 25 J= 1,MB
25    D(I)= D(I)+ Y(I,J)* E(J)
C     FORM C-YE     (F)
      DO 35 I= 1,MB
   35 F(I) = C(I) - D(I)
C     FORM G(C-YE)     (B)
      DO 45 I= 1,MN
      B(I,1) = 0.0
      DO 45 J= 1,MB
45    B(I,1) = B(I,1) + G(I,J)*F(J)
C     TRANSPOSE G     (T)
```

TABLE 3-1 *(continued)*

```
      DO 55 I= 1,MN
      DO 55 J= 1,MB
55    T(J,I) = G(I,J)
C     FORM GY    (R)
      DO 65 I= 1,MN
      DO 65 J= 1,MB
      R(I,J)= 0.0
      DO 65 L= 1,MB
65    R(I,J) = R(I,J) + G(I,L)*Y(L,J)
C     FORM GYT    (A)
      DO 75 I= 1,MN
      DO 75 J= 1,MN
      A(I,J)= 0.0
      DO 75 L= 1,MB
75    A(I,J)= A(I,J)+ R(I,L)*T(L,J)
      PRINT 60, ((A(I,J),I=1,MN),J=1,MN)
C     SOLVE (GYT)V= G(C-YE) FOR V
      CALL INVR(A,MN,B,1)
      PRINT 90, (B(I,1),I=1,MN)
   90 FORMAT (1H ,F12.8)
      CALL EXIT
      END
  2 4 2
B1    N(1,0)         .01
B2    N(2,1)         .000002
B3    N(2,0)         .0001
B4    N(0,0)         .000001    0.2
T1    B(1,2)-.008
T2    B(4,2) .008
```

by this program are $v_A = 0.0879042$ V and $v_B = -8.79024320$ V (output voltage, peak to peak).

The subroutine INVR (A, MN, B, 1) (see Appendix) not only inverts the matrix **A**, given the expression **AV** = **B**, but also multiplies the matrix **B** by $\mathbf{A}^{-1}$ to give $\mathbf{V} = \mathbf{A}^{-1}\ \mathbf{B}$ as the output. Thus the end result of the program is the vector of node voltage. Branch voltages or currents could be calculated by programming the appropriate equations from the flow chart of Fig. 3-22.

The program of Table 3-1 is directed to the formulation and solution of small circuits (10 branches or less), thus the program is not a general one. A more general program can be written to handle larger networks and to allow a more flexible input format. The input data for ECAP, for example, does not require information relating to the total number of nodes or branches nor is such a rigid data format required. Furthermore, the requirement of numbering the tree branches consecutively before proceeding to link branches as implied in this chapter need not be met for general programs such as ECAP. However, the program of Table 3-1 does demonstrate a basic approach to equation formulation and solution. The importance of topological methods is clearly visible in this example. We can note the ease with which the incidence matrix and the admittance matrix are constructed. Matrix manipulation is also handled with facility by the computer program. The most difficult and most time-consuming step is the solution of the equation involving inversion of a matrix. This is accomplished by calling the subroutine INVR to supply the necessary matrix inversion.

It should be pointed out that later programs requiring the tieset or cutset matrices can utilize information obtained from this program, since these matrices can both be found from a knowledge of the incidence matrix.

The Cutset Method

In using the node method we have chosen to base our calculations on the node-voltage vector **v**. Since all of these voltages are related to a reference node, this vector will contain $N - 1$ components, where N is the number of nodes in the circuit. Since there are $N - 1$ tree branches of a graph, we could base our calculations on the tree-branch voltage vector $\mathbf{e}_T$ which also has $N - 1$ independent voltages.

We first prove that all branch voltages can be expressed in terms of the tree-branch voltages as

$$\mathbf{e} = \mathbf{C}^t \mathbf{e}_T \tag{3-29}$$

To prove that Eq. 3-29 is correct we note that the vector **e** can be written as

$$\mathbf{e} = \begin{bmatrix} \mathbf{e}_T \\ \mathbf{e}_L \end{bmatrix}$$

where $\mathbf{e}_T$ is the tree-branch voltage vector and $\mathbf{e}_L$ is the link voltage vector. Equation 3-29 can be expressed as

$$\begin{bmatrix} \mathbf{e}_T \\ \mathbf{e}_L \end{bmatrix} = \begin{bmatrix} \mathbf{C}_T^t \\ \mathbf{C}_L^t \end{bmatrix} [\mathbf{e}_T] = \begin{bmatrix} \mathbf{C}_T^t \mathbf{e}_T \\ \mathbf{C}_L^t \mathbf{e}_T \end{bmatrix}$$

Since $\mathbf{C}_T^t$ is a unit matrix of dimension $N - 1$ and $\mathbf{e}_T$ has $N - 1$ components, we see that

$$\mathbf{C}_T^t \mathbf{e}_T = \mathbf{U} \mathbf{e}_T = \mathbf{e}_T$$

We must now show that

$$\mathbf{e}_L = \mathbf{C}_L^t \mathbf{e}_T$$

before accepting Eq. 3-29. From Eq. 3-2 we know that $\mathbf{Be} = 0$, thus we can write

$$[\mathbf{B}_T \ \mathbf{B}_L] \begin{bmatrix} \mathbf{e}_T \\ \mathbf{e}_L \end{bmatrix} = 0$$

which implies that

$$\mathbf{B}_L\,\mathbf{e}_L = -\,\mathbf{B}_T\,\mathbf{e}_T$$

Since $\mathbf{B}_L$ is a unit matrix and $\mathbf{C}_L^t = -\,\mathbf{B}_T$ (from Eq. 3-10), we see that

$$\mathbf{e}_L = -\,\mathbf{B}_T\,\mathbf{e}_T = \mathbf{C}_L^t\,\mathbf{e}_T$$

Thus Eq. 3-29 is true and can be used in succeeding calculations.

Returning to Eq. 3-25, we can write

$$\mathbf{I} - \mathbf{YE} + \mathbf{i} = \mathbf{Ye} \tag{3-30}$$

Using Eq. 3-29 and multiplying both sides of Eq. 3-30 by $\mathbf{C}$ we find that

$$\mathbf{C}(\mathbf{I} - \mathbf{YE}) + \mathbf{Ci} = \mathbf{CYC}^t\,\mathbf{e}_T$$

The tree-branch voltage vector is found to be

$$\mathbf{e}_T = [\mathbf{CYC}^t]^{-1}\,\mathbf{C}[\mathbf{I} - \mathbf{YE}] \tag{3-31}$$

Equation 3-29 is then applied to find $\mathbf{e}$.

Special Cases

There are some problems that require special attention when using the preceding methods of solution. One problem which has been discussed is that of a transadmittance or transimpedance that is proportional to the branch voltage or current rather than the element voltage or current. The circuit of Fig. 3-23 shows a case where the transadmittance is proportional to $e_g - e_k$ and

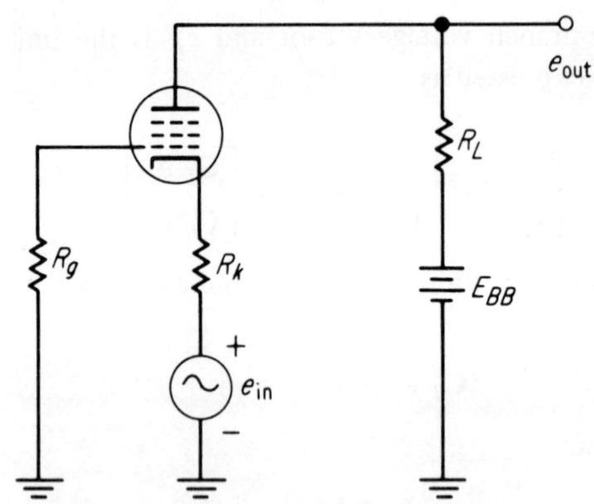

FIG. 3-23. Common-grid amplifier.

e_k is equal to the voltage across the cathode branch rather than the voltage across R_k.

The equivalent circuit in this case is shown in Fig. 3-24.

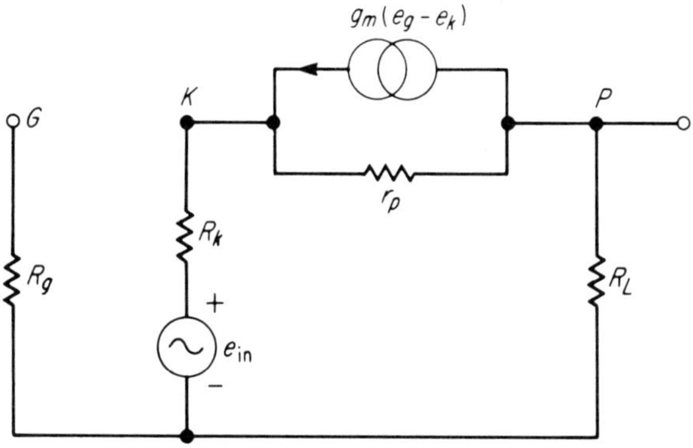

FIG. 3-24. Equivalent circuit of amplifier.

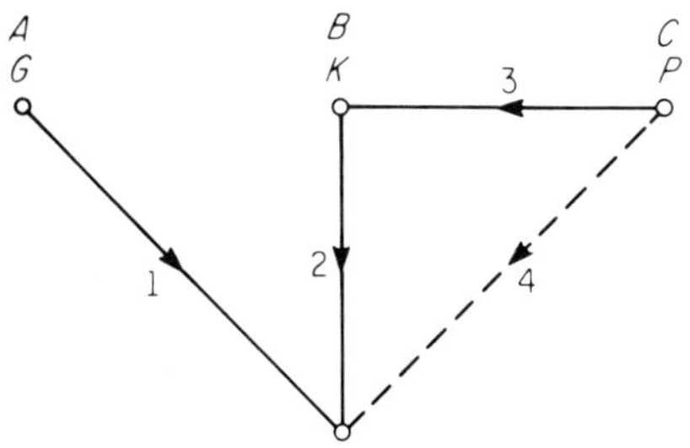

FIG. 3-25. Graph of amplifier circuit.

The graph of this circuit is shown in Fig. 3-25. The current of element 3 can be written as

$$J_3 = V_3/r_p + g_m V_1 - g_m e_2$$

We note that since the current is proportional to the negative of the branch voltage e_2, we cannot express this variation in terms of the normal **Y** matrix. To treat this problem, we return to Eq. 3-18, which states that

$$\mathbf{J} = \mathbf{Y}_S \mathbf{E} + \mathbf{Y}_B \mathbf{e} \tag{3-18}$$

By the principle of superposition we can break the current vector into two components; one due to the voltage source vector and one due to the branch voltage vector such that

$$\mathbf{J}_S = \mathbf{Y}_S \mathbf{E} \quad \text{and} \quad \mathbf{J}_B = \mathbf{Y}_B \mathbf{e}$$

The matrix $\mathbf{Y}_S$ equals $\mathbf{Y}_B$ unless we have modified $\mathbf{Y}_B$ to account for the transconductance as outlined previously.

Equation 3-18 can be written as

$$\mathbf{I} + \mathbf{i} = \mathbf{Y}_S \mathbf{E} + \mathbf{Y}_B \mathbf{e}$$

Using the fact that $\mathbf{Ai} = 0$ and $\mathbf{e} = \mathbf{A}^t\mathbf{v}$ allows us to write the node voltage as

$$\mathbf{v} = [\mathbf{A}\,\mathbf{Y}_B\,\mathbf{A}^t]^{-1}\,\mathbf{A}[\mathbf{I} - \mathbf{Y}_S\,\mathbf{E}] \tag{3-32}$$

An example will demonstrate the application of this equation.

EXAMPLE 3-4. Find the output voltage of the amplifier circuit of Fig. 3-23 if the input is a sinusoid.

Solution. The graph of the circuit is shown in Fig. 3-25. The **A** matrix is

$$\mathbf{A} = \begin{bmatrix} 1 & 0 & 0 & 0 \\ 0 & 1 & -1 & 0 \\ 0 & 0 & 1 & 1 \end{bmatrix}$$

In this case $\mathbf{Y}_S$ equals the **Y** matrix and is

$$\mathbf{Y}_S = \begin{bmatrix} 1/R_g & 0 & 0 & 0 \\ 0 & 1/R_k & 0 & 0 \\ g_m & 0 & 1/r_p & 0 \\ 0 & 0 & 0 & 1/R_L \end{bmatrix}$$

To find $\mathbf{Y}_B$ we simply add the transadmittance effects of the dependent current generator to the **Y** matrix to give

$$\mathbf{Y}_B = \mathbf{Y} + \mathbf{Y}' = \begin{bmatrix} 1/R_g & 0 & 0 & 0 \\ 0 & 1/R_k & 0 & 0 \\ g_m & -g_m & 1/r_p & 0 \\ 0 & 0 & 0 & 1/R_L \end{bmatrix}$$

Note that since the current generator is proportional to $g_m(E_1 + e_1) - g_m e_2$, the g_m term appears in the third row of the first column of both $\mathbf{Y}_S$ and $\mathbf{Y}_B$, but $-g_m$ only appears in the matrix $\mathbf{Y}_B$. The inverse of $[\mathbf{A}\,\mathbf{Y}_B\,\mathbf{A}^t]$ is found to be

$$\frac{1}{D}\begin{bmatrix} R_g D & 0 & 0 \\ g_m/R_L & (1/R_g r_p + 1/R_g R_L) & 1/R_g r_p \\ -g_m/R_k & (g_m/R_g + 1/R_g r_p) & (g_m/R_g + 1/R_g R_k + 1/R_g r_p) \end{bmatrix}$$

where $D = R_g R_k r_p R_L/(g_m r_p R_k + R_L + r_p + R_k)$. The source vector is

$$\mathbf{E} = \begin{bmatrix} 0 \\ -e_{\text{in}} \\ 0 \\ 0 \end{bmatrix}$$

and the node voltage given by Eq. 3-32 is

$$\mathbf{v} = \begin{bmatrix} 0 \\ \dfrac{e_{\text{in}}(R_L + r_p)}{g_m r_p R_k + R_L + r_p + R_k} \\ \dfrac{e_{\text{in}}(g_m r_p R_L + R_L)}{g_m r_p R_k + R_L + r_p + R_k} \end{bmatrix}$$

The output voltage corresponds to v_c and is

$$e_{\text{out}} = \frac{e_{\text{in}}(g_m r_p R_L + R_L)}{g_m r_p R_k + R_L + r_p + R_k}$$

The cutset method can be treated in the same way as the node method to give

$$\mathbf{v} = [\mathbf{C}\,\mathbf{Y}_B\,\mathbf{C}^t]^{-1}\,\mathbf{C}[\mathbf{I} - \mathbf{Y}_S\,\mathbf{E}] \tag{3-33}$$

A second problem arising is that of a circuit containing elements which cannot be included in the configuration of the standard branch as defined in Fig. 3-11. For example, the circuit of Fig. 3-26 contains a source V_{CC} that

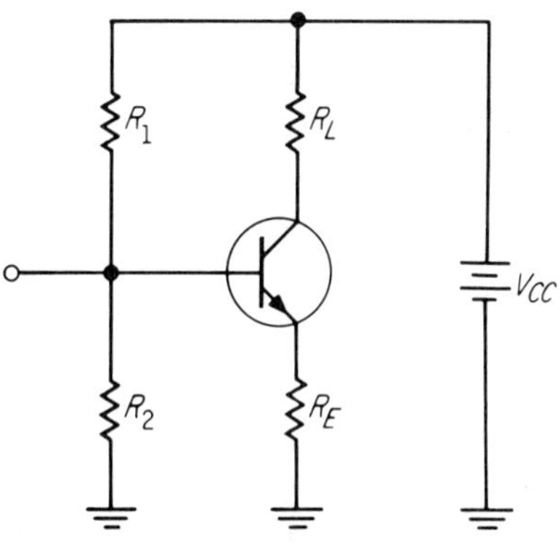

FIG. 3-26. Transistor circuit.

should be included in a branch. However, we must determine whether it should be included in the branch containing R_L or R_1. If we note that both R_1 and R_L are connected to a source of voltage V_{CC} we can conclude that the circuit operation would be unchanged if we used two separate sources as shown in Fig. 3-27. The circuit can now be analyzed in the usual way.

A similar problem occurs when a current source bridges a node as in Fig. 3-28. Again the configuration does not allow the current source to be associated with a particular branch. The circuit of Fig. 3-29 overcomes this problem while maintaining the same external current entering each node as in the circuit of Fig. 3-28. Note that node B still absorbs no net external current flow.

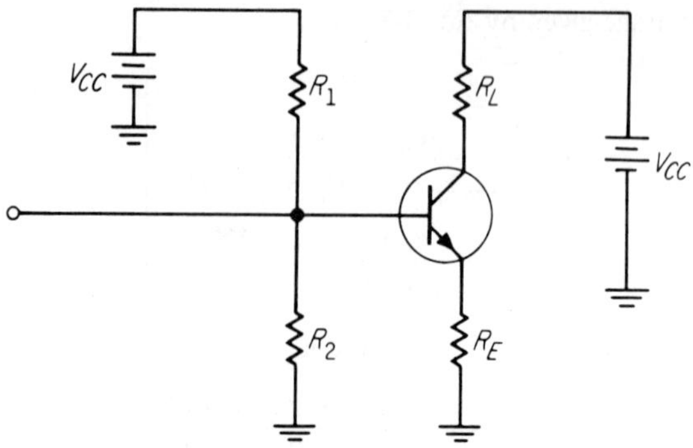

FIG. 3-27. Transistor circuit with separate sources.

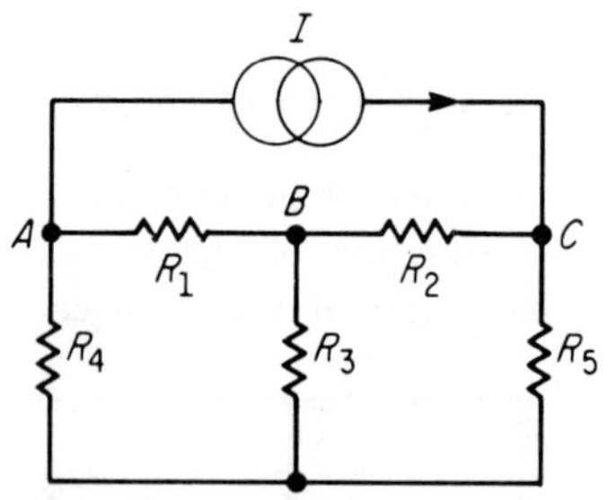

FIG. 3-28. Node bridging by a current source.

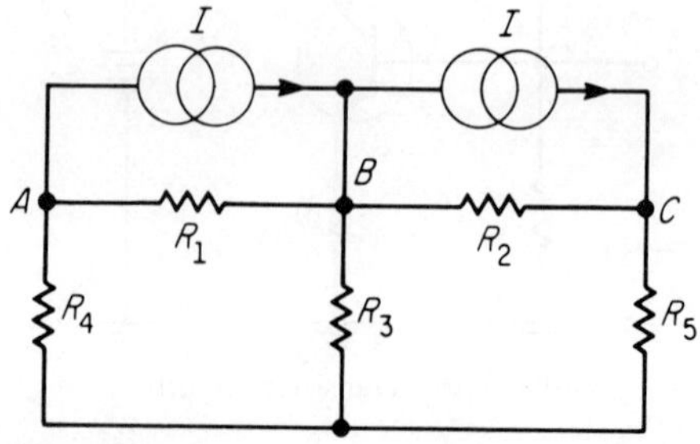

FIG. 3-29. Circuit with two separate current sources.

The circuit can now be divided into standard branches after this current source splitting has taken place.

Conversion Between Transadmittance and Transimpedance

It is possible to convert a transadmittance to a transimpedance, or vice-versa, if required. The tieset method requires that transimpedances be used while both the node and cutset methods use transadmittances. Let us return to the transistor amplifier of Fig. 3-17. The dependent voltage source of branch 5 is controlled by the element current in branch 4 and thus represents a transimpedance. A transadmittance, of course, is represented by a dependent current source controlled by an element voltage. The dependent voltage source can be converted to a current source equal to $\beta i_b r_c/r_c = \beta i_b$ with a shunt resistance r_c. We now have a current-controlled current source rather than a voltage-controlled current source. However, since we can write $i_b = e_{bb'}/r_{bb'} = V_4/r_{bb'}$, the dependent source current equals $\beta V_4/r_{bb'}$. The admittance matrix can be expressed as

$$\mathbf{Y} = \begin{bmatrix} 1/R_s & 0 & 0 & 0 & 0 \\ 0 & 1/r_d & 0 & 0 & 0 \\ 0 & 0 & 1/R_c & 0 & 0 \\ 0 & 0 & 0 & 1/r_{bb'} & 0 \\ 0 & 0 & 0 & -\beta/r_{bb'} & 1/r_c \end{bmatrix}$$

The negative sign associated with Y_{54} comes about because the current direction of the controlled source is opposite to that of the assumed current of branch 5 shown in Fig. 3-18.

The particular method of solution chosen determines whether the **Y** matrix or **Z** matrix is to be used. The Electronic Circuit Analysis Program (ECAP) available from International Business Machines Corporation uses the node method of solution, and thus accepts only controlled current sources as valid dependent elements. Dependent voltage sources must be converted to equivalent current sources to use this program.

The Ideal Transformer [4]

The ideal transformer shown in Fig. 3-30 exhibits the following relations with no frequency dependence:

$$v_2 = v_1/n \tag{3-34}$$

$$i_2 = -ni_1 \tag{3-35}$$

The ideal circuit shown can lead to problems of degeneracy under certain conditions [2]. To avoid these problems and to relate currents to voltages a series

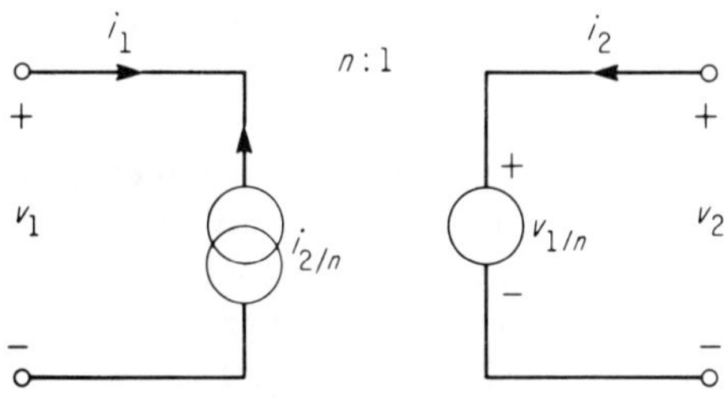

FIG. 3-30. Ideal transformer.

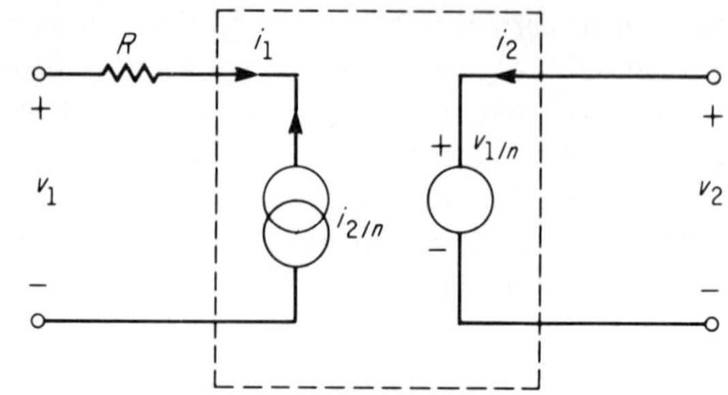

FIG. 3-31. Modified ideal transformer.

resistance R can be assumed in the primary circuit. The equations for the modified transformer of Fig. 3-31 then become

$$v_1 = i_1 R + v_2 n \tag{3-36}$$

$$i_2 = -ni_1 \tag{3-37}$$

Equations 3-36 and 3-37 can be solved for i_1 and i_2 to give

$$i_1 = v_1(1/R) - v_2(n/R) \tag{3-38}$$

$$i_2 = -v_1(n/R) + v_2(n^2/R) \tag{3-39}$$

These equations can easily be modeled by the network of Fig. 3-32. The current

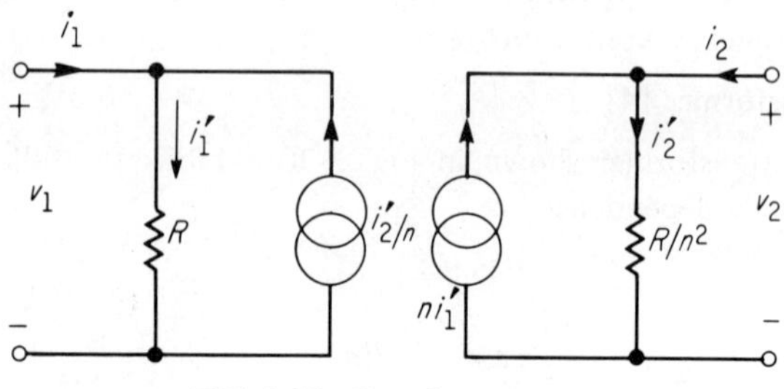

FIG. 3-32. Transformer model.

i_1' is equal to v_1/R resulting in a dependent secondary current source of nv_1/R. The current i_2' equals $v_2 n^2/R$ resulting in a dependent primary current source of nv_2/R. Equations 3-38 and 3-39 are then satisfied by this model. Since frequency effects have been neglected, the model of Fig. 3-32 is valid only in the midband of frequency. For modeling purposes the resistance R can be chosen to be small compared with other external primary resistances.

Earlier in this chapter it was noted that mutual impedances or admittances led to off-diagonal, but equal terms in the impedance or admittance matrix. In matrix form Eqs. 3-38 and 3-39 can be expressed as

$$\begin{bmatrix} i_1 \\ i_2 \end{bmatrix} = \begin{bmatrix} 1/R & -n/R \\ -n/R & n^2/R \end{bmatrix} \begin{bmatrix} v_1 \\ v_2 \end{bmatrix} \tag{3-40}$$

demonstrating the effect of mutual elements.

The resistance R inserted in series with the primary winding allows the practical equivalent circuit of Fig. 3-32 to be used. However, this element can lead to inaccuracies unless limited in value such that

$$R << n^2 R_L \tag{3-41}$$

where R_L is the load resistance connected to the secondary windings. An example will demonstrate the use of the transformer equivalent circuit and the choice of R.

EXAMPLE 3-5. A source with 1 $k\Omega$ output impedance and an open circuit output voltage of $e_{\text{in}} = 10 \sin \omega t$ is coupled to a 50-Ω load by means of a transformer with a turns ratio n of 10. Using the ideal transformer equivalent circuit calculate the voltage across the load and the primary current.

Solution. Figure 3-33 shows the ideal transformer equivalent circuits. Since a

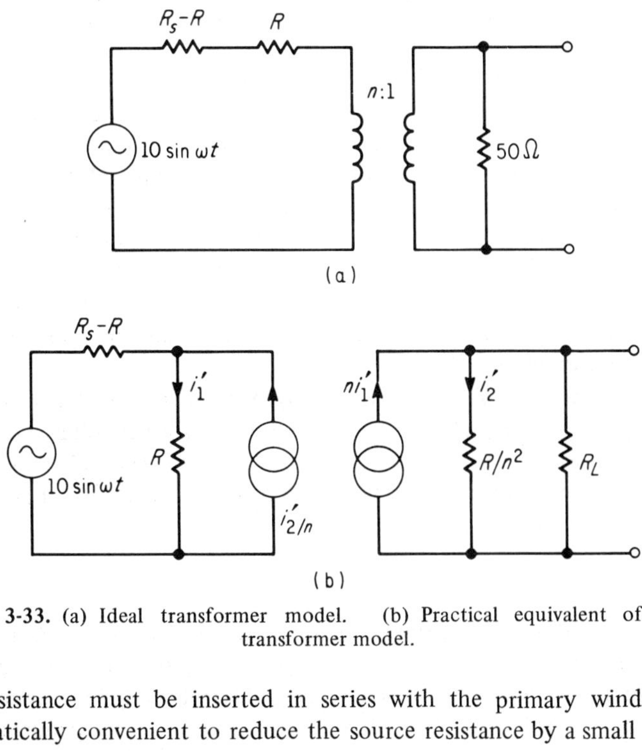

FIG. 3-33. (a) Ideal transformer model. (b) Practical equivalent of ideal transformer model.

series resistance must be inserted in series with the primary windings, it is mathematically convenient to reduce the source resistance by a small amount R

to preserve the correct series primary resistance. Thus R should be chosen to be small compared to R_s and compared to n^2R_L. After R is selected, the circuit of Fig. 3-33b can be applied to calculate the response. In this example a value of $R = 50\ \Omega$ is quite reasonable. The resulting circuit is shown in Fig. 3-34. The

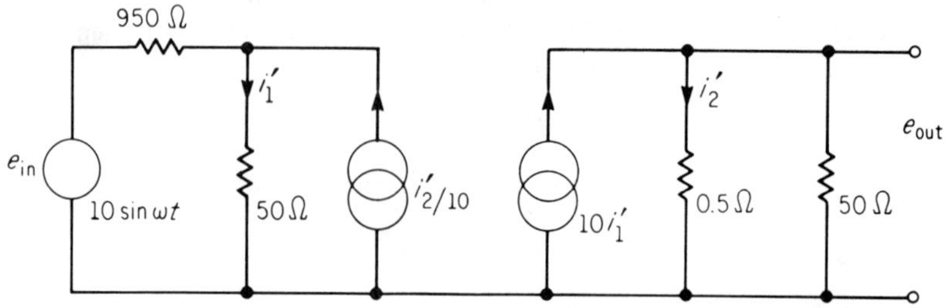

FIG. 3-34. Equivalent circuit for Example 3-5.

output voltage is easily found to be

$$e_{out} = e_{in}/12 = (5 \sin \omega t)/6$$

and the primary current is (5/3) sin ωt ma. The circuit of Fig. 3-33 can be used in automatic-analysis programs applying the methods developed in this chapter.

Discussion of Topological Methods

The methods outlined in the previous sections are merely formal application of procedures with which we are quite familiar. We could apply Ohm's and Kirchhoff's laws to analyze the simple circuits of Examples 3-2 to 3-4 without requiring the use of any topological matrices. However, if the circuit has only a few more nodes, manual analysis is time-consuming and subject to errors due to the increased complexity of the problem. The topological methods are easily extended to any practical number of nodes and are well suited to computer analysis because of the ease with which the digital computer stores and manipulates matrix information.

In any network problem there is a given number of voltages or currents that are linearly independent. The expression for any other voltage consists of a linear combination of the independent voltages. When we decide that the $N - 1$ node voltages are to form the voltage vector, we automatically obtain a set of $N - 1$ independent voltages. The equation

$$\mathbf{e} = \mathbf{A}^t \mathbf{v} \qquad [3\text{-}4]$$

can be thought of as a linear transformation to a new basis where $\mathbf{A}^t$ is the basis transformation matrix. The branch voltage is then expressible in terms of the node voltages and only $N - 1$ branch voltages will be linearly independent of the remaining branch voltages. Another way to obtain $N - 1$ independent voltages is to select the tree-branch voltages as the original vector. The equation

$$\mathbf{e} = \mathbf{C}^t \mathbf{e}_T \qquad [3\text{-}29]$$

is the transformation giving the branch voltage vector in this case.

In deriving the equations to solve the network problem, we apply a matrix that causes a variable to transform to the zero vector. For example, the matrix **B** multiplies both sides of Eq. 3-19, but $\mathbf{Be} = 0$ and thus one unknown variable is eliminated. In fact, this condition is simply the application of Kirchhoff's voltage law stating that the sum of the voltages around all tiesets must be zero.

Graph theory may not be a necessity for the circuit designer, but certainly this theory is prerequisite to the development of computer-aided design programs. The program of Table 3-1 utilizes topological methods to find the required output variables. However, only a minimal amount of input information is required. Generally the node numbers of each branch are specified along with element types and values. The computer then generates the required matrices and applies the appropriate solution steps.

The examples discussed in this chapter and the program of Table 3-1 were concerned with circuits consisting only of resistive elements to maintain simplicity. Capacitors and inductors can be included in the analysis assuming that the excitation frequency is known. We might note that a change of input frequency requires a complete reanalysis of the circuit to calculate the response. Thus, the methods described are not especially attractive when generating frequency responses, although they certainly can be used. Chapter 2 mentioned a more efficient approach to frequency analysis based on state space techniques and this method will be given a more detailed discussion in Chapter 5.

An impedance or admittance must be defined for each element when using this method. Since transient impedances are not defined, the methods described in this chapter are not applicable to the transient case. Chapter 5 will discuss the transient response of networks based on state variable analysis. We will find there that while the standard branch as defined in the present chapter is not used, the topological concepts discussed here will again be important.

BIBLIOGRAPHY

1. Branin, F. H., Jr., "Computer Methods of Network Analysis," *Proceedings of the Institute of Electrical and Electronics Engineers*, Vol. 55, No. 11 (November 1967).
2. Calahan, D. A., *Computer-Aided Network Design*, preliminary edition, McGraw-Hill Book Company, New York, 1968.
3. Cruz, J. B., Jr., and M. E. Van Valkenburg, *Introductory Signals and Circuits*, Blaisdell Publishing Company, Waltham, Mass., 1967.
4. Jensen, R. W., and M. D. Lieberman, *IBM Electronic Circuit Analysis Program*, Prentice-Hall, Inc., Englewood Cliffs, N.J., 1968.
5. Kuo, B. C., *Linear Networks and Systems*, McGraw-Hill Book Company, New York, 1967.

PROBLEMS

3-1 Draw the oriented graph of the circuit. Draw at least four trees of the graph, using one element per branch.

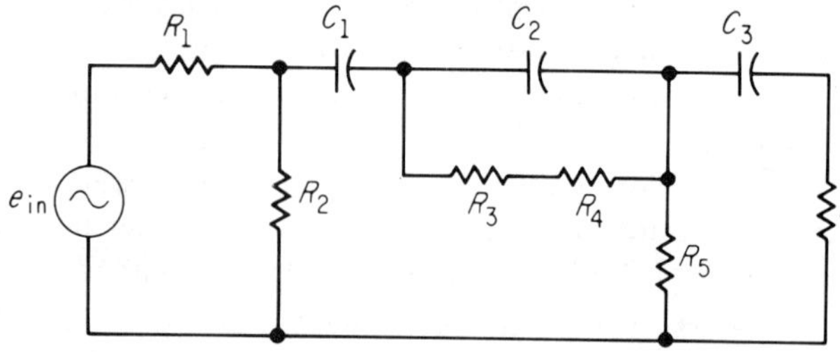

PROB. 3-1

3-2 Draw the oriented graph of the circuit. Draw at least four trees of the graph, using one element per branch.

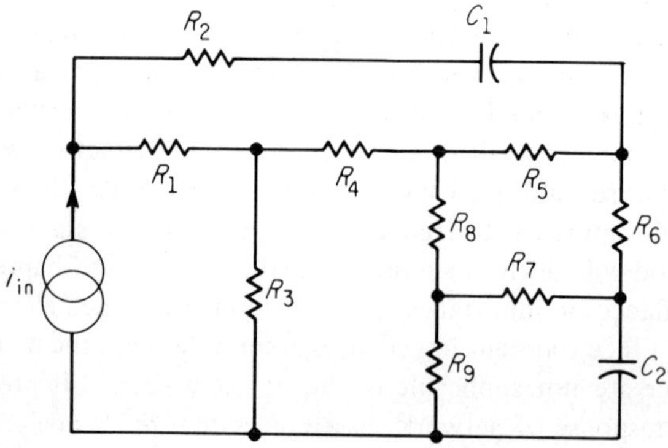

PROB. 3-2

3-3 Using the tree formed by branches 1, 2, 3, 4, and 5 construct the matrices **A**, **B**, and **C** from their respective definitions.

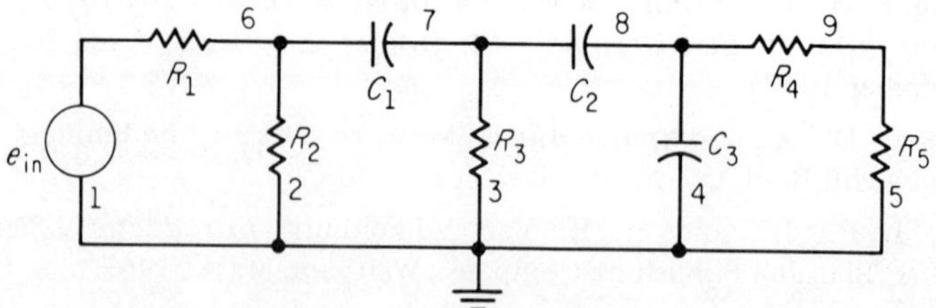

PROB. 3-3

3-4 From the **A** matrix found in Prob. 3-3, calculate the **B** and **C** matrices.

3-5 Using the tree formed by branches 1, 2, 3, 4 and 5 construct the matrices **A**, **B**, and **C** from their respective definitions.

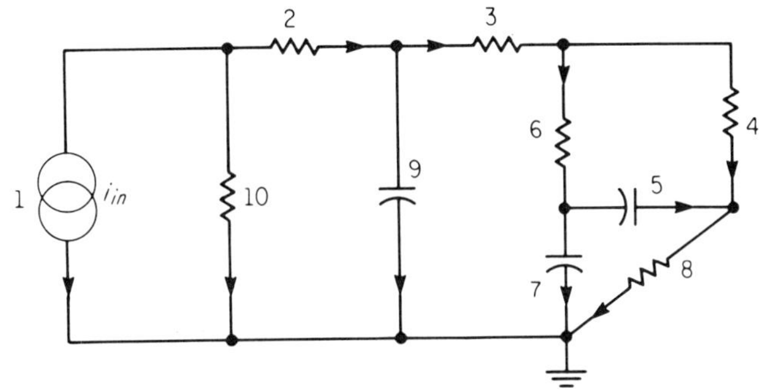

PROB. 3-5

3-6 From the **A** matrix found in Prob. 3-5, calculate the **B** and **C** matrices.

3-7 Using the tree formed by branches 1, 2, 3, 4, and 5 construct the matrices **A**, **B**, and **C** from their respective definitions.

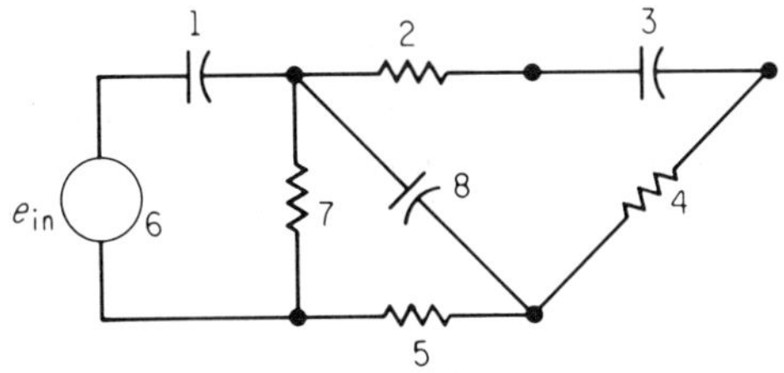

PROB. 3-7

3-8 From the **A** matrix found in Prob. 3-7, calculate the **B** and **C** matrices.

3-9 Find the branch-voltage vector by the node method using the standard branch. In this case $R_1 = 600\ \Omega$, $R_2 = 200\ \Omega$, $R_3 = 100\ \text{k}\Omega$, $R_4 = 2\ \text{k}\Omega$, $R_5 = 4\ \text{k}\Omega$, $R_6 = 2\ \text{k}\Omega$, $R_7 = 10\ \text{k}\Omega$ and $g_m = 40 \times 10^{-3}$ mho.

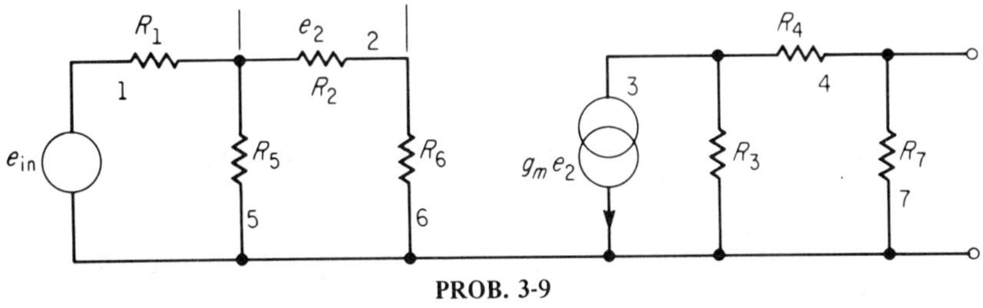

PROB. 3-9

3-10 Find the branch-current vector and branch-voltage vector of the circuit by the tieset method. The elements are $C_1 = 10\,\mu f$, $C_2 = 0.01\,\mu f$, $C_3 = 0.001\,\mu f$, $R_1 = 1\text{ k}\Omega$, $R_2 = 5\text{ k}\Omega$, and $R_3 = 20\text{ k}\Omega$. Assume $e_{in} = 12$ V peak at 1 kHz.

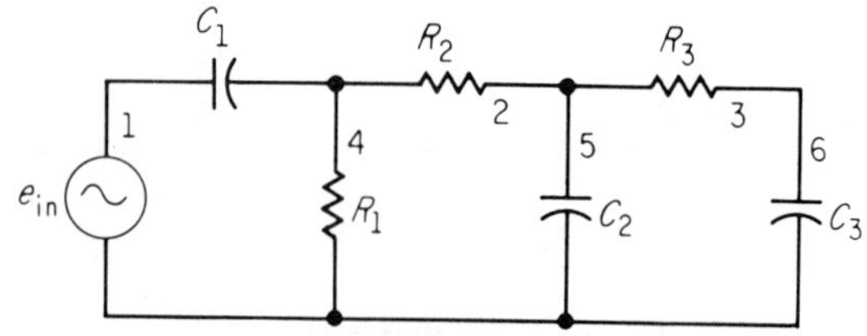

PROB. 3-10

3-11 Solve for branch currents and voltages of the circuit in Prob. 3-10 by the node method.

3-12 Solve for branch currents and node voltages of the circuit in Prob. 3-1 using the tieset method. The element values are $R_1 = 2$, $R_2 = 4$, $R_3 = 2$, $R_4 = 1$, $R_5 = 1$, $R_6 = 4$, $C_1 = 1$, $C_2 = 2$, and $C_3 = 8$, and $e_{in} = 10$ V peak at 10 Hz.

3-13 Solve for loop currents and branch voltages of the circuit in Prob. 3-1, using the node method and the element values given in Prob. 3-12.

3-14 Solve for branch currents and node voltages of the circuit in Prob. 3-2, using the tieset method. The element values are $R_1 = 4$, $R_2 = 2$, $R_3 = 1$, $R_4 = 4$, $R_5 = 6$, $R_6 = 1$, $R_7 = 10$, $R_8 = 3$, $R_9 = 10$, $C_1 = 2$, and $C_2 = 4$, and $i_{in} = 10 \sin 6.28t$ ma.

3-15 Solve for loop currents and branch voltages of the circuit in Prob. 3-2, using the node method and the element values given in Prob. 3-14.

3-16 Solve for branch currents and node voltages of the circuit in Prob. 3-3 using the tieset method. The element values are $R_1 = 1\text{ k}\Omega$, $R_2 = 1\text{ k}\Omega$, $R_3 = 1\text{ k}\Omega$, $R_4 = 1\text{ k}\Omega$, $R_5 = 10\text{ k}\Omega$, $C_1 = 0.01\,\mu f$, $C_2 = 0.001\,\mu f$, and $C_3 = 0.001\,\mu F$, and $e_{in} = 5$ V peak at 10 kHz.

3-17 Solve for loop currents and branch voltages of the circuit in Prob. 3-3 using the node method and the element values given in Prob. 3-16.

3-18 Work Example 3-2 using the node method.

3-19 Work Example 3-3 using the tieset method.

3-20 Work Example 3-3 using the cutset method.

3-21 Find the branch-voltage vector by an appropriate method.

3-22 (a) Draw the graph of the network, using the standard branch.
(b) Select an appropriate tree.
(c) Construct the tieset matrix for the graph, using the selected tree.
(d) Construct the element impedance matrix including the effects of the dependent current source.

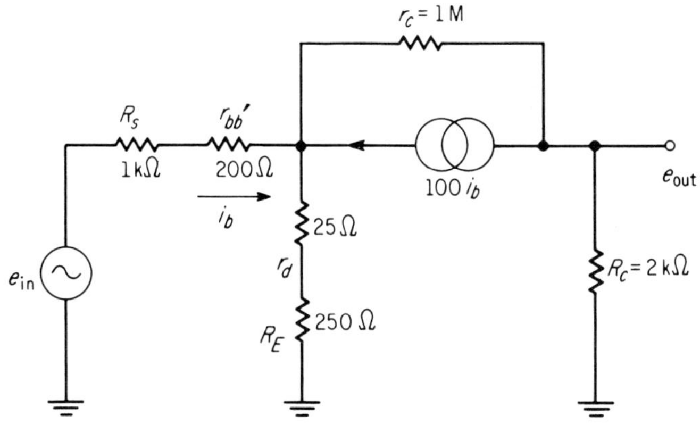

PROB. 3-21

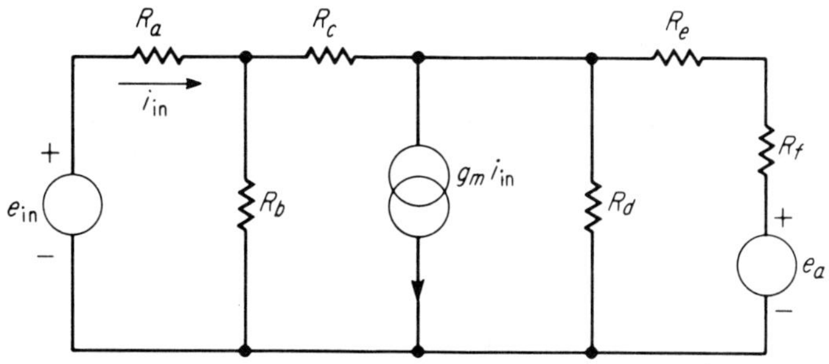

PROB. 3-22

3-23 Write the matrices $\mathbf{Y}_B$ and $\mathbf{Y}_S$ for the network and explain what these matrices represent.

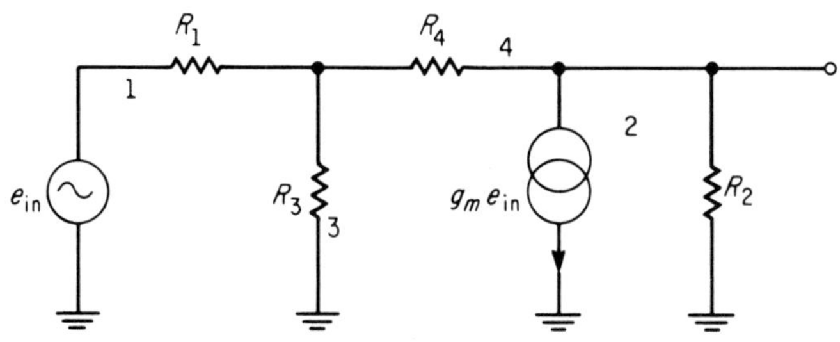

PROB. 3-23

3-24. Modify the program of Table 3-1 or write a program to solve Prob. 3-9.

3-25 Develop a program to find the node-voltage vector for the circuit of Prob. 3-22. The following values apply: $R_a = 1\ \text{k}\Omega$, $R_b = 4\ \text{k}\Omega$, $R_c =$

2 kΩ, $R_d = 30$ kΩ, $R_e = 5$ kΩ, $R_f = 8$ kΩ, $g_m = 8{,}000$ μmhos, $e_{in} =$ 2 sin 10,000t, and $e_a = 1$ sin 10,000t.

3-26 Repeat Prob. 3-25 for the circuit of Prob. 3-23. Assume $R_1 = 200\Omega$, $R_2 = 5$ kΩ, $R_3 = 2$ kΩ, $R_4 = 40$ kΩ, $g_m = 6{,}000$ μmhos and $e_{in} = 0.01$ sin 1000t.

3-27 Write a program to find the node-voltage vector of a circuit containing

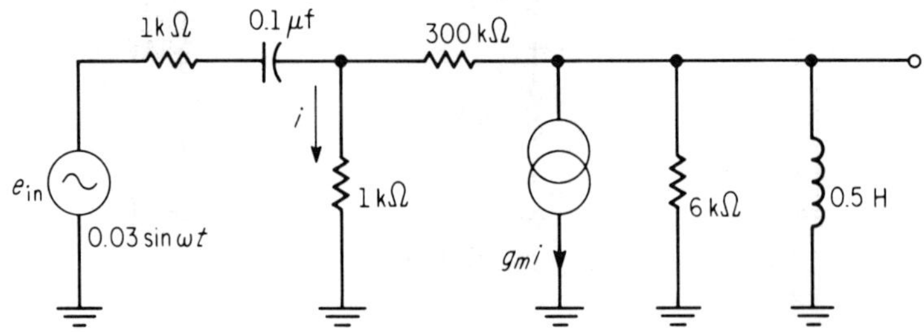

PROB. 3-27

reactive elements and up to 10 branches. Test your program by analyzing the circuit shown at $\omega = 6{,}280$ rad/sec. Assume $g_m = 0.04$ mho.

chapter **4**

THE STATE VARIABLE

We have considered methods of formulating and solving network equations by the cutset, node, and tieset methods. In reviewing the necessary equations for these methods one realizes that a branch impedance or admittance matrix is applied in each case. Thus, these methods are applicable only to the d-c or a-c network wherein branch impedances can be expressed. Networks accepting transient inputs, as typified by logic circuits or multivibrators, make up an important class of circuits to which the techniques of analysis thus far discussed do not apply. We must therefore consider methods that can be utilized to formulate and solve equations relating to networks with transient excitation.

One extremely useful technique is the state-variable method. With this approach equations can be formulated and solved for the transient as well as the a-c case. The basic idea is to express the required information relating to network or system response in terms of a first-order differential (matrix) equation. If the state vector is appropriately defined, it is always possible to write the equation in the proper form for linear networks. Once the state equation is formulated, well-known mathematical techniques can be applied to the solution. This chapter discusses the formulation of state equations for linear networks, while Chapter 5 considers the machine solution of the state equation.

A. STATE EQUATIONS

The state variable or state vector for a system is that vector whose linearly independent components can describe the present state or any possible future state of the system. If the vector components are properly chosen, the state vector $\mathbf{x}$ satisfies the standard state equation [8]

$$\dot{\mathbf{x}} = \mathbf{A}\mathbf{x} + \mathbf{B}\mathbf{u} \tag{4-1}$$

where $\mathbf{x}$ is the state vector and $\mathbf{u}$ is the input vector. The output equation is

$$\mathbf{y} = \mathbf{C}\mathbf{x} + \mathbf{D}\mathbf{u} \tag{4-2}$$

where $\mathbf{y}$ is the output vector.

We must note here that the matrices $\mathbf{A}$, $\mathbf{B}$, $\mathbf{C}$, and $\mathbf{D}$ are not the topological matrices as defined in Chapter 3. This ambiguity is introduced in an effort to conform to standard notation for the state equations. In a later section of this chapter it will be necessary to use the topological matrices. To avoid confusion these matrices will be expressed in terms of the tree-branch and link-branch submatrices. For example, the incidence matrix will be written as

$$[A_T \; A_L]$$

The simplest choice of state vector is one having components that describe the energy stored by different elements of the system. In network problems, the components are normally chosen to be capacitor voltages and inductor currents allowing independently specifiable initial conditions [5]. Equations 4-1 and 4-2 can be solved in general terms so that any system obeying the state equation can be described by a state vector at any instant of time. We must note that many specific problems of interest can often be solved more rapidly by straightforward methods. For example, if the input current of the network of Fig. 4-1 is required

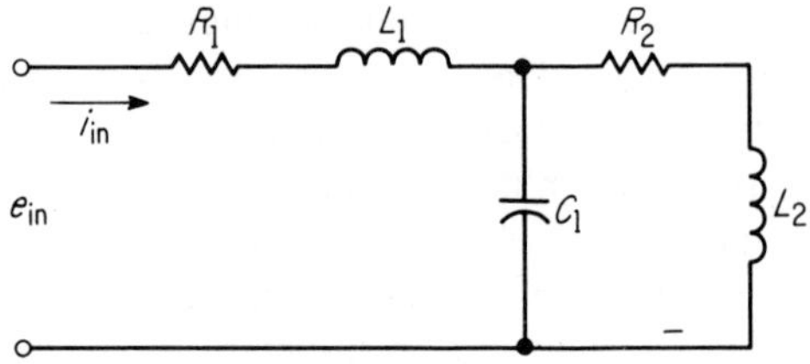

FIG. 4-1. Electrical network.

for a given excitation, the network can be reduced to an equivalent complex impedance in the Laplace domain. The Laplace variable $I_{in}(s)$ can be found and transformed to the time domain by standard methods to evaluate i_{in}. In this case, the state vector and state equations yield excess information and would probably require more effort to solve. While there may be advantages in using the state concept when many variables are required or a complex system is to be analyzed, the real justification for using this approach is found in the generality and orderliness of the method. The method is amenable to digital (or analog) computer solution and a given program can be applied to many different analysis problems. As programs become more specific they lose flexibility and can be quite useless when the problem is slightly modified.

As a simple example of the state-vector approach we shall write the state equations for the circuit of Fig. 4-2. The two first-order differential equations

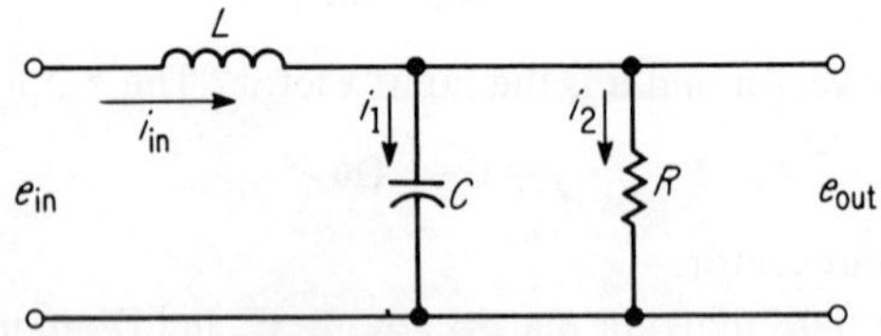

FIG. 4-2. *RLC* network.

governing the network response are

$$e_{in} = L\frac{di_{in}}{dt} + e_{out}$$

$$C\frac{de_{out}}{dt} = i_{in} - e_{out}/R$$

where i_{in} equals the inductor current and e_{out} equals the capacitor voltage. If the components of the state vector $\mathbf{x}$ are chosen as i_{in} and e_{out}, the state equation can be expressed in standard form as

$$\dot{\mathbf{x}} = \frac{d}{dt}\begin{bmatrix} i_{in} \\ e_{out} \end{bmatrix} = \begin{bmatrix} 0 & -1/L \\ 1/C & -1/RC \end{bmatrix} \begin{bmatrix} i_{in} \\ e_{out} \end{bmatrix} + \begin{bmatrix} 1/L \\ 0 \end{bmatrix} [e_{in}]$$

In this case, though not in general, one of the components of $\mathbf{x}$ is the output voltage. The above equation corresponds to Eq. 4-1, while Eq. 4-2 will be

$$\mathbf{y} = [e_{out}] = [0 \quad 1] \begin{bmatrix} i_{in} \\ e_{out} \end{bmatrix}$$

The matrix $\mathbf{D}$ is zero in this equation, but this occurs only when e_{out} can be written as a component or a linear combination of components of the state vector. Suppose that the desired output signal now becomes the inductor voltage, that is,

$$e_{out} = L\frac{di_{in}}{dt} = e_{in} - v_c$$

where v_c is the capacitor voltage.

If $\mathbf{x}$ is expressed as $\mathbf{x} = \begin{bmatrix} i_{in} \\ v_c \end{bmatrix}$ the output can be written as

$$\mathbf{y} = [e_{out}] = [0 \quad -1] \begin{bmatrix} i_{in} \\ v_c \end{bmatrix} + [1]\ [e_{in}]$$

and $\mathbf{D}$ is nonzero.

EXAMPLE 4-1. Write the state equation and output equation for the network of Fig. 4.3.

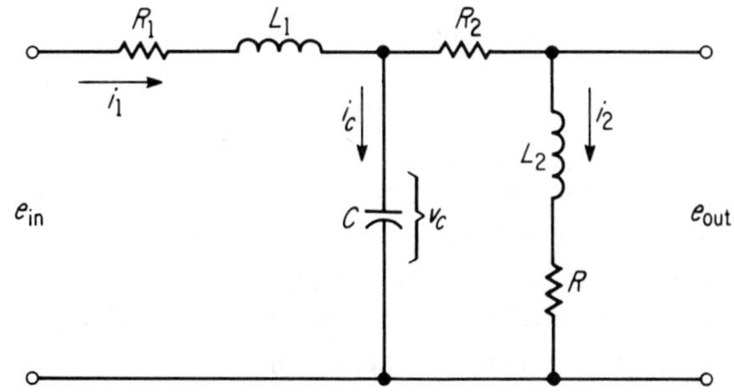

FIG. 4-3. Network for Example 4-1.

Solution. The set of first-order differential equations describing this network is

$$e_{in} = R_1 i_1 + L_1 \frac{di_1}{dt} + v_c$$

$$v_c = (R_2 + R_3)\, i_2 + L_2 \frac{di_2}{dt}$$

$$C \frac{dv_c}{dt} = i_1 - i_2$$

The state vector is chosen to have components i_1, i_2, and v_c. The state equation corresponding to Eq. 4-1 is

$$\dot{\mathbf{x}} = \begin{bmatrix} -R_1/L_1 & 0 & -1/L_1 \\ 0 & -(R_2 + R_3)/L_2 & 1/L_2 \\ 1/C & -1/C & 0 \end{bmatrix} \begin{bmatrix} i_1 \\ i_2 \\ v_c \end{bmatrix} + \begin{bmatrix} 1/L_1 \\ 0 \\ 0 \end{bmatrix} [e_{in}]$$

The output equation can be written by noting that

$$e_{out} = R_3 i_2 + L_2 \frac{di_2}{dt}$$

$$= v_c - R_2 i_2$$

Equation 4-2 becomes

$$\mathbf{y} = [e_{out}] = [0 \quad -R_2 \quad 1] \begin{bmatrix} i_1 \\ i_2 \\ v_c \end{bmatrix}$$

and again the matrix **D** is zero.

We can now discuss the reasons for choosing inductor currents and capacitor voltages as components of the state vector. The voltage around any loop is made up of source voltages, inductor voltages, resistor voltages, and capacitor voltages. Equation 4-1 allows any source voltages to be accounted for by means of the matrix **B**. The inductor voltage can be expressed in terms of a first derivative of the inductor current (derivative of a state variable). The resistor voltage can be expressed either in terms of inductor current or capacitor voltage. The current into a given node can be expressed in terms of source current, inductor current, resistor current, and capacitor current. Since capacitor current is expressible as a first derivative of capacitor voltage, this variable can be written in terms of a state variable. Thus both of Kirchhoff's equations can be written in terms of input variables, inductor current and capacitor voltages, and the first derivatives of inductor currents and capacitor voltages. Choosing the state variables in this manner then guarantees that Eq. 4-1 can be written.

There are other possible choices for the state vector, and these possibilities must be explored since the state concept can be used in control systems and areas wherein the choice of a state vector is not so obvious. Before leaving networks however, let us return to the circuit of Fig. 4-2 to examine the analysis in more detail. To find the output voltage classically we would write

$$e_{in} = L\frac{di_{in}}{dt} + e_{out}$$

$$i_{in} = \frac{e_{out}}{R} + C\frac{de_{out}}{dt}$$

Eliminating di_{in}/dt from the first equation gives

$$e_{in} = \frac{L}{R}\frac{de_{out}}{dt} + LC\frac{d^2 e_{out}}{dt^2} + e_{out} \tag{4-3}$$

Notice that a second derivative of e_{out} appears in this equation. If the state vector is chosen as

$$\mathbf{x} = \begin{bmatrix} e_{out} \\ \dfrac{de_{out}}{dt} \end{bmatrix}$$

the equation can be written as the first-order differential equation

$$\begin{bmatrix} \dfrac{de_{out}}{dt} \\ \dfrac{d^2 e_{out}}{dt^2} \end{bmatrix} = \begin{bmatrix} 0 & 1 \\ -1/LC & -1/RC \end{bmatrix} \begin{bmatrix} e_{out} \\ \dfrac{de_{out}}{dt} \end{bmatrix} + \begin{bmatrix} 0 \\ 1/LC \end{bmatrix} e_{in} \tag{4-4}$$

or

$$\dot{\mathbf{x}} = \mathbf{Ax} + \mathbf{Bu}$$

With this choice of state vector, the matrices **A** and **B** are different from those resulting when **x** was chosen as

$$\mathbf{x} = \begin{bmatrix} i_{in} \\ e_{out} \end{bmatrix}$$

Equation 4-3 is a second-order differential equation whereas Eq. 4-4 represents two first-order differential equations. By carefully defining the state vector, an nth order linear differential equation can always be reduced to a set of n first order differential equations. This allows the state vector concept to be applied to any system that can be described by a linear differential equation. In general, the components of the state vector are chosen to be the dependent variable and all derivatives of this variable up to the (nth-1).

Suppose that a fourth-order equation describes some physical system (a complex system to require a fourth-order equation)

$$M\frac{d^4r}{dt^4} + N\frac{d^3r}{dt^3} + P\frac{d^2r}{dt^2} + Q\frac{dr}{dt} + Rr = v(t)$$

where $v(t)$ is the input excitation. We can write the four equations

$$\frac{d(r)}{dt} = \frac{dr}{dt}$$

$$\frac{d\left(\frac{dr}{dt}\right)}{dt} = \frac{d^2r}{dt^2}$$

$$\frac{d\left(\frac{d^2r}{dt^2}\right)}{dt} = \frac{d^3r}{dt^3}$$

$$\frac{d^4r}{dt^4} = v(t)/M - (N/M)\frac{d^3r}{dt^3} - (P/M)\frac{d^2r}{dt^2} - (Q/M)\frac{dr}{dt} - (R/M)r$$

Choosing the state vector as

$$\mathbf{x} = \begin{bmatrix} r \\ \frac{dr}{dt} \\ \frac{d^2r}{dt^2} \\ \frac{d^3r}{dt^3} \end{bmatrix}$$

allows the fourth-order equation to be expressed as

$$\dot{\mathbf{x}} = \mathbf{A}\mathbf{x} + \mathbf{B}\mathbf{u}$$

where

$$\mathbf{A} = \begin{bmatrix} 0 & 1 & 0 & 0 \\ 0 & 0 & 1 & 0 \\ 0 & 0 & 0 & 1 \\ -R/M & -Q/M & -P/M & -N/M \end{bmatrix}, \quad \mathbf{B} = \begin{bmatrix} 0 \\ 0 \\ 0 \\ 1/M \end{bmatrix}$$

and

$$\mathbf{u} = [v(t)]$$

EXAMPLE 4-2. Write the state equation for the control system of Fig. 4.4. The

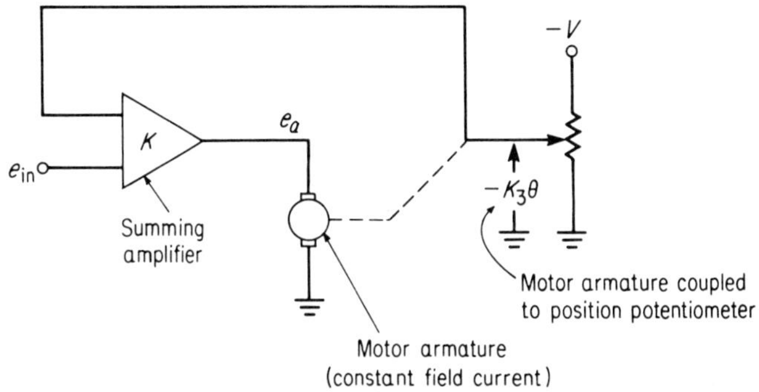

FIG. 4-4. Position-control system.

voltage applied to the armature is dropped across the armature resistance and the back emf of the motor, that is

$$e_a = R_a i_a + K_1 \frac{d\theta}{dt} \tag{4-5}$$

where θ is the shaft position and K_1 is a motor constant. The torque developed by the motor is directly proportional to armature current and this torque is absorbed by frictional effects of the motor and acceleration of the shaft. The equation for torque is

$$T = K_2 i_a = J\frac{d^2\theta}{dt^2} + B\frac{d\theta}{dt} \tag{4-6}$$

where K_2 is a constant, J is the inertia of the shaft (plus the load being driven), and B is a constant accounting for frictional effects.

Solution. Noting that $e_a = (e_{in} - K_3\theta)K$ and applying Eqs. 4-5 and 4-6 results in

$$e_{in} = \frac{R_a J}{KK_2}\frac{d^2\theta}{dt^2} + \left(\frac{R_a B}{KK_2} + \frac{K_1}{K}\right)\frac{d\theta}{dt} + K_3\theta \tag{4-7}$$

The state vector $\mathbf{x}$ is

$$\mathbf{x} = \begin{bmatrix} \theta \\ \dfrac{d\theta}{dt} \end{bmatrix}$$

and

$$\frac{d}{dt}\begin{bmatrix} \theta \\ \dfrac{d\theta}{dt} \end{bmatrix} = \begin{bmatrix} 0 & 1 \\ -KK_2K_3/R_aJ & -(B/J + K_1K_2/R_aJ) \end{bmatrix}\begin{bmatrix} \theta \\ \dfrac{d\theta}{dt} \end{bmatrix} + \begin{bmatrix} 0 \\ KK_2/R_aJ \end{bmatrix}[e_{in}]$$

The output quantity is θ, thus

$$[\theta] = [1 \quad 0] \begin{bmatrix} \theta \\ \dfrac{d\theta}{dt} \end{bmatrix} = [1 \quad 0]\,\mathbf{x}$$

We see that state equations can be written for systems or networks by following certain prescribed rules. It was previously mentioned that the state-equation approach can be used by the digital computer in computer-aided design. Perhaps it is now appropriate to expand this point.

In Chapter 3 topology was considered as an aid to equation formulation. The computer considers the topological information and formulates the analysis equations based on some network-analysis method. The node, cutset, and tieset methods of Chapter 3 require that impedance or admittance of each element be known. Thus the formulated equations are only valid for d-c analysis or a-c analysis at fixed frequencies. ECAP uses this approach for d-c and a-c analyses, but bases transient analysis on a different method [9].

The state equation contains no impedance or admittance terms and expresses the time dependence of the network response in more fundamental terms. A transient input voltage, expressed as a function of time, can be directly substituted into the state equation before solution. Transient problems can then be treated by application and solution of the state equation. Machine solution of this equation can be done by integration or other algorithms for solving a first-order differential equation. The SCEPTRE [7] program is a good example of a commercial network-analysis program that applies the state-variable approach. There are others including ECAP that use state equations or modified state equations for transient analysis. However, the formulation in the following sections applies almost directly to that utilized in SCEPTRE.

B. NETWORK STATE EQUATIONS

The simple networks of the preceding section allowed the state equations to be written by inspection. With networks that are only slightly more complex, the inspection method can become very difficult to apply. A general procedure for writing the state equations using network graphs will be developed in this section. This method is similar to that outlined by Calahan [3] and is based on earlier work by Bashkow [1] and Bryant [2].

The network of Fig. 4-5 is a rather simple circuit for which the state equation cannot be written by inspection. This network is described by the equations

$$e_{\text{in}} = L\frac{di_L}{dt} + R_1 C_1 \frac{dv_{c1}}{dt} + v_{c1}$$

$$R_1 C_1 \frac{dv_{c1}}{dt} + v_{c1} = R_2 C_2 \frac{dv_{c2}}{dt} + v_{c2}$$

$$C_1 \frac{dv_{c1}}{dt} + C_2 \frac{dv_{c2}}{dt} = i_L$$

The first two equations result from summing voltages around two loops while the third equation sums the current at a node. We note that all three equations contain two derivative terms. Each derivative must be expressed in terms of the capacitor voltages and inductor current rather than derivatives of these variables, thus some manipulation is necessary before the standard form of the state equation can be written. The second and third equations can be solved simultaneously to find dv_{c1}/dt and dv_{c2}/dt in terms of the proper variables. As the complexity of the circuit increases it becomes very difficult to write the state equation.

When the resistive voltages can be expressed directly in terms of inductor currents or capacitive voltages, the state equations can be written immediately. The circuit of Fig. 4-5 demonstrates that this is not always possible, and this type of problem necessitates the development of suitable new techniques.

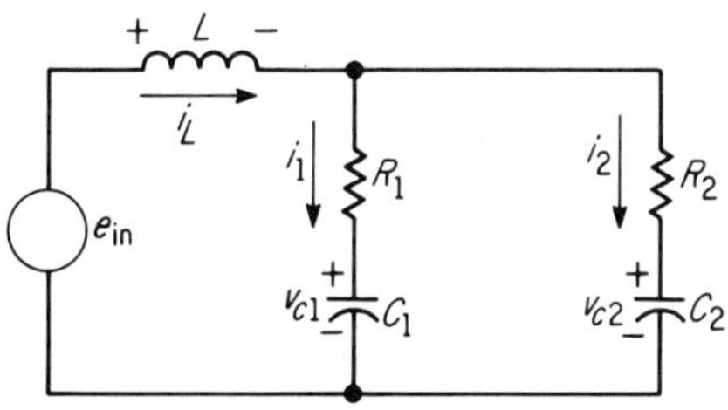

FIG. 4-5. *RLC* network.

Returning to the expressions describing the network of Fig. 4-5, we note that $i_1 = C_1 \dfrac{dv_{c1}}{dt}$ and $i_2 = C_2 \dfrac{dv_{c2}}{dt}$. Using these relationships we can write the three equations such that only one derivative of a state variable appears in each equation. We now have

$$\frac{di_L}{dt} = (1/L)\, e_{\text{in}} - (R_1/L)\, i_1 - (1/L)\, v_{c1}$$

$$\frac{dv_{c1}}{dt} = (1/C_1) i_L - (1/C_1) i_2$$

$$\frac{dv_{c2}}{dt} = (1/C_2)\, i_L - (1/C_2)\, i_1$$

In matrix form,

$$\begin{bmatrix} \dfrac{di_L}{dt} \\ \dfrac{dv_{c1}}{dt} \\ \dfrac{dv_{c2}}{dt} \end{bmatrix} = \begin{bmatrix} 0 & -1/L & 0 \\ 1/C_1 & 0 & 0 \\ 1/C_2 & 0 & 0 \end{bmatrix} \begin{bmatrix} i_L \\ v_{c1} \\ v_{c2} \end{bmatrix} + \begin{bmatrix} -R_1/L & 0 \\ 0 & -1/C_1 \\ -1/C_2 & 0 \end{bmatrix} \begin{bmatrix} i_1 \\ i_2 \end{bmatrix} + \begin{bmatrix} 1/L \\ 0 \\ 0 \end{bmatrix} [e_{in}]$$

or

$$\dot{\mathbf{x}} = \mathbf{A}'\mathbf{x} + \mathbf{B}'\mathbf{u} + \mathbf{C}'\mathbf{i}_R \tag{4-8}$$

The resistive currents i_1 and i_2 are now found in terms of the state vector (and source vector if necessary). This gives

$$i_R = \begin{bmatrix} i_1 \\ i_2 \end{bmatrix} = 1/(R_1 + R_2) \begin{bmatrix} R_2 & -1 & 1 \\ R_1 & 1 & -1 \end{bmatrix} \begin{bmatrix} i_L \\ v_{c1} \\ v_{c2} \end{bmatrix} = \mathbf{T}\mathbf{x} \tag{4-9}$$

The expression for $\dot{\mathbf{x}}$ in terms of $\mathbf{x}$ and $\mathbf{u}$ is

$$\dot{\mathbf{x}} = (\mathbf{A}' + \mathbf{C}'\mathbf{T})\,\mathbf{x} + \mathbf{B}'\mathbf{u}$$

which is the normal form of the equation. The matrix $\mathbf{T}$ was obtained by solving for i_1 and i_2 in terms of the state variables. This can also be difficult when more variables are involved. In this case the matrix $\mathbf{B}'$ equals $\mathbf{B}$.

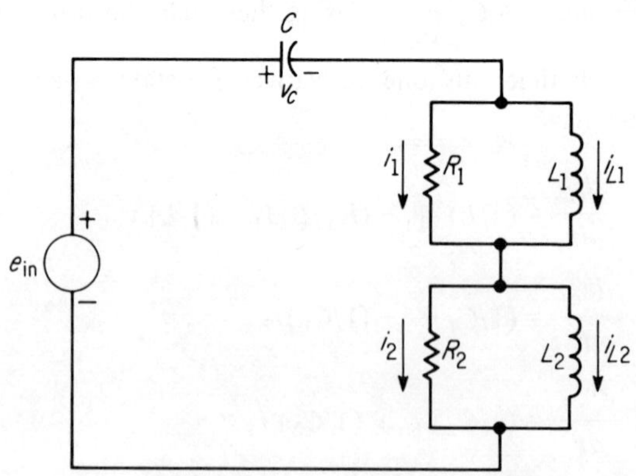

FIG. 4-6. *RLC* network.

The same kind of problem is encountered when resistive voltages must be expressed in terms of state variables in order to write the state equations. Figure 4-6 shows a network of this type.

From the equations

$$e_{\text{in}} = v_c + v_{R_1} + v_{R_2}$$

$$L_1 \frac{di_{L1}}{dt} = v_{R_1}$$

$$L_2 \frac{di_{L2}}{dt} = v_{R_2}$$

$$C \frac{dv_c}{dt} = v_{R_1}/R_1 + i_{L_1}$$

$$C \frac{dv_c}{dt} = v_{R_2}/R_2 + i_{L_2}$$

we can write

$$\frac{d}{dt}\begin{bmatrix} v_c \\ i_{L_1} \\ i_{L_2} \end{bmatrix} = \begin{bmatrix} 0 & 1/C & 0 \\ 0 & 0 & 0 \\ 0 & 0 & 0 \end{bmatrix} \begin{bmatrix} v_c \\ i_{L_1} \\ i_{L_2} \end{bmatrix} + \begin{bmatrix} 1/CR_1 & 0 \\ 1/L_1 & 0 \\ 0 & 1/L_2 \end{bmatrix} \begin{bmatrix} v_{R_1} \\ v_{R_2} \end{bmatrix}$$

The voltages v_{R_1} and v_{R_2} can be expressed in terms of the state vector and e_{in} as

$$\begin{bmatrix} v_{R_1} \\ v_{R_2} \end{bmatrix} = \frac{1}{R_1 + R_2} \begin{bmatrix} -R_1 & -R_1R_2 & R_1R_2 \\ -R_2 & R_1R_2 & -R_1R_2 \end{bmatrix} \begin{bmatrix} v_c \\ i_{L_1} \\ i_{L_2} \end{bmatrix} + \frac{1}{R_1 + R_2} \begin{bmatrix} R_1 \\ R_2 \end{bmatrix} e_{\text{in}}$$

Combining these equations gives

$$\dot{\mathbf{x}} = \frac{1}{R_1 + R_2} \begin{bmatrix} -1/C & R_1/C & R_2/C \\ -R_1/L_1 & -R_1R_2/L_1 & R_1R_2/L_1 \\ -R_2/L_2 & R_1R_2/L_2 & -R_1R_2/L_2 \end{bmatrix} \mathbf{x} + \frac{1}{R_1 + R_2} \begin{bmatrix} 1/C \\ R_1/L_1 \\ R_2/L_2 \end{bmatrix} e_{\text{in}}$$

In the networks of Figs. 4-5 and 4-6 the state equations could not be written directly. An intermediate vector, consisting of either the resistive currents or resistive voltages, had to be used. This vector then had to be found in terms of the state vector and source vector. A more orderly procedure can be used to express this intermediate vector as a function of the appropriate variables.

Returning to Eq. 3-1, we see that

$$[\mathbf{A}_T \quad \mathbf{A}_L] \begin{bmatrix} \mathbf{i}_T \\ \mathbf{i}_L \end{bmatrix} = 0 \qquad [3\text{-}1]$$

where $\mathbf{i}_T$ is the column vector of tree-branch currents and $\mathbf{i}_L$ is the column vector of link-branch currents. Solving for $\mathbf{i}_T$ in terms of $\mathbf{i}_L$ gives

$$\mathbf{i}_T = \mathbf{A}_T^{-1} \mathbf{A}_L \, \mathbf{i}_L \tag{4-10}$$

Equation 4-10 indicates that the tree-branch currents can be written in terms of the link-branch currents. Starting with Eq. 3-2

$$[\mathbf{B}_T \quad \mathbf{B}_L] \begin{bmatrix} \mathbf{e}_T \\ \mathbf{e}_L \end{bmatrix} = 0 \qquad [3\text{-}2]$$

allows us to express the tree-branch voltages in terms of the link-branch voltages as

$$\mathbf{e}_T = -\mathbf{B}_T^{-1} \mathbf{B}_L \, \mathbf{e}_L \tag{4-11}$$

We can also write $\mathbf{e}_L$ in terms of $\mathbf{e}_T$ as

$$\mathbf{e}_L = -\mathbf{B}_L^{-1} \, \mathbf{B}_T \, \mathbf{e}_T \tag{4-12}$$

Equation 4-10 can be simplified by applying Eq. 3-11 to give

$$\mathbf{i}_T = -\mathbf{C}_L \, \mathbf{i}_L \tag{4-13}$$

Equation 4-12 reduces to

$$\mathbf{e}_L = -\mathbf{B}_T \, \mathbf{e}_T \tag{4-14}$$

upon noting that $\mathbf{B}_L$ is the unit matrix as discussed in connection with the formation of the tieset matrix in Chapter 3.

For most networks a tree of the graph can be so chosen to include all capacitors, all voltage sources, and some resistors. The links then consist of all inductances, the remaining resistors not included in the tree, and all current sources. The tree resulting from this particular method of selection is called a proper tree. We will see later that occasionally we must include inductors in the tree or delete capacitors from the tree, but in many cases a proper tree can be formed as prescribed above.

Since capacitor currents will be tree-branch currents, we can express capacitor currents $\left(\text{and hence } \dfrac{dv_c}{dt}\right)$ in terms of inductor link currents, resistive link currents, and independent source currents. The inductor voltages $\left(\text{and hence } \dfrac{di_L}{dt}\right)$ can be expressed in terms of capacitor tree voltages, resistive tree voltages, and independent source voltages. The derivatives of the state variables (dv_c/dt and di_L/dt) can then be written as functions of inductor link currents and capacitor tree voltages, resistive link currents and resistive tree voltages, and independent current sources and voltage sources. This matrix equation resembles the state equation

$$\dot{\mathbf{x}} = \mathbf{A}\mathbf{x} + \mathbf{B}\mathbf{u}$$

but contains the resistive link currents and tree voltages as additional variables. By applying Ohm's law and solving for these variables in terms of the state vector and source vector, these resistive variables can be eliminated. The state equation is then expressible in terms of the state vector and source vector. Before developing the general theory, a specific example will be considered to demonstrate the method.

EXAMPLE 4-3. Write the state equation for the network of Fig. 4-6 applying Eqs. 4-13 and 4-14.

Solution. The oriented network graph and the proper tree of the graph are shown in Fig. 4-7. The tree-branch currents can be written in terms of the inductive link and resistive link currents as

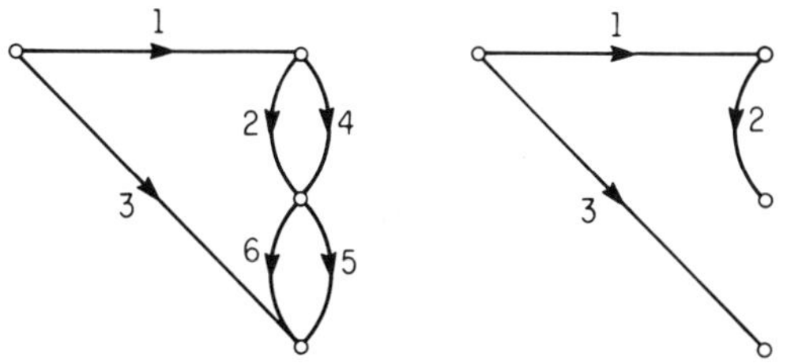

FIG. 4-7. Oriented graph and proper tree of the circuit of Fig. 4-6.

$$\begin{bmatrix} i_1 \\ i_2 \\ i_3 \end{bmatrix} = \begin{bmatrix} 0 & 1 & 1 \\ -1 & 1 & 1 \\ 0 & -1 & -1 \end{bmatrix} \begin{bmatrix} i_4 \\ i_5 \\ i_6 \end{bmatrix} = \begin{bmatrix} 0 & 1 \\ -1 & 1 \\ 0 & -1 \end{bmatrix} \begin{bmatrix} i_4 \\ i_5 \end{bmatrix} + \begin{bmatrix} 1 \\ 1 \\ -1 \end{bmatrix} [i_6]$$

The link voltages are expressed in terms of tree-branch voltages as

$$\begin{bmatrix} v_4 \\ v_5 \\ v_6 \end{bmatrix} = \begin{bmatrix} 0 & 1 & 0 \\ -1 & -1 & 1 \\ -1 & -1 & 1 \end{bmatrix} \begin{bmatrix} v_1 \\ v_2 \\ v_3 \end{bmatrix} = \begin{bmatrix} 0 \\ -1 \\ -1 \end{bmatrix} [v_1] + \begin{bmatrix} 1 \\ -1 \\ -1 \end{bmatrix} [v_2] + \begin{bmatrix} 0 \\ 1 \\ 1 \end{bmatrix} [v_3]$$

Note that the tree-branch voltage vector is resolved into capacitor voltages, resistor voltages, and source voltages. We now have expressions for capacitor current i_1 and inductor voltages v_4 and v_5. From the previous two equations we express these variables as

$$\begin{bmatrix} i_1 \\ v_4 \\ v_5 \end{bmatrix} = \begin{bmatrix} 0 & 0 & 1 \\ 0 & 0 & 0 \\ -1 & 0 & 0 \end{bmatrix} \begin{bmatrix} v_1 \\ i_4 \\ i_5 \end{bmatrix} + \begin{bmatrix} 1 & 0 \\ 0 & 1 \\ 0 & -1 \end{bmatrix} \begin{bmatrix} i_6 \\ v_2 \end{bmatrix} + \begin{bmatrix} 0 \\ 0 \\ 1 \end{bmatrix} [v_3]$$

The current i_1 is equal to $C dv_1/dt$ while v_4 equals $L_1 \, di_4/dt$ and v_5 equals $L_2 \, di_5/dt$. Thus the preceding equation is very similar to the state equation of interest. The major difference is that a vector associated with the resistive branches appears in the equation along with the state vector and source vector. We must eliminate this vector from the equation by expressing it in terms of the state vector and source vector.

The first two equations of this example can again be used to derive expressions for the resistive variables i_2 and v_6, that is

$$\begin{bmatrix} i_2 \\ v_6 \end{bmatrix} = \begin{bmatrix} 0 & -1 & 1 \\ -1 & 0 & 0 \end{bmatrix} \begin{bmatrix} v_1 \\ i_4 \\ i_5 \end{bmatrix} + \begin{bmatrix} 1 & 0 \\ 0 & -1 \end{bmatrix} \begin{bmatrix} i_6 \\ v_2 \end{bmatrix} + \begin{bmatrix} 0 \\ 1 \end{bmatrix} [v_3]$$

Since branches 2 and 6 are resistive branches we can express Ohm's law for these branches as

$$\begin{bmatrix} i_2 \\ v_6 \end{bmatrix} = \begin{bmatrix} 0 & 1/R_1 \\ R_2 & 0 \end{bmatrix} \begin{bmatrix} i_6 \\ v_2 \end{bmatrix}$$

Solving these equations simultaneously gives

$$\begin{bmatrix} i_6 \\ v_2 \end{bmatrix} = \frac{1}{R_1 + R_2} \begin{bmatrix} -1 & R_1 & -R_1 \\ R_1 & -R_1 R_2 & R_1 R_2 \end{bmatrix} \begin{bmatrix} v_1 \\ i_4 \\ i_5 \end{bmatrix} + \frac{1}{R_1 + R_2} \begin{bmatrix} 1 \\ R_1 \end{bmatrix} [v_3]$$

This value is now substituted into the equation for i_1, v_4, and v_5 to give

$$\begin{bmatrix} C dv_1/dt \\ L_1 \, di_4/dt \\ L_2 \, di_5/dt \end{bmatrix} = \frac{1}{R_1 + R_2} \begin{bmatrix} -1 & R_1 & R_2 \\ -R_1 & -R_1 R_2 & R_1 R_2 \\ -R_2 & R_1 R_2 & -R_1 R_2 \end{bmatrix} \begin{bmatrix} v_1 \\ i_4 \\ i_5 \end{bmatrix} + \frac{1}{R_1 + R_2} \begin{bmatrix} 1 \\ R_1 \\ R_2 \end{bmatrix} [v_3]$$

This equation can be compared to that derived earlier for this circuit.

In comparing the effort required to obtain the state equation in this example to the previous derivation, it is obvious that more steps are used in Example 4-3. There are two important justifications for proceeding as outlined in the example. This is a general method which we shall develop more fully in succeeding sections. The alternate derivation requires the solution of simultaneous equations to eliminate an unwanted vector. As the magnitude of the problem increases, the effort required to use this method increases rapidly. The general method of the example utilizes matrices in eliminating a vector, and of course matrix theory is very efficiently applied to this type of problem even for a large number of variables. Using the general method also allows application of the digital computer in formulating the state equations. Only element values and topological information need be entered, and the computer can rapidly derive the state equation for the network.

Formal Procedure for Obtaining the State Equation

Let us assume that we have a network with m capacitors all of which can be included in a proper tree, n tree resistors, p voltage sources, q link inductors, r link resistors, and s current sources. The network graph is drawn numbering the capacitive tree branches from 1 to m. The n tree resistors are assigned the branch numbers $m+1$ to $m+n$ and the voltage source branches are assigned numbers $m+n+1$ to $m+n+p$. The inductive link branches are assigned numbers $m+n+p+1$ to $m+n+p+q$, the resistive link branches are assigned numbers $m+n+p+q+1$ to $m+n+p+q+r$, and the current source branches are assigned $m+n+p+q+r+1$ to $m+n+p+q+r+s$. The portion of the cutset matrix corresponding to link branches ($\mathbf{C}_L$) and the portion of the tieset matrix corresponding to the proper tree ($\mathbf{B}_T$) are evaluated. Equation 4-13 can be written as

$$\mathbf{i}_T = \begin{bmatrix} \mathbf{i}_{TC} \\ \mathbf{i}_{TR} \\ \mathbf{i}_{TG} \end{bmatrix} = -\mathbf{C}_L \begin{bmatrix} \mathbf{i}_{LL} \\ \mathbf{i}_{LR} \\ \mathbf{i}_{LG} \end{bmatrix} = -\mathbf{C}_L \mathbf{i}_L \tag{4-15}$$

where $\mathbf{i}_{TC}$ is the subvector of capacitor currents, $\mathbf{i}_{TR}$ is the subvector of resistive tree-branch currents, $\mathbf{i}_{TG}$ is the subvector of voltage source tree-branch currents $\mathbf{i}_{LL}$ is the subvector of inductor currents, $\mathbf{i}_{LR}$ is the subvector of resistive link-branch currents, and $\mathbf{i}_{LG}$ is the subvector of current source link-branch currents. The matrix $\mathbf{C}_L$ can be partitioned to express $\mathbf{i}_T$ as

$$\begin{bmatrix} \mathbf{i}_{TC} \\ \mathbf{i}_{TR} \\ \mathbf{i}_{TG} \end{bmatrix} = -\begin{bmatrix} \mathbf{C}_{LLC} & \mathbf{C}_{LRC} & \mathbf{C}_{LGC} \\ \mathbf{C}_{LLR} & \mathbf{C}_{LRR} & \mathbf{C}_{LGR} \\ \mathbf{C}_{LLG} & \mathbf{C}_{LRG} & \mathbf{C}_{LGG} \end{bmatrix} \begin{bmatrix} \mathbf{i}_{LL} \\ \mathbf{i}_{LR} \\ \mathbf{i}_{LG} \end{bmatrix} \tag{4-16}$$

The dimensions of the submatrices are

$$\begin{array}{lll} \mathbf{C}_{LLC} \;\; m \times q, & \mathbf{C}_{LRC} \;\; m \times r, & \mathbf{C}_{LGC} \;\; m \times s \\ \mathbf{C}_{LLR} \;\; n \times q, & \mathbf{C}_{LRR} \;\; n \times r, & \mathbf{C}_{LGR} \;\; n \times s \\ \mathbf{C}_{LLG} \;\; p \times q, & \mathbf{C}_{LRG} \;\; p \times r, & \mathbf{C}_{LGG} \;\; p \times s \end{array}$$

Equation 4-16 is equivalent to

$$\begin{bmatrix} \mathbf{i}_{TC} \\ \mathbf{i}_{TR} \\ \mathbf{i}_{TG} \end{bmatrix} = -\begin{bmatrix} \mathbf{C}_{LLC} \\ \mathbf{C}_{LLR} \\ \mathbf{C}_{LLG} \end{bmatrix} \mathbf{i}_{LL} - \begin{bmatrix} \mathbf{C}_{LRC} \\ \mathbf{C}_{LRR} \\ \mathbf{C}_{LRG} \end{bmatrix} \mathbf{i}_{LR} - \begin{bmatrix} \mathbf{C}_{LGC} \\ \mathbf{C}_{LGR} \\ \mathbf{C}_{LGG} \end{bmatrix} \mathbf{i}_{LG} \tag{4-17}$$

From this equation it is easy to express $\mathbf{i}_{TC}$ and $\mathbf{i}_{TR}$ in terms of $\mathbf{i}_{LL}$, $\mathbf{i}_{LR}$, and $\mathbf{i}_{LG}$ as

$$\mathbf{i}_{TC} = -\mathbf{C}_{LLC}\,\mathbf{i}_{LL} - \mathbf{C}_{LRC}\,\mathbf{i}_{LR} - \mathbf{C}_{LGC}\,\mathbf{i}_{LG} \tag{4-18}$$

and

$$\mathbf{i}_{TR} = -\mathbf{C}_{LLR}\,\mathbf{i}_{LL} - \mathbf{C}_{LRR}\,\mathbf{i}_{LR} - \mathbf{C}_{LGR}\,\mathbf{i}_{LG} \tag{4-19}$$

We proceed in a similar fashion to write the link-branch voltages as

$$\mathbf{e}_L = \begin{bmatrix} \mathbf{e}_{LL} \\ \mathbf{e}_{LR} \\ \mathbf{e}_{LG} \end{bmatrix} = -\mathbf{B}_T \begin{bmatrix} \mathbf{e}_{TC} \\ \mathbf{e}_{TR} \\ \mathbf{e}_{TG} \end{bmatrix} = -\mathbf{B}_T\,\mathbf{e}_T \tag{4-20}$$

where $\mathbf{e}_{LL}$ is the subvector of inductor voltages, $\mathbf{e}_{LR}$ is the subvector of resistive link-branch voltages, $\mathbf{e}_{LG}$ is the subvector of current source link-branch voltages, $\mathbf{e}_{TC}$ is the subvector of capacitor voltages, $\mathbf{e}_{TR}$ is the subvector of resistive tree-branch voltages, and $\mathbf{e}_{TG}$ is the subvector of voltage source tree branch voltages. The matrix $\mathbf{B}_T$ can be partitioned to express $\mathbf{e}_L$ as

$$\begin{bmatrix} \mathbf{e}_{LL} \\ \mathbf{e}_{LR} \\ \mathbf{e}_{LG} \end{bmatrix} = -\begin{bmatrix} \mathbf{B}_{TCL} & \mathbf{B}_{TRL} & \mathbf{B}_{TGL} \\ \mathbf{B}_{TCR} & \mathbf{B}_{TRR} & \mathbf{B}_{TGR} \\ \mathbf{B}_{TCG} & \mathbf{B}_{TRG} & \mathbf{B}_{TGG} \end{bmatrix} \begin{bmatrix} \mathbf{e}_{TC} \\ \mathbf{e}_{TR} \\ \mathbf{e}_{TG} \end{bmatrix} \tag{4-21}$$

The dimensions of the submatrices are

$$\begin{array}{lll} \mathbf{B}_{TCL}\ \ q \times m, & \mathbf{B}_{TRL}\ \ q \times n, & \mathbf{B}_{TGL}\ \ q \times p \\ \mathbf{B}_{TCR}\ \ r \times m, & \mathbf{B}_{TRR}\ \ r \times n, & \mathbf{B}_{TGR}\ \ r \times p \\ \mathbf{B}_{TCG}\ \ s \times m, & \mathbf{B}_{TRG}\ \ s \times n, & \mathbf{B}_{TGG}\ \ s \times p \end{array}$$

Equation 4-21 is equivalent to

$$\begin{bmatrix} \mathbf{e}_{LL} \\ \mathbf{e}_{LR} \\ \mathbf{e}_{LG} \end{bmatrix} = -\begin{bmatrix} \mathbf{B}_{TCL} \\ \mathbf{B}_{TCR} \\ \mathbf{B}_{TCG} \end{bmatrix} \mathbf{e}_{TC} - \begin{bmatrix} \mathbf{B}_{TRL} \\ \mathbf{B}_{TRR} \\ \mathbf{B}_{TRG} \end{bmatrix} \mathbf{e}_{TR} - \begin{bmatrix} \mathbf{B}_{TGL} \\ \mathbf{B}_{TGR} \\ \mathbf{B}_{TGG} \end{bmatrix} \mathbf{e}_{TG} \tag{4-22}$$

From this equation $\mathbf{e}_{LL}$ and $\mathbf{e}_{LR}$ can be expressed as

$$\mathbf{e}_{LL} = -\mathbf{B}_{TCL}\,\mathbf{e}_{TC} - \mathbf{B}_{TRL}\,\mathbf{e}_{TR} - \mathbf{B}_{TGL}\,\mathbf{e}_{TG} \tag{4-23}$$

and

$$\mathbf{e}_{LR} = -\mathbf{B}_{TCR}\,\mathbf{e}_{TC} - \mathbf{B}_{TRR}\,\mathbf{e}_{TR} - \mathbf{B}_{TGR}\,\mathbf{e}_{TG} \tag{4-24}$$

Equations 4-18 and 4-23 are now combined to give

$$\begin{bmatrix} \mathbf{i}_{TC} \\ \mathbf{e}_{LL} \end{bmatrix} = -\begin{bmatrix} \mathbf{0} & \mathbf{C}_{LLC} \\ \mathbf{B}_{TCL} & \mathbf{0} \end{bmatrix} \begin{bmatrix} \mathbf{e}_{TC} \\ \mathbf{i}_{LL} \end{bmatrix} - \begin{bmatrix} \mathbf{0} & \mathbf{C}_{LRC} \\ \mathbf{B}_{TRL} & \mathbf{0} \end{bmatrix} \begin{bmatrix} \mathbf{e}_{TR} \\ \mathbf{i}_{LR} \end{bmatrix} - \begin{bmatrix} \mathbf{0} & \mathbf{C}_{LGC} \\ \mathbf{B}_{TGL} & \mathbf{0} \end{bmatrix} \begin{bmatrix} \mathbf{e}_{TG} \\ \mathbf{i}_{LG} \end{bmatrix} \tag{4-25}$$

Note that the left-hand side of Eq. 4-25 contains capacitor currents which are proportional to the first derivative of capacitor voltages. The inductor voltages which are proportional to the first derivatives of inductor current are also present. These derivatives are expressed in terms of capacitor voltages and inductor currents, source voltages and source currents, and resistive tree-branch voltages and resistive link-branch currents. We must now express $\mathbf{e}_{TR}$ and $\mathbf{i}_{LR}$ as functions of $\mathbf{e}_{TC}$, $\mathbf{i}_{LL}$, $\mathbf{e}_{TG}$, and $\mathbf{i}_{LG}$ in order to write the desired state equation.

Combining Eqs. 4-19 and 4-24 we have

$$\begin{bmatrix} \mathbf{e}_{LR} \\ \mathbf{i}_{TR} \end{bmatrix} = -\begin{bmatrix} \mathbf{B}_{TCR} & \mathbf{0} \\ \mathbf{0} & \mathbf{C}_{LLR} \end{bmatrix} \begin{bmatrix} \mathbf{e}_{TC} \\ \mathbf{i}_{LL} \end{bmatrix} - \begin{bmatrix} \mathbf{B}_{TRR} & \mathbf{0} \\ \mathbf{0} & \mathbf{C}_{LRR} \end{bmatrix} \begin{bmatrix} \mathbf{e}_{TR} \\ \mathbf{i}_{LR} \end{bmatrix} - \begin{bmatrix} \mathbf{B}_{TGR} & \mathbf{0} \\ \mathbf{0} & \mathbf{C}_{LGR} \end{bmatrix} \begin{bmatrix} \mathbf{e}_{TG} \\ \mathbf{i}_{LG} \end{bmatrix} \tag{4-26}$$

The resistive voltages and currents are related by

$$\begin{bmatrix} \mathbf{e}_{LR} \\ \mathbf{i}_{TR} \end{bmatrix} = \begin{bmatrix} \mathbf{0} & \mathbf{R}_{L} \\ \mathbf{G}_{T} & \mathbf{0} \end{bmatrix} \begin{bmatrix} \mathbf{e}_{TR} \\ \mathbf{i}_{LR} \end{bmatrix} \tag{4-27}$$

where $\mathbf{R}_L$ is an $r \times r$ matrix whose diagonal elements are link-branch resistances and $\mathbf{G}_T$ is an $n \times n$ matrix whose diagonal elements are tree-branch conductances. The off-diagonal elements are zero.

Equations 4-26 and 4-27 can be solved to express $\mathbf{e}_{TR}$ and $\mathbf{i}_{LR}$ as

$$\begin{bmatrix} \mathbf{e}_{TR} \\ \mathbf{i}_{LR} \end{bmatrix} = -\begin{bmatrix} \mathbf{B}_{TRR} & \mathbf{R}_{L} \\ \mathbf{G}_{T} & \mathbf{C}_{LRR} \end{bmatrix}^{-1} \begin{bmatrix} \mathbf{B}_{TCR} & \mathbf{0} \\ \mathbf{0} & \mathbf{C}_{LLR} \end{bmatrix} \begin{bmatrix} \mathbf{e}_{TC} \\ \mathbf{i}_{LL} \end{bmatrix} - \begin{bmatrix} \mathbf{B}_{TRR} & \mathbf{R}_{L} \\ \mathbf{G}_{T} & \mathbf{C}_{LRR} \end{bmatrix}^{-1} \begin{bmatrix} \mathbf{B}_{TGR} & \mathbf{0} \\ \mathbf{0} & \mathbf{C}_{LGR} \end{bmatrix} \begin{bmatrix} \mathbf{e}_{TG} \\ \mathbf{i}_{LG} \end{bmatrix} \tag{4-28}$$

Using Eq. 4-28 we can eliminate the variables $\mathbf{e}_{TR}$ and $\mathbf{i}_{LR}$ from Eq. 4-25 to yield

$$\begin{bmatrix} \mathbf{i}_{TC} \\ \mathbf{e}_{LL} \end{bmatrix} = \left\{ \begin{bmatrix} \mathbf{0} & \mathbf{C}_{LRC} \\ \mathbf{B}_{TRL} & \mathbf{0} \end{bmatrix} \left(\begin{bmatrix} \mathbf{B}_{TRR} & \mathbf{R}_{L} \\ \mathbf{G}_{T} & \mathbf{C}_{LRR} \end{bmatrix}^{-1} \right) \begin{bmatrix} \mathbf{B}_{TCR} & \mathbf{0} \\ \mathbf{0} & \mathbf{C}_{LLR} \end{bmatrix} - \begin{bmatrix} \mathbf{0} & \mathbf{C}_{LLC} \\ \mathbf{B}_{TCL} & \mathbf{0} \end{bmatrix} \right\} \begin{bmatrix} \mathbf{e}_{TC} \\ \mathbf{i}_{LL} \end{bmatrix} + \left\{ \begin{bmatrix} \mathbf{0} & \mathbf{C}_{LRC} \\ \mathbf{B}_{TRL} & \mathbf{0} \end{bmatrix} \left(\begin{bmatrix} \mathbf{B}_{TRR} & \mathbf{R}_{L} \\ \mathbf{G}_{T} & \mathbf{C}_{LRR} \end{bmatrix}^{-1} \right) \right.$$

$$\times \begin{bmatrix} \mathbf{B}_{TGR} & \mathbf{0} \\ \mathbf{0} & \mathbf{C}_{LGR} \end{bmatrix} - \begin{bmatrix} \mathbf{0} & \mathbf{C}_{LGC} \\ \mathbf{B}_{TGL} & \mathbf{0} \end{bmatrix} \Bigg\} \begin{bmatrix} \mathbf{e}_{TG} \\ \mathbf{i}_{LG} \end{bmatrix} \tag{4-29}$$

Since

$$\begin{bmatrix} \mathbf{i}_{TC} \\ \mathbf{e}_{LL} \end{bmatrix} = \begin{bmatrix} C_1 \dfrac{dv_1}{dt} \\ C_2 \dfrac{dv_2}{dt} \\ \cdot \\ \cdot \\ \cdot \\ C_m \dfrac{dv_m}{dt} \\ L_1 \dfrac{di_{L1}}{dt} \\ L_2 \dfrac{di_{L2}}{dt} \\ \cdot \\ \cdot \\ \cdot \\ L_q \dfrac{di_{Lq}}{dt} \end{bmatrix}$$

Eq. 4-29 represents the state equation.

It should be pointed out that the zero elements shown in the matrices are really zero submatrices with dimensions determined by the nonzero submatrices.

Although there is a formidable number of submatrices required in the preceding derivation we should review the steps required to appreciate the advantages. The initial step is to construct a proper tree and number the branches as prescribed. The second step is to obtain the topological matrices $\mathbf{C}_L$ and $\mathbf{B}_T$. Given a minimal amount of input information, the digital computer can perform these first two steps with ease. The matrices $\mathbf{C}_L$ and $\mathbf{B}_T$ are then partitioned into submatrices. The matrices $\mathbf{R}_L$ and $\mathbf{G}_T$ are generated and $\mathbf{e}_{TR}$ and $\mathbf{i}_{LR}$ are found by inverting the appropriate matrix (Eq. 4-28). The state equation is then obtained simply by matrix manipulation. The prime requisite in applying this method is ability to manipulate matrices and bookkeeping ability, both attributes possessed by the digital computer. As shown later the computer not only generates the state equations, but can efficiently solve these equations. Thus the previous theory is directed mainly at computer-aided design applications.

While this method can be useful in a number of applications there are several problems that can arise. We must first consider the modifications of theory necessary to accommodate dependent sources. We will then consider the difficulties that can arise when no proper tree of the network can be formed. This situation occurs when all branches of a loop contain capacitors or when only inductors are connected to a given node.

State Equation for Networks with Dependent Sources

The procedure for obtaining the state equation when dependent current and/or voltage sources are present is similar to the previous theory except that the elimination of an additional unknown vector is required. The tree-current vector is partitioned into four submatrices relating to capacitor, resistor, independent voltage source, and dependent voltage source currents. The link-current vector is partitioned into submatrices relating to inductor, resistor, independent current source, and dependent current source currents. We can then write

$$\begin{bmatrix} \mathbf{i}_{TC} \\ \mathbf{i}_{TR} \\ \mathbf{i}_{TG} \\ \mathbf{i}_{TD} \end{bmatrix} = - \begin{bmatrix} \mathbf{C}_{LLC} & \mathbf{C}_{LRC} & \mathbf{C}_{LGC} & \mathbf{C}_{LDC} \\ \mathbf{C}_{LLR} & \mathbf{C}_{LRR} & \mathbf{C}_{LGR} & \mathbf{C}_{LDR} \\ \mathbf{C}_{LLG} & \mathbf{C}_{LRG} & \mathbf{C}_{LGG} & \mathbf{C}_{LDG} \\ \mathbf{C}_{LLD} & \mathbf{C}_{LRD} & \mathbf{C}_{LGD} & \mathbf{C}_{LDD} \end{bmatrix} \begin{bmatrix} \mathbf{i}_{LL} \\ \mathbf{i}_{LR} \\ \mathbf{i}_{LG} \\ \mathbf{i}_{LD} \end{bmatrix} \tag{4-30}$$

where $\mathbf{i}_{TD}$ is the subvector of dependent voltage source (tree) currents and $\mathbf{i}_{LD}$ is the subvector of dependent current source (link) currents. All other variables have been defined previously. If there are u dependent voltage sources and v dependent current sources the dimensions of the previously undefined submatrices are

$$\mathbf{C}_{LDC}\ m \times v, \quad \mathbf{C}_{LDR}\ n \times v, \quad \mathbf{C}_{LDG}\ p \times v, \quad \mathbf{C}_{LDD}\ u \times v$$
$$\mathbf{C}_{LLD}\ u \times q, \quad \mathbf{C}_{LRD}\ u \times r, \quad \mathbf{C}_{LGD}\ u \times s$$

The link-branch voltages are expressed in terms of the tree branch voltages as

$$\begin{bmatrix} \mathbf{e}_{LL} \\ \mathbf{e}_{LR} \\ \mathbf{e}_{LG} \\ \mathbf{e}_{LD} \end{bmatrix} = - \begin{bmatrix} \mathbf{B}_{TCL} & \mathbf{B}_{TRL} & \mathbf{B}_{TGL} & \mathbf{B}_{TDL} \\ \mathbf{B}_{TCR} & \mathbf{B}_{TRR} & \mathbf{B}_{TGR} & \mathbf{B}_{TDR} \\ \mathbf{B}_{TCG} & \mathbf{B}_{TRG} & \mathbf{B}_{TGG} & \mathbf{B}_{TDG} \\ \mathbf{B}_{TCD} & \mathbf{B}_{TRD} & \mathbf{B}_{TGD} & \mathbf{B}_{TDD} \end{bmatrix} \begin{bmatrix} \mathbf{e}_{TC} \\ \mathbf{e}_{TR} \\ \mathbf{e}_{TG} \\ \mathbf{e}_{TD} \end{bmatrix} \tag{4-31}$$

where $\mathbf{e}_{LD}$ is the subvector of voltages across the dependent current sources and $\mathbf{e}_{TD}$ is the subvector of dependent voltage source voltages. The dimensions of the new submatrices are

$$\mathbf{B}_{TDL}\ q \times u, \quad \mathbf{B}_{TDR}\ r \times u, \quad \mathbf{B}_{TDG}\ s \times u, \quad \mathbf{B}_{TDD}\ v \times u$$
$$\mathbf{B}_{TCD}\ v \times m, \quad \mathbf{B}_{TRD}\ v \times n, \quad \mathbf{B}_{TGD}\ v \times p$$

The equation for capacitor currents and inductor voltages can be written as

$$\begin{bmatrix} \mathbf{i}_{TC} \\ \mathbf{e}_{LL} \end{bmatrix} = -\begin{bmatrix} \mathbf{0} & \mathbf{C}_{LLC} \\ \mathbf{B}_{TCL} & \mathbf{0} \end{bmatrix} \begin{bmatrix} \mathbf{e}_{TC} \\ \mathbf{i}_{LL} \end{bmatrix} - \begin{bmatrix} \mathbf{0} & \mathbf{C}_{LRC} \\ \mathbf{B}_{TRL} & \mathbf{0} \end{bmatrix} \begin{bmatrix} \mathbf{e}_{TR} \\ \mathbf{i}_{LR} \end{bmatrix}$$
$$- \begin{bmatrix} \mathbf{0} & \mathbf{C}_{LGC} \\ \mathbf{B}_{TGL} & \mathbf{0} \end{bmatrix} \begin{bmatrix} \mathbf{e}_{TG} \\ \mathbf{i}_{LG} \end{bmatrix} - \begin{bmatrix} \mathbf{0} & \mathbf{C}_{LDC} \\ \mathbf{B}_{TDL} & \mathbf{0} \end{bmatrix} \begin{bmatrix} \mathbf{e}_{TD} \\ \mathbf{i}_{LD} \end{bmatrix} \qquad (4\text{-}32)$$

The resistive tree-branch currents and resistive link voltages are

$$\begin{bmatrix} \mathbf{e}_{LR} \\ \mathbf{i}_{TR} \end{bmatrix} = -\begin{bmatrix} \mathbf{B}_{TCR} & \mathbf{0} \\ \mathbf{0} & \mathbf{C}_{LLR} \end{bmatrix} \begin{bmatrix} \mathbf{e}_{TC} \\ \mathbf{i}_{LL} \end{bmatrix} - \begin{bmatrix} \mathbf{B}_{TRR} & \mathbf{0} \\ \mathbf{0} & \mathbf{C}_{LRR} \end{bmatrix} \begin{bmatrix} \mathbf{e}_{TR} \\ \mathbf{i}_{LR} \end{bmatrix}$$
$$- \begin{bmatrix} \mathbf{B}_{TGR} & \mathbf{0} \\ \mathbf{0} & \mathbf{C}_{LGR} \end{bmatrix} \begin{bmatrix} \mathbf{e}_{TG} \\ \mathbf{i}_{LG} \end{bmatrix} - \begin{bmatrix} \mathbf{B}_{TDR} & \mathbf{0} \\ \mathbf{0} & \mathbf{C}_{LDR} \end{bmatrix} \begin{bmatrix} \mathbf{e}_{TD} \\ \mathbf{i}_{LD} \end{bmatrix} \qquad (4\text{-}33)$$

The state equation can be derived from Eq. 4-32 if the vectors relating to resistive elements and dependent sources are eliminated. The dependent source vector will depend on other network variables and it is possible to express this dependence as

$$\begin{bmatrix} \mathbf{e}_{TD} \\ \mathbf{i}_{LD} \end{bmatrix} = [\mathbf{G}_1] \begin{bmatrix} \mathbf{e}_{TC} \\ \mathbf{i}_{LL} \end{bmatrix} + [\mathbf{G}_2] \begin{bmatrix} \mathbf{e}_{TR} \\ \mathbf{i}_{LR} \end{bmatrix} + [\mathbf{G}_3] \begin{bmatrix} \mathbf{e}_{TG} \\ \mathbf{i}_{LG} \end{bmatrix} \qquad (4\text{-}34)$$

Equation 4-34 is substituted into Eqs. 4-32 and 4-33 to eliminate the dependent source vector. Ohm's law for the resistive branches is then applied to eliminate the resistive element vector as previously done, and the state equation can then be expressed as a function of the correct variables (capacitor voltages, inductor currents, and independent source variables). A simple example will demonstrate the procedure outlined above.

EXAMPLE 4-4. Write the state equation for the transistor equivalent circuit shown in Fig. 4-8.

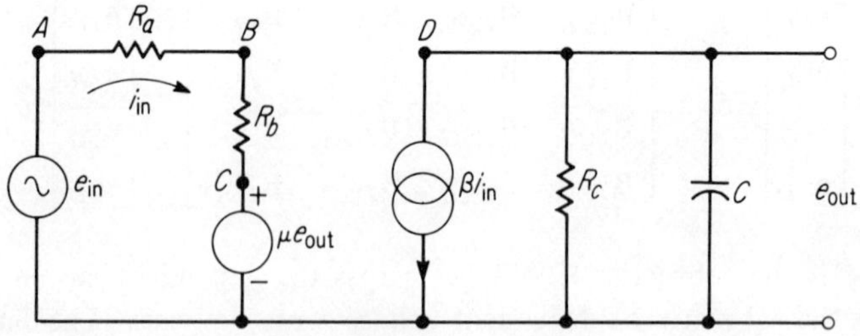

FIG. 4-8. Transistor equivalent circuit.

Solution. The network graph and proper tree are shown in Fig. 4-9. Note that the tree branches are numbered so that the capacitive branch is number one,

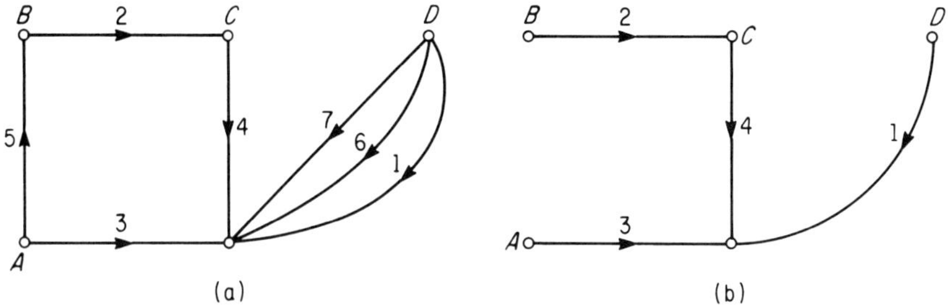

FIG. 4-9. (a) Network graph, and (b) proper tree of the circuit shown in Fig. 4-8.

the resistive branch is number 2, the independent source is number 3, and the dependent source is number 4. The matrix $\mathbf{C}_L$ is

$$\mathbf{C}_L = \begin{bmatrix} 0 & 1 & 1 \\ -1 & 0 & 0 \\ 1 & 0 & 0 \\ -1 & 0 & 0 \end{bmatrix}$$

and

$$\mathbf{B}_T = \begin{bmatrix} 0 & 1 & -1 & 1 \\ -1 & 0 & 0 & 0 \\ -1 & 0 & 0 & 0 \end{bmatrix}$$

After applying Eqs. 4-13 and 4-14 and suitably partitioning the matrices $\mathbf{C}_L$ and $\mathbf{B}_T$ we arrive at the equations

$$[\mathbf{i}_{TC}] = [0]\,[\mathbf{e}_{TC}] - [0 \;\; 0 \;\; 1] \begin{bmatrix} \mathbf{e}_{TR} \\ \mathbf{i}_{LR} \end{bmatrix} - [0]\,[\mathbf{e}_{TG}] - [0 \;\; 1] \begin{bmatrix} \mathbf{e}_{TD} \\ \mathbf{i}_{LD} \end{bmatrix}$$

and

$$\begin{bmatrix} \mathbf{e}_{LR} \\ \mathbf{i}_{TR} \end{bmatrix} = - \begin{bmatrix} 0 \\ -1 \\ 0 \end{bmatrix} [\mathbf{e}_{TC}] - \begin{bmatrix} 1 & 0 & 0 \\ 0 & 0 & 0 \\ 0 & -1 & 0 \end{bmatrix} \begin{bmatrix} \mathbf{e}_{TR} \\ \mathbf{i}_{LR} \end{bmatrix} - \begin{bmatrix} -1 \\ 0 \\ 0 \end{bmatrix} [\mathbf{e}_{TG}] - \begin{bmatrix} 1 & 0 \\ 0 & 0 \\ 0 & 0 \end{bmatrix} \begin{bmatrix} \mathbf{e}_{TD} \\ \mathbf{i}_{LD} \end{bmatrix}$$

All subvectors above are of dimension 1×1 except $\mathbf{e}_{LR}$ and $\mathbf{i}_{LR}$, which are 2×1 subvectors corresponding to the variables associated with the two resistive links (branches 5 and 6). We will first eliminate the vector involving the dependent sources from both equations. To do this we write

$$\begin{bmatrix} \mathbf{e}_{TD} \\ \mathbf{i}_{LD} \end{bmatrix} = \begin{bmatrix} \mu \\ 0 \end{bmatrix} [\mathbf{e}_{TC}] + \begin{bmatrix} 0 & 0 & 0 \\ 0 & \beta & 0 \end{bmatrix} \begin{bmatrix} \mathbf{e}_{TR} \\ \mathbf{i}_{LR} \end{bmatrix}$$

Substituting this vector into the equations results in

$$[\mathbf{i}_{TC}] = [0]\,[\mathbf{e}_{TC}] - [0 \;\; \beta \;\; 1] \begin{bmatrix} \mathbf{e}_{TR} \\ \mathbf{i}_{LR} \end{bmatrix} - [0]\,[\mathbf{e}_{TG}]$$

and

$$\begin{bmatrix} \mathbf{e}_{LR} \\ \mathbf{i}_{TR} \end{bmatrix} = - \begin{bmatrix} \mu \\ -1 \\ 0 \end{bmatrix} [\mathbf{e}_{TC}] - \begin{bmatrix} 1 & 0 & 0 \\ 0 & 0 & 0 \\ 0 & -1 & 0 \end{bmatrix} \begin{bmatrix} \mathbf{e}_{TR} \\ \mathbf{i}_{LR} \end{bmatrix} - \begin{bmatrix} -1 \\ 0 \\ 0 \end{bmatrix} [\mathbf{e}_{TG}]$$

The vectors involving resistive variables are eliminated by combining the last equation with Ohm's law for the resistive branches

$$\begin{bmatrix} \mathbf{e}_{LR} \\ \mathbf{i}_{TR} \end{bmatrix} = \begin{bmatrix} 0 & R_a & 0 \\ 0 & 0 & R_c \\ 1/R_b & 0 & 0 \end{bmatrix} \begin{bmatrix} \mathbf{e}_{TR} \\ \mathbf{i}_{LR} \end{bmatrix}$$

The vector $\begin{bmatrix} \mathbf{e}_{TR} \\ \mathbf{i}_{LR} \end{bmatrix}$ is found to be

$$\begin{bmatrix} \mathbf{e}_{TR} \\ \mathbf{i}_{LR} \end{bmatrix} = - \begin{bmatrix} \dfrac{\mu R_b}{R_a + R_b} \\[2ex] \dfrac{\mu}{R_a + R_b} \\[2ex] -1/R_c \end{bmatrix} [\mathbf{e}_{TC}] - \begin{bmatrix} \dfrac{-R_b}{R_a + R_b} \\[2ex] \dfrac{-1}{R_a + R_b} \\[2ex] 0 \end{bmatrix} [\mathbf{e}_{TG}]$$

When this equation is used in the equation for capacitor current the result is

$$[\mathbf{i}_{TC}] = \left[\frac{\beta\mu}{R_a + R_b} - \frac{1}{R_c} \right] [\mathbf{e}_{TC}] - \left[\frac{\beta}{R_a + R_b} \right] [\mathbf{e}_{TG}]$$

Since $\mathbf{i}_{TC}$ in this case equals $C\dfrac{dv_c}{dt}$, $\mathbf{e}_{TC} = v_c$, and $\mathbf{e}_{TG} = e_{\text{in}}$, we can write the state equation as

$$\frac{d}{dt}[v_c] = \frac{1}{C} \left[\frac{\beta\mu}{R_a + R_b} - \frac{1}{R_c} \right] [v_c] - \frac{1}{C} \left[\frac{\beta}{R_a + R_b} \right] [e_{\text{in}}]$$

A difficulty that remains to be considered involves the occurrence of excess elements. Occasionally a network will occur such that the subgraph containing all capacitors and voltage sources will contain a loop, and hence will not qualify as a proper tree. Another network may contain a node formed only by inductors. In this case any tree of the graph must include at least one inductor. Examples of these are shown in Fig. 4-10.

The network of Fig. 4-10a corresponds to the high-frequency, hybrid π equivalent circuit of a transistor with stray capacitance appearing across the output terminals. The three inductors of Fig. 4-10b might represent the

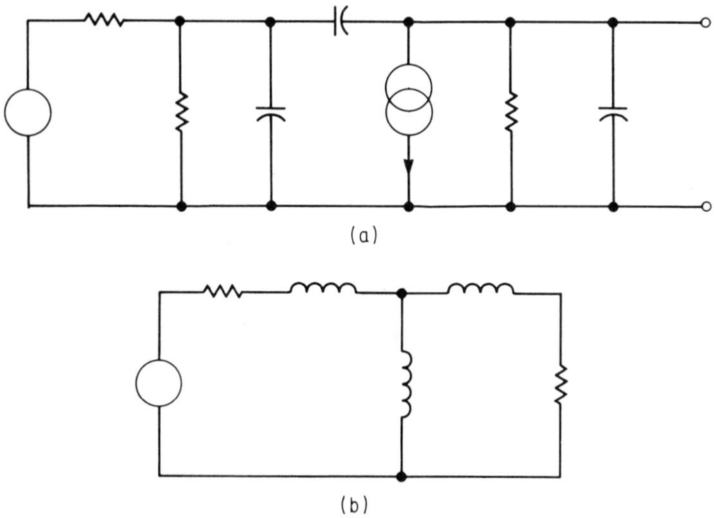

FIG. 4-10. Networks with excess elements. (a) Capacitive loop. (b) Inductive node.

equivalent circuit of a transformer. In writing the state equation for the network of Fig. 4-10a, only two capacitor voltages need be known to determine the voltage across the third. Thus we need only consider two capacitor voltages in the state vector. The third capacitor is called an *excess element.* An analogous situation occurs at an inductive node. Two inductor currents determine the third inductor current; therefore only two inductor currents need be included in the state vector and the third inductor is an excess element.

In constructing a modified proper tree, excess capacitors are considered as links and excess inductors are considered as tree branches. The excess-capacitor variables are included in the resistive link subvectors while excess-inductor variables are included in the resistive tree branch subvectors. The relationship between voltage and current across these excess elements is represented by a first derivative, and this fact will sometimes result in a state equation that includes not only the input source variables but derivatives of source variables as well. The following two examples will demonstrate the procedure used when excess elements are present.

EXAMPLE 4-5. Write the state equation for the circuit of Fig. 4-11.

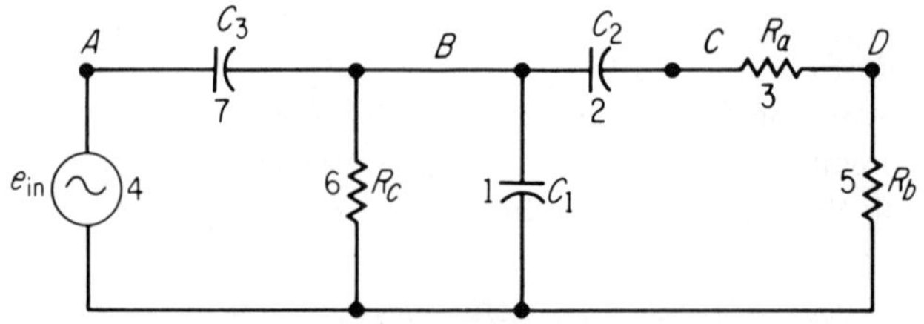

FIG. 4-11. Network with an excess capacitor.

Solution. A modified proper tree is drawn in Fig. 4-12. In this situation the capacitors C_1 and C_2 will be included in the tree, while C_3 is a link branch. The capacitive tree subvector $\mathbf{i}_{TC}$ is made up of current through C_1 and C_2 while the

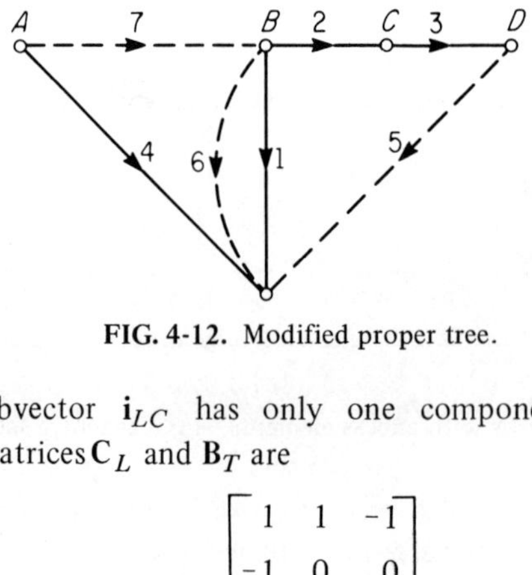

FIG. 4-12. Modified proper tree.

capacitive link subvector $\mathbf{i}_{LC}$ has only one component, namely current through C_3. The matrices $\mathbf{C}_L$ and $\mathbf{B}_T$ are

$$\mathbf{C}_L = \begin{bmatrix} 1 & 1 & -1 \\ -1 & 0 & 0 \\ -1 & 0 & 0 \\ 0 & 0 & 1 \end{bmatrix}$$

and

$$\mathbf{B}_T = \begin{bmatrix} -1 & 1 & 1 & 0 \\ -1 & 0 & 0 & 0 \\ 1 & 0 & 0 & -1 \end{bmatrix}$$

The application of Eqs. 4-13 and 4-14 result in

$$[\mathbf{i}_{TC}] = \begin{bmatrix} 0 & -1 & -1 & 1 \\ 0 & 1 & 0 & 0 \end{bmatrix} \begin{bmatrix} \mathbf{e}_{TR} \\ \mathbf{i}_{LR} \\ \mathbf{i}_{LC} \end{bmatrix}$$

and

$$\begin{bmatrix} \mathbf{e}_{LR} \\ \mathbf{i}_{TR} \\ \mathbf{e}_{LC} \end{bmatrix} = \begin{bmatrix} 1 & -1 \\ 1 & 0 \\ 0 & 0 \\ -1 & 0 \end{bmatrix} [\mathbf{e}_{TC}] + \begin{bmatrix} -1 & 0 & 0 & 0 \\ 0 & 0 & 0 & 0 \\ 0 & 1 & 0 & 0 \\ 0 & 0 & 0 & 0 \end{bmatrix} \begin{bmatrix} \mathbf{e}_{TR} \\ \mathbf{i}_{LR} \\ \mathbf{i}_{LC} \end{bmatrix} + \begin{bmatrix} 0 \\ 0 \\ 0 \\ 1 \end{bmatrix} [\mathbf{e}_{TG}]$$

The *V-I* relationships for the resistive branches and the excess capacitor can be expressed as

$$\begin{bmatrix} \mathbf{e}_{TR} \\ \mathbf{i}_{LR} \\ \mathbf{i}_{LC} \end{bmatrix} = \begin{bmatrix} 0 & 0 & R_a & 0 \\ 1/R_b & 0 & 0 & 0 \\ 0 & 1/R_c & 0 & 0 \\ 0 & 0 & 0 & C_3 \dfrac{d}{dt} \end{bmatrix} \begin{bmatrix} \mathbf{e}_{LR} \\ \mathbf{i}_{TR} \\ \mathbf{e}_{LC} \end{bmatrix}$$

Combining the last two equations and solving for the desired variables gives

$$\begin{bmatrix} \mathbf{e}_{TR} \\ \mathbf{i}_{LR} \\ \mathbf{i}_{LC} \end{bmatrix} = \begin{bmatrix} \dfrac{R_a}{R_a + R_b} & -\dfrac{R_a}{R_a + R_b} \\ \dfrac{1}{R_a + R_b} & -\dfrac{1}{R_a + R_b} \\ \dfrac{1}{R_c} & 0 \\ -C_3 \dfrac{d}{dt} & 0 \end{bmatrix} [\mathbf{e}_{TC}] + \begin{bmatrix} 0 \\ 0 \\ 0 \\ C_3 \dfrac{d}{dt} \end{bmatrix} [\mathbf{e}_{TG}]$$

The capacitor tree current can now be written as

$$[\mathbf{i}_{TC}] = \begin{bmatrix} \left(-\dfrac{(R_a + R_b + R_c)}{(R_a + R_b) R_c} - C_3 \dfrac{d}{dt} \right) & \dfrac{1}{R_a + R_b} \\ \dfrac{1}{R_a + R_b} & \dfrac{-1}{R_a + R_b} \end{bmatrix} [\mathbf{e}_{TC}] + \begin{bmatrix} C_3 \dfrac{d}{dt} \\ 0 \end{bmatrix} [\mathbf{e}_{TG}]$$

We note that there is a derivative of a state variable on the right-hand side of the equation. This term can be transposed to the other side and the state equation written:

$$\frac{d}{dt} \begin{bmatrix} v_{c1} \\ v_{c2} \end{bmatrix} = \begin{bmatrix} \dfrac{-(R_a + R_b + R_c)}{(R_a + R_b) R_c (C_1 + C_3)} & \dfrac{1}{(R_a + R_b)(C_1 + C_3)} \\ \dfrac{1}{(R_a + R_b) C_2} & \dfrac{-1}{(R_a + R_b) C_2} \end{bmatrix} \begin{bmatrix} v_{c1} \\ v_{c2} \end{bmatrix}$$

$$+ \begin{bmatrix} \left(\dfrac{C_3}{C_1 + C_3}\right) \dfrac{d}{dt} \\ 0 \end{bmatrix} [\mathbf{e}_{\text{in}}]$$

In general, the derivative of e_{in} can be found from a knowledge of e_{in}.

The matter of an excess inductor is treated in much the same way as the excess capacitor. However, the inductor forms a tree branch and is initially included in the tree branch vector.

The general procedure for writing the state equation is summarized below.

1. Draw the network graph and select a proper tree. The proper tree should include all capacitors if possible, all voltage sources, both independent and dependent, and any necessary resistors or excess inductors.
2. Number the tree branches consecutively starting with capacitors, resistors, excess inductors, independent voltage sources, and dependent voltage sources.
3. Number the link branches consecutively starting with inductors, resistors, excess capacitors, independent current sources, and dependent current sources.

   The tree and link vectors now appear as

$$[\mathbf{i}_T] = \begin{bmatrix} \mathbf{i}_{TC} \\ \mathbf{i}_{TR} \\ \mathbf{i}_{TL} \\ \mathbf{i}_{TG} \\ \mathbf{i}_{TD} \end{bmatrix}, \qquad [\mathbf{i}_L] = \begin{bmatrix} \mathbf{i}_{LL} \\ \mathbf{i}_{LR} \\ \mathbf{i}_{LC} \\ \mathbf{i}_{LG} \\ \mathbf{i}_{LD} \end{bmatrix}$$

$$\mathbf{e}_T = \begin{bmatrix} \mathbf{e}_{TC} \\ \mathbf{e}_{TR} \\ \mathbf{e}_{TL} \\ \mathbf{e}_{TG} \\ \mathbf{e}_{TD} \end{bmatrix}, \qquad \mathbf{e}_L = \begin{bmatrix} \mathbf{e}_{LL} \\ \mathbf{e}_{LR} \\ \mathbf{e}_{LC} \\ \mathbf{e}_{LG} \\ \mathbf{e}_{LD} \end{bmatrix}$$

4. Determine the topological matrices $\mathbf{C}_L$ and $\mathbf{B}_T$.
5. Using the topological matrices, write the tree-current vector in terms of the link-current vector and write the link-voltage vector in terms of the tree-voltage vector.
6. Partition the matrices $\mathbf{C}_L$ and $\mathbf{B}_T$ such that the appropriate equations for currents and voltages can be written in terms of the correct vectors as indicated below. The vector

$$\begin{bmatrix} \mathbf{i}_{TC} \\ \mathbf{e}_{LL} \end{bmatrix}$$

should be written in terms of

$$\begin{bmatrix} \mathbf{e}_{TC} \\ \mathbf{i}_{LL} \end{bmatrix}, \quad \begin{bmatrix} \mathbf{e}_{TR} \\ \mathbf{i}_{LR} \\ \mathbf{e}_{TL} \\ \mathbf{i}_{LC} \end{bmatrix}, \quad \begin{bmatrix} \mathbf{e}_{TG} \\ \mathbf{i}_{LG} \end{bmatrix}, \quad \text{and} \quad \begin{bmatrix} \mathbf{e}_{TD} \\ \mathbf{i}_{LD} \end{bmatrix}$$

The vector

$$\begin{bmatrix} \mathbf{e}_{LR} \\ \mathbf{i}_{TR} \\ \mathbf{e}_{LC} \\ \mathbf{i}_{TL} \end{bmatrix}$$

should be written in terms of the same variables used to express capacitor currents and inductor voltages.

7. Eliminate the dependent source vector by expressing this vector in terms of the other variables.
8. Write the matrix equation expressing the V-I relationships of the resistive branches and the excess branches. This will be of the form

$$\begin{bmatrix} \mathbf{e}_{TR} \\ \mathbf{i}_{LR} \\ \mathbf{e}_{TL} \\ \mathbf{i}_{LC} \end{bmatrix} = \begin{bmatrix} 0 & \mathbf{R}_T & 0 & 0 \\ \mathbf{G}_L & 0 & 0 & 0 \\ 0 & 0 & 0 & \mathbf{L}\dfrac{d}{dt} \\ 0 & 0 & \mathbf{C}\dfrac{d}{dt} & 0 \end{bmatrix} \begin{bmatrix} \mathbf{e}_{LR} \\ \mathbf{i}_{TR} \\ \mathbf{e}_{LC} \\ \mathbf{i}_{TL} \end{bmatrix}$$

9. Using the above equation eliminate the vector

$$\begin{bmatrix} \mathbf{e}_{TR} \\ \mathbf{i}_{LR} \\ \mathbf{e}_{TL} \\ \mathbf{i}_{LC} \end{bmatrix}$$

from the equation for capacitor tree-branch currents and inductor link voltages.

The equation resulting from this last step can easily be written as the state equation by dividing each row of the matrices by the correct constant.

The Output Equation

The output equation depends not only on the state vector and the source vector, but also on the particular choice of output variables. In the simplest case only a single output voltage or current may be required. However, any combination of voltages or currents may be required in a given problem. To write the output in general terms we have

$$\mathbf{y} = [\mathbf{Q}^1] \begin{bmatrix} \mathbf{e} \\ \mathbf{i} \end{bmatrix} \tag{4-35}$$

where $\mathbf{e}$ and $\mathbf{i}$ are branch variables. The matrix $\mathbf{Q}^1$ determines which of the variables are chosen as output variables. Partitioning the branch variables gives

$$\mathbf{y} = [\mathbf{Q}^1] \begin{bmatrix} \mathbf{e}_T \\ \mathbf{e}_L \\ \mathbf{i}_T \\ \mathbf{i}_L \end{bmatrix} = [\mathbf{Q}^1] \begin{bmatrix} \mathbf{e}_T \\ -\mathbf{B}_T\mathbf{e}_T \\ -\mathbf{C}_L\mathbf{i}_L \\ \mathbf{i}_L \end{bmatrix} = [\mathbf{Q}] \begin{bmatrix} \mathbf{e}_T \\ \mathbf{i}_L \end{bmatrix} \tag{4-36}$$

Again, the tree-branch and link variables can be partitioned to give

$$\mathbf{y} = [\mathbf{Q}_1\ \mathbf{Q}_2\ \mathbf{Q}_3\ \mathbf{Q}_4] \begin{bmatrix} \mathbf{e}_{TC} \\ \mathbf{i}_{LL} \\ \mathbf{e}_{TR} \\ \mathbf{i}_{LR} \\ \mathbf{e}_{TL} \\ \mathbf{i}_{LC} \\ \mathbf{e}_{TG} \\ \mathbf{i}_{LG} \\ \mathbf{e}_{TD} \\ \mathbf{i}_{LD} \end{bmatrix}$$

$$= [\mathbf{Q}_1] \begin{bmatrix} \mathbf{e}_{TC} \\ \mathbf{i}_{LL} \end{bmatrix} + [\mathbf{Q}_2] \begin{bmatrix} \mathbf{e}_{TR} \\ \mathbf{i}_{LR} \\ \mathbf{e}_{TL} \\ \mathbf{i}_{LC} \end{bmatrix} + [\mathbf{Q}_3] \begin{bmatrix} \mathbf{e}_{TG} \\ \mathbf{i}_{LG} \end{bmatrix} + [\mathbf{Q}_4] \begin{bmatrix} \mathbf{e}_{TD} \\ \mathbf{i}_{LD} \end{bmatrix} \tag{4-37}$$

As in the procedure for generating the state equation, the vectors

$$\begin{bmatrix} \mathbf{e}_{TD} \\ \mathbf{i}_{LD} \end{bmatrix} \quad \text{and} \quad \begin{bmatrix} \mathbf{e}_{TR} \\ \mathbf{i}_{LR} \\ \mathbf{e}_{TL} \\ \mathbf{i}_{LC} \end{bmatrix}$$

can be eliminated from Eq. 4-37. The remaining two variables in terms of which $\mathbf{y}$ is expressed are the appropriate ones.

We might note at this point that the derivation of the output equation could be combined with that of the state equation as done by Calahan [3]. This approach results in greater computational efficiency since only a single equation need be treated.

C. SELECTION OF A PROPER TREE BY SCEPTRE

There are several other approaches to the formulation of the state equation. Some of these methods, such as the application of Lagrange equations, can be more efficient for manual formulation. The preceding method however, is attractive for machine formulation of equations. We must emphasize that the most complex aspect of this formulation is simply that of detailed bookkeeping. While it may be difficult for the human mind to apply the necessary bookkeeping to effectively use the method, the computer has no problem in this regard. This method can then form the basis for equation formulation and thus allows the engineer to use the computer as a design aid.

At the beginning of this section we referred to a program for automatic network analysis called SCEPTRE [7]. Since this program utilizes a state-variable approach we will discuss it in more detail at this point.

In order to formulate the equations as previously outlined the computer must select a proper tree of the network. This selection is based on the topological information specified by the designer. The input specifications must include nodal connections of each branch in terms of a FROM node and a TO node. The type of element and element value must also be listed. In SCEPTRE the computer first generates a branch list such that all voltage sources appear first, capacitors next, resistors next, inductors next, and then current sources. The node numbers assigned to the first element of the list are compared to all other element node numbers. Any element with the same node numbers as this first reference element will be classified as a link branch and removed from the list. Any element with a node number corresponding to the TO node of the reference element will have this node number replaced by the FROM element of the reference element. After the first element is compared to all others of the list the second element, which may have had a node number changed, becomes the reference element. Node numbers of all succeeding elements are compared to those of the new reference element. Again link branches are removed when both node numbers correspond to the reference element nodes. Branches with the same node number as the TO node of the reference element have this node number changed to the FROM node number. This process is repeated until all elements of the list have either been identified as links or have served as reference elements. We note that elements that appear early in the list such as voltage sources or capacitors have little probability of being classified as links. Elements appearing later such as inductors or current sources will tend to be identified as links since more elements for the tree have already been chosen before these elements are considered in the program.

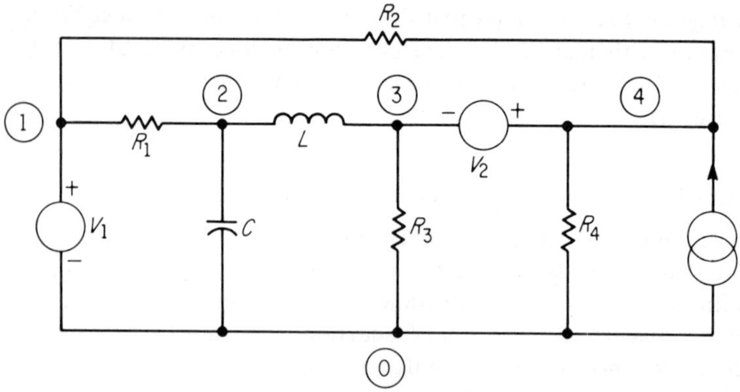

FIG. 4-13. Network to demonstrate proper tree selection.

A simple example will demonstrate this procedure. Consider the circuit of Fig. 4-13. The designer must supply the necessary nodal information for the computer to generate the following list:

FROM	Element	TO
0	V_1	1
3	V_2	4
0	C	2
1	R_1	2
1	R_2	4
3	R_3	0
4	R_4	0
2	L	3
0	I	4

The element V_1 is chosen as the reference element. No other element has two corresponding node numbers, thus no branches are identified as links at this point. The TO node of this element is node 1 and both R_1 and R_2 are connected to this node. Both of these elements must have node number 1 changed to the number of the FROM node of V_1 which is zero. The modified list appears below.

FROM	Element	TO
0	V_1	1
3	V_2	4
0	C	2
0	R_1	2
0	R_2	4
3	R_3	0
4	R_4	0
2	L	3
0	I	4

The second element now becomes the reference element and node numbers of all succeeding elements are compared to those of V_2. Again no link branches are found, however R_2, R_4, and I contain the TO node number of the reference element. Changing these numbers to the FROM node number 3 results in a new list.

FROM	Element	TO
0	V_1	1
3	V_2	4
0	C	2
0	R_1	2
0	R_2	3
3	R_3	0
3	R_4	0
2	L	3
0	I	3

The third element now becomes the reference element and it is noted that R_1 has the same node numbers, thus R_1 is a link branch and is removed from the list. The TO node of the reference element is number 2. Element L connects to this node and consequently the list becomes

FROM	Element	TO	Links
0	V_1	1	R_1
3	V_2	4	
0	C	2	
0	R_2	3	
3	R_3	0	
3	R_4	0	
0	L	3	
0	I	3	

Element R_2 now becomes the reference element and all succeeding elements must be links since they each have corresponding node numbers. The final list is then

FROM	Element	TO	Links
0	V_1	1	R_1, R_3, R_4, L, I
3	V_2	4	
0	C	2	
0	R_2	3	

The remaining list elements are tree branches, and these branches are seen to consist of voltage sources, a capacitor, and a resistor while the links consist of resistors, an inductor, and a current source.

The computer has now identified the elements of the proper tree and can apply the formulation method outlined earlier in this section. While capacitor loop or inductive node networks can be treated as outlined, there are certain restrictions on acceptable network topologies. The networks must not have a loop containing only capacitors plus a dependent voltage source nor a node with only inductors plus a dependent current source.

Fig. 4-14 shows a flow chart of the operations required to select a proper tree and construct an incidence matrix with lower-numbered columns that are

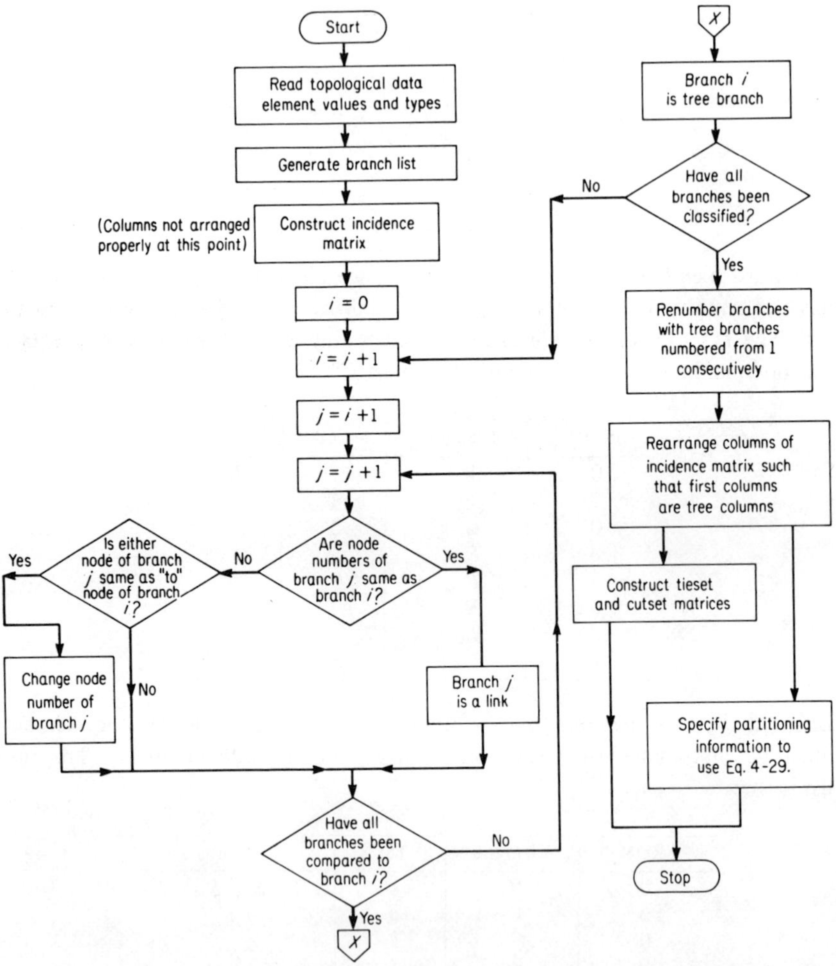

FIG. 4-14. Flow chart for selection of a proper tree.

associated with tree branches. Once $\mathbf{A}_T$ and $\mathbf{A}_L$ have been constructed, the tieset and cutset matrices **B** and **C** can be found. Equation 4-29, which corresponds rather closely to the desired state equation, requires the topological matrices $\mathbf{B}_T$ and $\mathbf{C}_L$. Thus, in selecting a proper tree these matrices and other necessary information to use Eq. 4-29 are generated. The formulation of the state equation is then completed by matrix manipulation based on Eq. 4-29.

The output information of Fig. 4-14 includes the number of each different element type so that the submatrices of Eq. 4-29 can easily be constructed. Table 4-1 shows the program that implements the flow chart of Fig. 4-14. This program is written in Fortran II and was executed on the IBM 1620. It was assumed that the branch list was already known for the circuit of Fig. 4-13. The output consists of the incidence matrix of the circuit with branches numbered as shown in the first list relating to this circuit. The properly arranged incidence matrix is also found along with the numbers of tree branches, link branches, voltage sources, capacitors, tree resistors, link resistors, inductors, and current sources. This program is not designed to consider networks having excess elements or dependent sources.

TABLE 4-1. Program to Find Proper Tree.

```
      DIMENSION A(10,10),D(10,10),NF(10),NT(10),V(10),JT(10),MT(10)
      READ 109,MB,MN
C     MB IS NUMBER OF BRANCHES
C     MN IS NUMBER OF NODES (EXCEPT REFERENCE)
109   FORMAT (2I2)
      READ 99,(NF(I),NT(I),V(I),JT(I),MT(I),I=1,MB)
C     I IS BRANCH NUMBER, NF IS FROM NODE, NT IS TO NODE
C     V IS ELEMENT VALUE, JT IS ELEMENT TYPE
C     MT TELLS TREE OR LINK
C     WHEN MT IS 0 BRANCH IS TREE, OTHER NUMBER BRANCH IS
C     LINK
C     MT(I) HAS ALL ZERO ELEMENTS INITIALLY
99    FORMAT (2I4,F10.6,2I4)
C     CONSTRUCT INCIDENCE MATRIX D
      DO 20 I = 1,MN
      DO 20 J = 1,MB
20    D(I,J) = 0.
      DO 8 I =1,MB
      IF (NF(I))4,4,3
3     IP = NF(I)
      D(IP,I) = 1.
4     IF (NT(I)) 8,8,7
7     IQ = NT(I)
      D (IQ,I) = -1.
8     CONTINUE
      PRINT 209, ((D(I,J),J=1,MB),I=1,MN)
209   FORMAT (1H , 9F4.0)
C     SELECT TREE BRANCHES AND LINKS
      KV = 0
C     KV IS TREE VOLTAGE SOURCE NUMBER
      KC =0
C     KC IS TREE CAPACITOR NUMBER
      KR = 0
```

TABLE 4-1 (*continued*)

```
C       KR IS TREE RESISTOR NUMBER
        IR = 0
C       IR IS LINK RESISTOR NUMBER
        IL = 0
C       IL IS LINK INDUCTOR NUMBER
        IC = 0
C       IC IS LINK CURRENT SOURCE NUMBER
        IK =0
C       IK IS TREE BRANCH NUMBER
        K = 0
C       K IS LINK BRANCH NUMBER
        I = 0
51      I = I + 1
        IF(I-MB)200,200,103
200     J = I
        IF (MT(I))51,53,51
53      IF(I-MB)52,102,102
52      J = J + 1
        IF(MT(J))52,61,52
61      IF(NT(I)-NT(J))33,5,33
5       IF(NF(I)-NF(J))43,6,43
43      NT(J) = NF(I)
220     IF(J-MB)52,102,102
6       K = K + 1
        MT(J) = MN + K
        IMNK = MN+K
        DO 70 L = 1,MN
70      A(L,IMNK)= D(L,J)
        IF (JT(J) - 4)71,72,73
71      IR = IR + 1
        GO TO 74
72      IL = IL + 1
        GO TO 74
73      IC = IC + 1
74      IF(J-MB)52,102,103
33      IF(NT(I)-NF(J))220,63,220
63      IF(NF(I)-NT(J))83,6,83
83      NF(J) = NF(I)
        GO TO 220
102     IK = IK + 1
        DO 100 M = 1,MN
100     A(M,IK) = D(M,I)
        IF(JT(I)-2) 333,334,335
333     KV = KV + 1
        GO TO 51
334     KC = KC + 1
        GO TO 51
335     KR = KR + 1
        GO TO 51
103     CONTINUE
        PRINT 309,IK,K,KV,KC,KR,IR,IL,IC
```

TABLE 4-1 *(continued)*

```
309   FORMAT (1H ,8I4)
      PRINT 409, ((A(I,J),J=1,MB),I=1,MN)
C     A(I,J) IS INCIDENCE MATRIX WITH FIRST MN COLUMNS
C     CORRESPONDING TO TREE BRANCHES
409   FORMAT(1H ,9F4.0)
      CALL EXIT
      END
 -1.  0.  0.  1.  1.  0.  0.  0.  0.
  0.  0. -1. -1.  0.  0.  0.  1.  0.
  0.  1.  0.  0.  0.  1.  0. -1.  0.
  0. -1.  0.  0. -1.  0.  1.  0. -1.
   4   5   2   1   1   3   1   1
 -1.  0.  0.  1.  1.  0.  0.  0.  0.
  0.  0. -1.  0. -1.  0.  0.  1.  0.
  0.  1.  0.  0.  0.  1.  0. -1.  0.
  0. -1.  0. -1.  0.  0.  1.  0. -1.
```

This program demonstrates the practical application of graph theory to the machine formulation of the state equation. Chapter 3 showed how to automatically form the basic equations necessary to apply the node method. It should be obvious that graph theory is important to the field of computer-aided network design.

D. SYSTEM STATE EQUATIONS

In the first section of this chapter it was shown that the state equation for a mechanical system could be derived from the differential equation describing the system. A second method of importance is that of constructing electrical analogs of the system [4] and then proceeding as outlined in the previous section.

Three important variables in mechanical systems are inertial force, damping force, and restoring force. If x is used to denote displacement the equations can be written as

$$f_i = m \frac{d^2 x}{dt^2}$$

$$f_b = b \frac{dx}{dt}$$

and

$$f_k = kx$$

The first equation is a statement of Newton's second law, the second is the law of viscous friction, and the third is Hooke's law. The constants m, b, and k are mass, coefficient of viscous friction, and spring constant respectively. If we substitute velocity r for dx/dt in these equations we find that

$$f_i = m\frac{dr}{dt} \tag{4-38}$$

$$f_b = br \tag{4-39}$$

and

$$f_k = k\int r\,dt \tag{4-40}$$

Two sets of electrical equations describing the *V-I* relationships of resistors, capacitors, and inductors are very similar to Eqs. 4-38 to 4-40. We can write

$$v_L = L\frac{di_L}{dt} \tag{4-41}$$

$$v_R = Ri_R \tag{4-42}$$

and

$$v_C = \frac{1}{C}\int i_C\,dt \tag{4-43}$$

A second set of similar equations is

$$i_C = C\frac{dv_C}{dt} \tag{4-44}$$

$$i_R = \left(\frac{1}{R}\right) v_R \tag{4-45}$$

and

$$i_L = \frac{1}{L}\int v_L\,dt \tag{4-46}$$

Thus if we associate voltage with force and current with velocity, Eqs. 4-41 to 4-43 are analogous to the force equations. If current is associated with force and voltage with velocity Eqs. 4-44 to 4-46 are analogous to the force equations. The mechanical system of Fig. 4-15a is analogous to the electrical circuits of Figs. 4-15b and 4-15c, provided the correct variables are associated together.

For the circuit of Fig. 4-15b the voltage and current are related by

$$v(t) = \frac{1}{C}\int i\,dt + L\frac{di}{dt} + R_i$$

Making the appropriate substitutions leads to

$$f(t) = k\int r\,dt + m\frac{dr}{dt} + br$$

which is the correct force equation. For the second analog we can write

$$i(t) = \frac{1}{L}\int v\,dt + C\frac{dv}{dt} + \frac{v}{R}$$

Replacing current by force, voltage by velocity, and using the appropriate constants result in the same force equation.

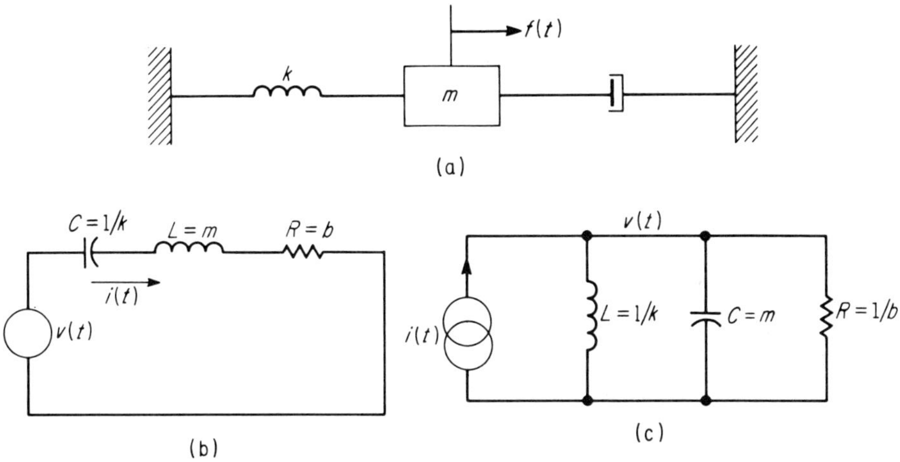

FIG. 4-15. (a) Mechanical system. (b) Electrical analog. (c) Second analog.

The equations governing the simple mechanical system of Fig. 4-15 are easy enough to write without resorting to analogs. However, the analysis of more complex systems can often be simplified by using analogs. Consider the system of Fig. 4-16a and the electrical analog of Fig. 4-16b. The state equations can

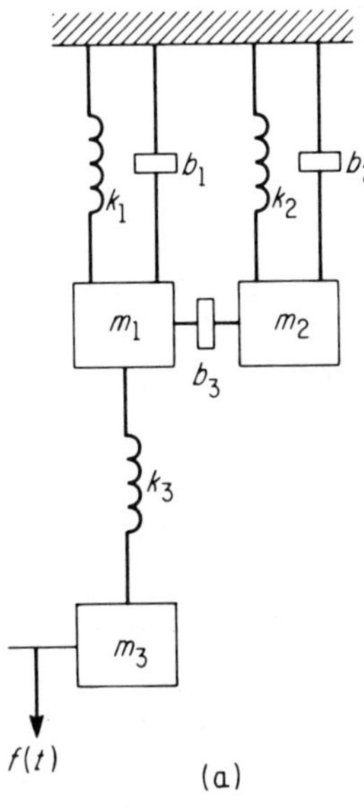

FIG. 4-16. (a) Complex mechanical system.

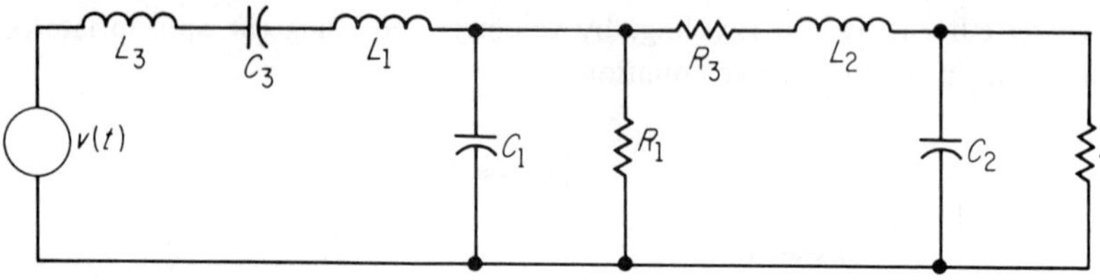

$$L_1 = m_1 \quad C_1 = 1/k_1 \quad R_1 = b_1$$
$$L_2 = m_2 \quad C_2 = 1/k_2 \quad R_2 = b_2$$
$$L_3 = m_3 \quad C_3 = 1/k_3 \quad R_3 = b_3$$

(b)

FIG. 4-16. (b) Electrical analog.

now be written using the techniques discussed in conjunction with network problems.

BIBLIOGRAPHY

1. Bashkow, T., "The A Matrix: A New Network Description," *IRE Trans. on Circuit Theory*, Vol. CT-4 (September 1957), pp. 117-119.
2. Bryant, P. R., "The Explicit Form of Bashkow's A Matrix," *IRE Trans. on Circuit Theory*, Vol. CT-9 (1962), pp. 303-306.
3. Calahan, D. A., *Computer-Aided Network Design*, preliminary edition, McGraw-Hill Book Company, New York, 1968.
4. Chen, C. F., and I. J. Haas, *Elements of Control Systems Analysis*, Prentice-Hall, Inc., Englewood Cliffs, N.J., 1967.
5. Kuh, E. S., and R. A. Rohrer, "The State-Variable Approach to Network Analysis," *Proc. IEEE*, Vol. 53 (July 1965), pp. 672-686.
6. Kuo, B. C., *Linear Networks and Systems*, McGraw-Hill Book Company, New York, 1967.
7. Sedore, S. R., "SCEPTRE: A Program for Automatic Network Analysis," *IBM Journal*, Vol. 11 (November 1967), pp. 627-637.
8. Zadeh, L. A., and C. A. Desoer, *Linear System Theory*, McGraw-Hill Book Company, New York, 1963.
9. 1620 Electronic Circuit Analysis Program (ECAP), (1620-EE-02X) User's Manual, IBM Publication H20-0170-1, IBM Corporation, White Plains, N.Y.

PROBLEMS

4-1 to 4-8 Write the state equations for the networks shown, using capacitor voltages and inductor currents as the components of the state vector. Write the output equations. Base your formulation on $\mathbf{B}_T$ and $\mathbf{C}_L$.

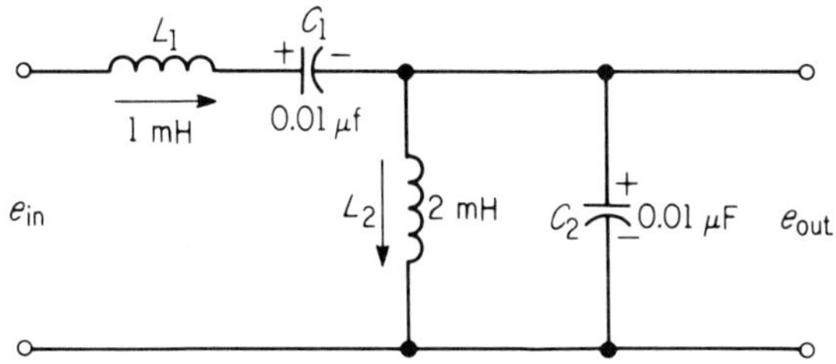

PROB. 4-1

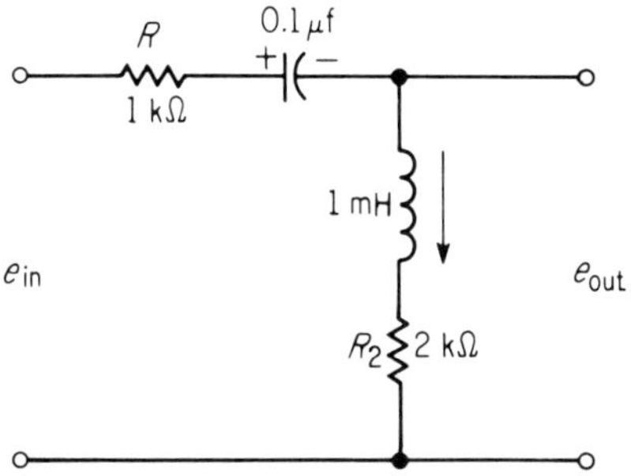

PROB. 4-2

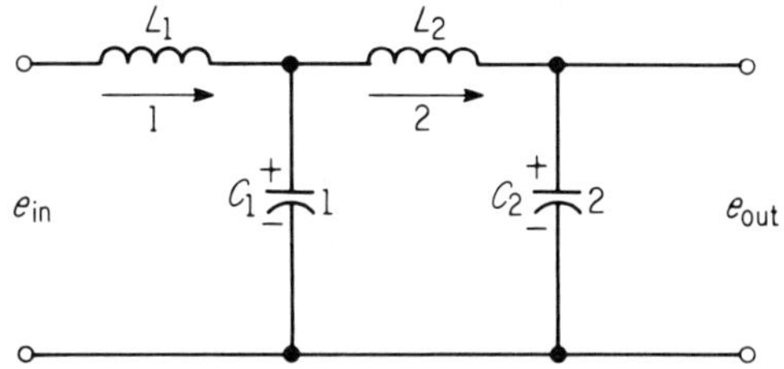

PROB. 4-3

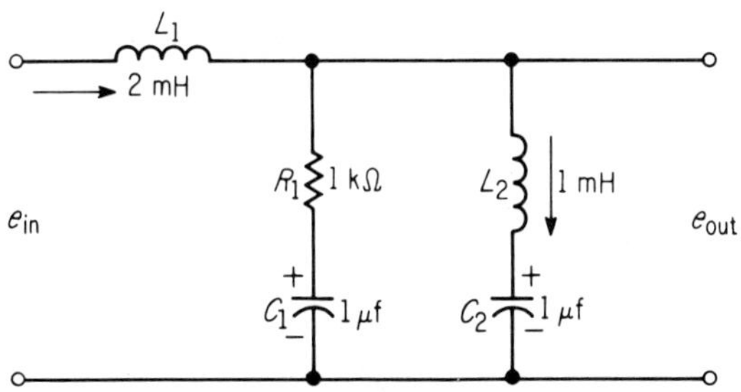

PROB. 4-4

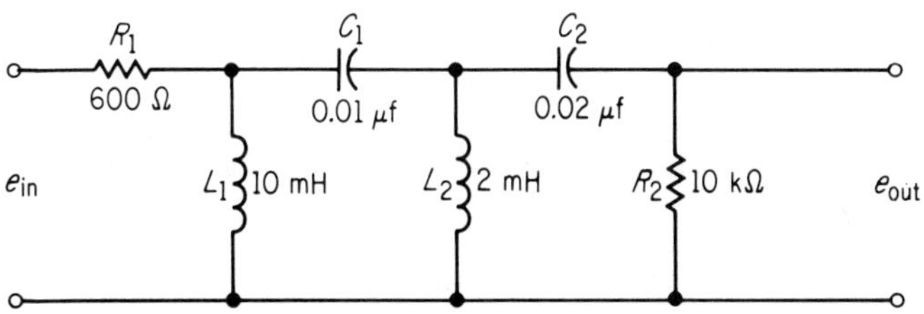

PROB. 4-5

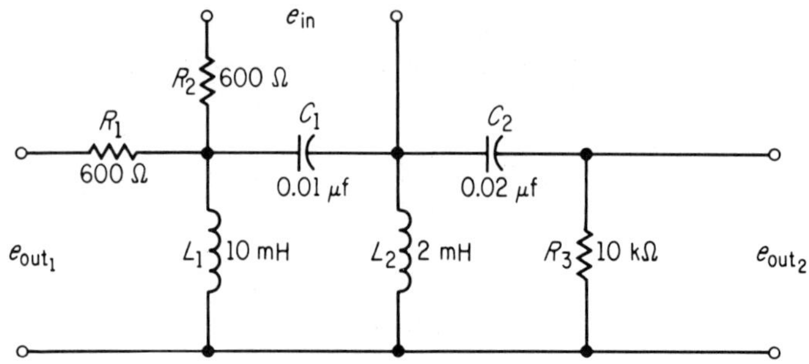

PROB. 4-6

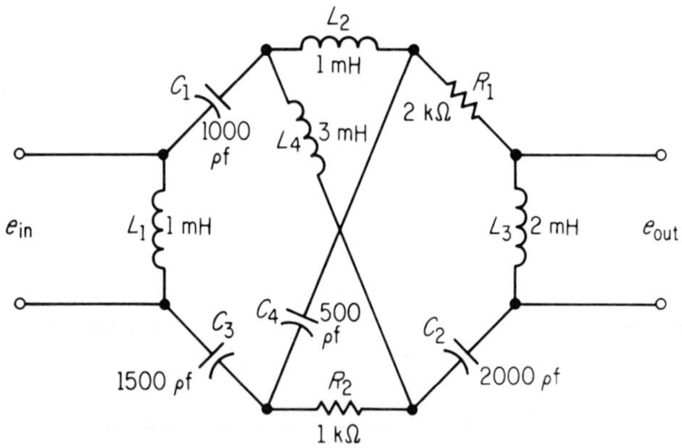

PROB. 4-7

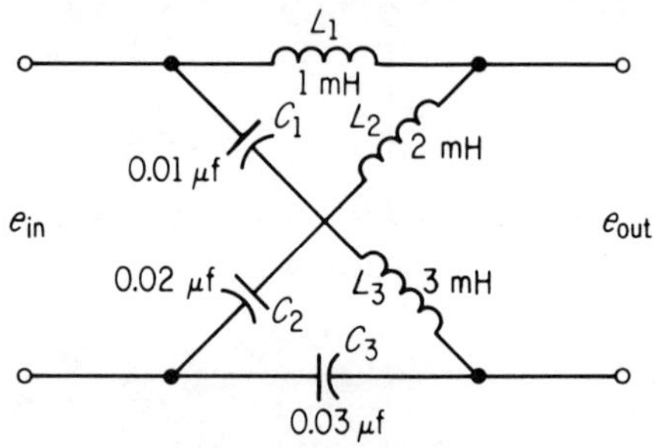

PROB. 4-8

4-9 Express the 5th-order differential equation as a state equation:

$$10\frac{d^5x}{dt^5} - 2\frac{d^4x}{dt^4} + \frac{d^3x}{dt} + \frac{d^2x}{dt} + x(t) = v(t)$$

4-10 Express as a state equation

$$\frac{d^4x}{dt^4} + 2\frac{d^2x}{dt^2} + \frac{dx}{dt} = v(t)$$

4-11 Express as a differential equation

$$\dot{\mathbf{x}} = \begin{bmatrix} 0 & 1 & 0 & 0 \\ 0 & 0 & 1 & 0 \\ 0 & 0 & 0 & 1 \\ 10 & -2 & 1 & -4 \end{bmatrix} \begin{bmatrix} x \\ \dfrac{dx}{dt} \\ \dfrac{d^2x}{dt^2} \\ \dfrac{d^3x}{dt^3} \end{bmatrix} + \begin{bmatrix} 0 \\ 0 \\ 0 \\ 3 \end{bmatrix} [v(t)]$$

4-12 to 4-14 Formulate the state equation for the networks shown using the matrices $\mathbf{B}_T$ and $\mathbf{C}_L$. Use capacitor voltages and inductor currents as components of the state equation.

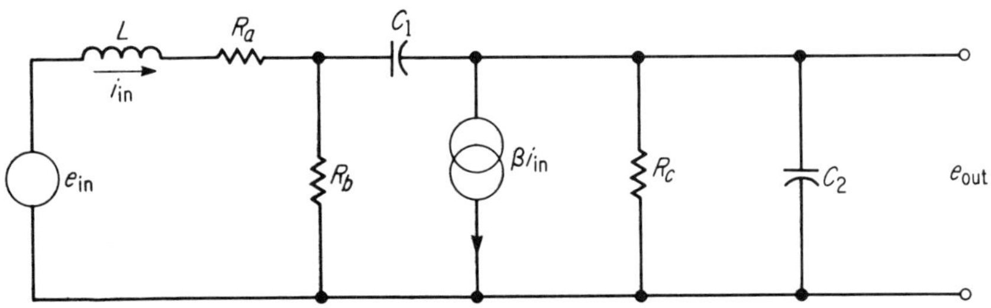

PROB. 4-12

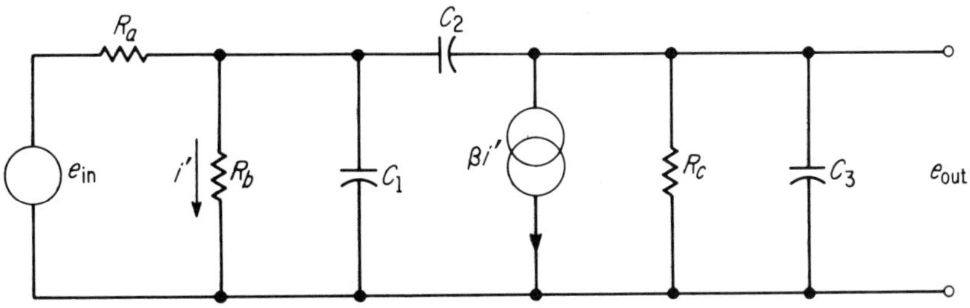

PROB. 4-13

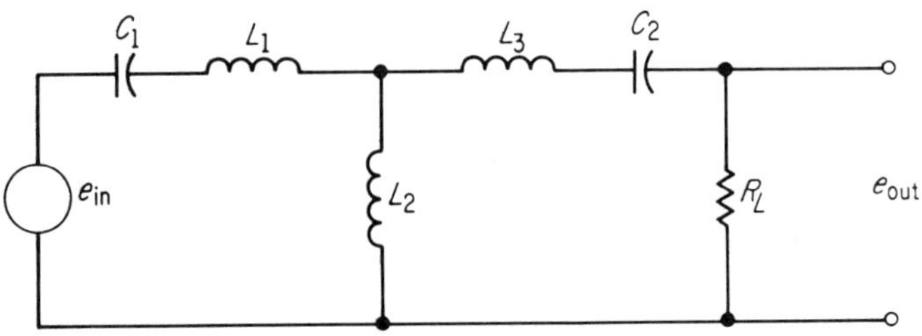

PROB. 4-14

4-15 Write a program to formulate the state equation for networks with dependent sources. Formulate the state equation for the network of Prob. 4-12 if e_{in} changes from 0 to 10 volts at $t = 0$. The element values are $L = 1$ mH, $R_a = 1$ kΩ, $R_b = 2$ kΩ, $R_c = 3$ kΩ, $C_1 = 100$ pf, $C_2 = 0.2$ μf, and $\beta = 100$.

chapter 5

NUMERICAL SOLUTION OF THE STATE EQUATION

Chapters 2 and 4 indicated that the state-variable concept is quite important in network analysis, especially when machine analysis is to be undertaken. Chapter 3 introduced topology theory and the standard branch and demonstrated methods of applying these ideas to machine formulation of equations. Machine solution of the network problem based on the node, cutset, and tieset methods was also discussed. It was pointed out that these methods require the existence of an impedance or admittance for each branch. Thus these methods are not applicable to transient analysis. The state variable approach is commonly used for the transient case. Chapter 4 considered methods of automatic formulation of the state equation by digital computer. A scheme for automatic formulation of equations by computer was also demonstrated for a simple network. We must now consider the solution of the state equation using a computer.

This chapter considers several computer methods of solving the state equation given the matrices **A** and **B** and the initial vector $\mathbf{x}_0$. There is no single method that is conclusively the most effective to apply. Each method has advantages and disadvantages. Numerical integration is very useful in many cases, but becomes inefficient for certain networks as discussed later. Even within a given method of solution there are several alternate possibilities, each leading to different efficiencies for a given problem. Both ECAP and SCEPTRE use numerical integration to solve transient problems. ECAP uses an integration subroutine that is inherently inaccurate unless the problem is subdivided into very small time intervals compared to circuit time constants [5]. Thus many calculations must be made to carry a problem to completion. SCEPTRE allows a choice of integration methods [4], all of which are more accurate than that of ECAP. While SCEPTRE integrations are generally more efficient than those performed by ECAP, circuits with widely differing time constants require excessive computer time by either program. Other methods of solution may be more applicable for circuits of this nature.

A. ZERO-INPUT RESPONSE

In practice, the networks to be analyzed will normally be excited by an input signal. The problem then consists of solving the nonhomogeneous state vector. Methods of solution are discussed later in this chapter. While circuits with initial charges or currents, but no inputs, are rarely analyzed in actual practice, it is instructive to consider the zero-input response. The methods developed for analysis can in general be extended to cover the more useful finite input case. Thus some detail will now be devoted to obtaining the zero-input response of a network.

In Chapter 2 the solution of the homogeneous state equation was considered. The solution to the equation

$$\dot{\mathbf{x}} = \mathbf{A}\mathbf{x} \tag{2-36}$$

was shown to be

$$\mathbf{x} = e^{\mathbf{A}t}\,\mathbf{x}_0 \tag{5-1}$$

where

$$e^{\mathbf{A}t} = \sum_{n=0}^{\infty} \frac{1}{n!}\mathbf{A}^n t^n \tag{2-42}$$

Equation 2-36 corresponds to the state equation with no input present. The response then depends on the initial state vector $\mathbf{x}_0$. If $\mathbf{x}_0$ is the zero vector, implying an initially relaxed system, the state vector is identically zero for all time.

In ordinary differential equations, a solution such as Eq. 5-1 is the end result of the problem. Equation 5-1, however, is a matrix equation and requires the expansion of the exponential function of a matrix to complete the solution. There are at least three methods of obtaining the value of $\mathbf{x}$ as a function of time. The first is simply to evaluate all elements of the exponential matrix series expansion as demonstrated in Chapter 2. A second approach is to find the eigenvectors of $\mathbf{A}$ which will diagonalize both $\mathbf{A}$ and the exponential matrix. If $\mathbf{P}$ is the matrix having eigenvectors of $\mathbf{A}$ as its columns, then

$$e^{\Lambda t} = \mathbf{P}^{-1}\, e^{\mathbf{A}t}\, \mathbf{P} \tag{5-2}$$

and

$$e^{\mathbf{A}t} = \mathbf{P}e^{\Lambda t}\,\mathbf{P}^{-1} \tag{5-3}$$

where Λ is a diagonal matrix whose elements correspond to the eigenvalues of $\mathbf{A}$. Once the eigenvalues and eigenvectors are known, the calculation of $e^{\Lambda t}$ is much easier for large matrices than the calculation of $e^{\mathbf{A}t}$.

A third approach to finding $\mathbf{x}$ is to integrate Eq. 2-36. This method of solution is more appropriate to machine calculation than hand calculation. Integration is an efficient method of solution if the eigenvalues are not widely separated. When the eigenvalues are widely separated, the efficiency of integration by the digital computer is very poor [1]. A more detailed evaluation of the three approaches will be given later in this chapter. Let us consider these methods by means of a simple example.

EXAMPLE 5-1. Evaluate the zero input response of the network of Fig. 5-1 if $e_{out}(0) = v_c(0) = 10$ and $i_{in}(0) = i_L(0) = 0$.

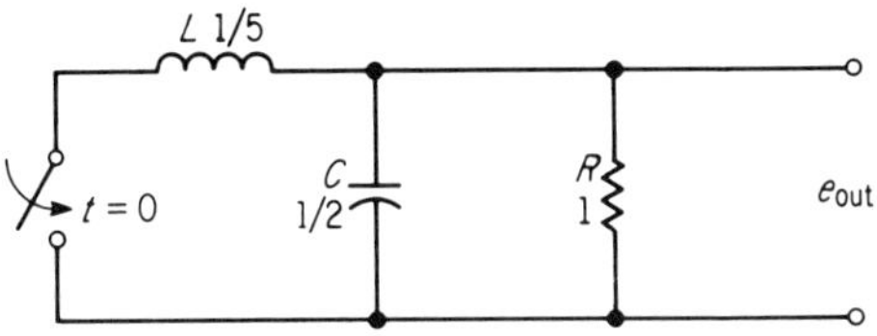

FIG. 5-1. *RLC* network.

Solution. The state equation for this network has been determined in Sec. A of Chapter 4 to be

$$\dot{\mathbf{x}} = \begin{bmatrix} 0 & -1/L \\ 1/C & -1/RC \end{bmatrix} \begin{bmatrix} i_{in} \\ e_{out} \end{bmatrix} + \begin{bmatrix} 1/L \\ 0 \end{bmatrix} [e_{in}]$$

where $\mathbf{x} = \begin{bmatrix} i_{in} \\ e_{out} \end{bmatrix}$. Substituting the correct element values and taking $[e_{in}] = 0$, we have

$$\dot{\mathbf{x}} = \begin{bmatrix} 0 & -5 \\ 2 & -2 \end{bmatrix} \begin{bmatrix} i_{in} \\ e_{out} \end{bmatrix}$$

The solution of this equation is

$$\mathbf{x} = \mathbf{e}^{\mathbf{A}t}\,\mathbf{x}_0$$

where

$$\mathbf{x}_0 = \begin{bmatrix} 0 \\ 10 \end{bmatrix}$$

The method used in Example 2-11 can be applied to evaluate this function by hand calculation. The eigenvalues of **A** are $\lambda_1 = -1 - j3$ and $\lambda_2 = -1 + j3$. Since **A** is a 2 X 2 matrix, we know that

$$\mathbf{F}(\mathbf{A}) = K_0\mathbf{U} + K_1\mathbf{A}$$

and

$$f(\lambda) = K_0 + K_1\lambda$$

Using the two eigenvalues in the previous equation gives

$$e^{-t}e^{-j3t} = K_0 - K_1 - j3K_1$$

and

$$e^{-t}e^{+j3t} = K_0 - K_1 + j3K_1$$

Solving for K_0 and K_1 results in

$$K_0 = e^{-t} \cos 3t + \frac{e^{-t}}{3} \sin 3t$$

and

$$K_1 = \frac{e^{-t}}{3} \sin 3t$$

The solution is then

$$\begin{bmatrix} i_{\text{in}} \\ e_{\text{out}} \end{bmatrix} = \frac{e^{-t}}{3} \begin{bmatrix} (3 \cos 3t + \sin 3t & -5 \sin 3t \\ 2 \sin 3t & (3 \cos 3t - \sin 3t) \end{bmatrix} \begin{bmatrix} 0 \\ 10 \end{bmatrix}$$

$$= \begin{bmatrix} (-50\, e^{-t}/3) \sin 3t \\ 10\, e^{-t} \cos 3t - (10\, e^{-t}/3) \sin 3t \end{bmatrix}$$

This answer can be checked by considering the diagonalized matrix **A**. Two possible eigenvectors corresponding to the eigenvalues of **A** are

$$\mathbf{u}(-1 - j3) = \begin{bmatrix} 1 \\ 1/5 + j3/5 \end{bmatrix}$$

and

$$\mathbf{u}(-1 + j3) = \begin{bmatrix} 1 \\ 1/5 - j3/5 \end{bmatrix}$$

The matrix **P** is

$$\mathbf{P} = \begin{bmatrix} 1 & 1 \\ 1/5 + j3/5 & 1/5 - j3/5 \end{bmatrix}$$

while

$$\mathbf{P}^{-1} = -j\frac{5}{6} \begin{bmatrix} -1/5 + j3/5 & 1 \\ 1/5 + j3/5 & -1 \end{bmatrix}$$

The function $\mathbf{e}^{\Lambda t}$ is now equal to

$$\exp\left(\begin{bmatrix} -1 - j3 & 0 \\ 0 & -1 + j3 \end{bmatrix} t\right)$$

Based on the power-series expansion for the exponential, it is easy to show (see Eq. 2-66) that

$$\exp \begin{bmatrix} \lambda_1 t & 0 & \cdots & 0 \\ \cdot & \lambda_2 t & & \cdot \\ \cdot & & & \cdot \\ \cdot & & \cdot & \cdot \\ 0 & & \cdots & \lambda_n t \end{bmatrix} = \begin{bmatrix} e^{\lambda_1 t} & 0 & \cdots & 0 \\ \cdot & e^{\lambda_2 t} & & \cdot \\ \cdot & & \cdot\ \cdot & \cdot \\ 0 & & \cdots & e^{\lambda_n t} \end{bmatrix}$$

Thus

$$\mathbf{e}^{\Lambda t} = \begin{bmatrix} e^{-t}\, e^{-j3t} & 0 \\ 0 & e^{-t}\, e^{j3t} \end{bmatrix}$$

We now evaluate $\mathbf{e}^{\mathbf{A}t}$ from Eq. 5-3 as

$$\mathbf{e}^{\mathbf{A}t} = \mathbf{P}\mathbf{e}^{\Lambda t}\,\mathbf{P}^{-1} \tag{5-3}$$

$$= \frac{e^{-t}}{3}\begin{bmatrix} 3\cos 3t + \sin 3t & -5\sin 3t \\ 2\sin 3t & 3\cos 3t - \sin 3t \end{bmatrix}$$

Multiplying $\mathbf{x}_0$ by this matrix yields the same result as obtained previously.

In the case of a 2×2 matrix, the above two methods require roughly the same amount of calculation to arrive at a solution. Let us consider the steps required by each method to estimate the time requirements for larger matrices. Both the remainder method and the eigenvector method depend on the eigenvalues of **A** and thus these eigenvalues must be evaluated. One equation corresponding to each individual eigenvalue is generated in the remainder method. These n equations (for an nth-order matrix) must then be solved simultaneously for n coefficients ($K_0, K_1, \ldots, K_{n-1}$). When the coefficients are found, the quantity $\mathbf{e}^{\mathbf{A}t}$ can be expressed as a finite power series involving powers of **A** up to the $(n-1)$th power. Thus, all powers of **A** up to the $(n-1)$th must be evaluated. The power series terms are then summed to obtain the final result.

Use of the eigenvector method requires the selection of n eigenvectors. This amounts to solving n equations for $n-1$ unknowns ($2n$ equations for $2(n-1)$ unknowns if the eigenvectors are complex) to find each eigenvector. The inverse of the eigenvector matrix **P** is found and then Eq. 5-3 can be applied to yield the final result. While the eigenvector method is more complex than the remainder method in general, there are many practical cases where this is not so. Often, the matrix **A** will be a sparse matrix having many zero off-diagonal elements. This simplifies both the eigenvector selection and the inversion of **P** and may lead to a more efficient method than the remainder method. Example 5-2 will demonstrate the effect that zero elements of **A** have on the solution.

EXAMPLE 5-2. Solve the state equation for the circuit of Fig. 5-2 if the initial capacitor voltages are $v_{c_1}(0) = 0$, $v_{c_2}(0) = 4$, $v_{c_3}(0) = 0$, $v_{c_4}(0) = 6$, and $v_{c_5}(0) = 10$. All switches are closed at $t = 0$.

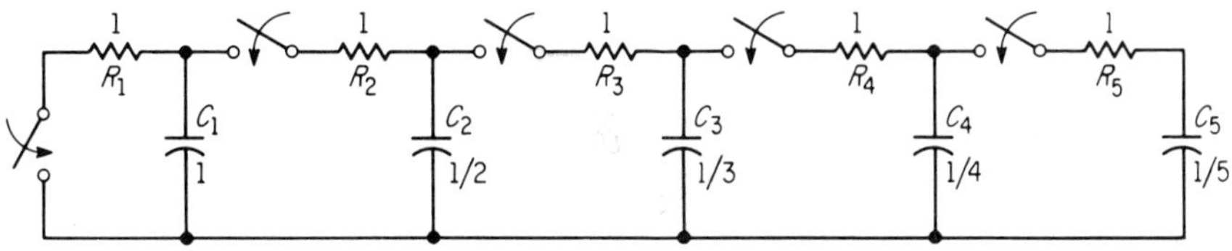

FIG. 5-2. *RC* network.

Solution. The state equation for the network is

$$\frac{d}{dt}\begin{bmatrix} v_{c_1} \\ v_{c_2} \\ v_{c_3} \\ v_{c_4} \\ v_{c_5} \end{bmatrix} = \begin{bmatrix} -\frac{1}{C_1}\left(\frac{1}{R_1}+\frac{1}{R_2}\right) & \frac{1}{C_1 R_2} & 0 & 0 & 0 \\ \frac{1}{C_2 R_2} & -\frac{1}{C_2}\left(\frac{1}{R_2}+\frac{1}{R_3}\right) & \frac{1}{C_2 R_3} & 0 & 0 \\ 0 & \frac{1}{C_3 R_3} & -\frac{1}{C_3}\left(\frac{1}{R_3}+\frac{1}{R_4}\right) & \frac{1}{C_3 R_4} & 0 \\ 0 & 0 & \frac{1}{C_4 R_4} & -\frac{1}{C_4}\left(\frac{1}{R_4}+\frac{1}{R_5}\right) & \frac{1}{C_4 R_5} \\ 0 & 0 & 0 & \frac{1}{C_5 R_5} & -\frac{1}{C_5 R_5} \end{bmatrix}$$

$$\times \begin{bmatrix} v_{c_1} \\ v_{c_2} \\ v_{c_3} \\ v_{c_4} \\ v_{c_5} \end{bmatrix} = \begin{bmatrix} -2 & 1 & 0 & 0 & 0 \\ 2 & -4 & 2 & 0 & 0 \\ 0 & 3 & -6 & 3 & 0 \\ 0 & 0 & 4 & -8 & 4 \\ 0 & 0 & 0 & 5 & -5 \end{bmatrix} \begin{bmatrix} v_{c_1} \\ v_{c_2} \\ v_{c_3} \\ v_{c_4} \\ v_{c_5} \end{bmatrix}$$

There are standard subroutines available to solve for eigenvalues and eigenvectors although these subroutines may be somewhat inefficient if designed for general problem solution. Special subroutines can be written to minimize computer time for particular types of networks. The eigenvalues for this example were evaluated by computer to be

$$\begin{aligned} \lambda_1 &= -0.263564 \\ \lambda_2 &= -1.413400 \\ \lambda_3 &= -3.596400 \\ \lambda_4 &= -7.085790 \\ \lambda_5 &= -12.64080 \end{aligned}$$

The eigenvectors for an RC ladder network can be found easily by solving the equations

$$\begin{aligned} (-2-\lambda)u_1 + u_2 &= 0 \\ 2u_1 + (-4-\lambda)u_2 + 2u_3 &= 0 \\ 3u_2 + (-6-\lambda)u_3 + 3u_4 &= 0 \end{aligned}$$

$$\begin{aligned} 4u_3 + (-8 - \lambda)u_4 + 4u_5 \qquad &= 0 \\ 5u_4 + (-5 - \lambda)u_5 &= 0 \end{aligned}$$

Since the value of one component of each eigenvector is arbitrary, only four equations need be solved. Furthermore, if the eigenvector subroutine is written to take advantage of the zero coefficients of the equations, the eigenvectors can be found more efficiently. For this example the eigenvectors were found to be

$$\begin{bmatrix} 1.000000 \\ 1.736440 \\ 2.244040 \\ 2.554500 \\ 2.696640 \end{bmatrix}, \begin{bmatrix} 1.000000 \\ 0.586000 \\ -0.241350 \\ -0.955592 \\ -1.332170 \end{bmatrix}, \begin{bmatrix} 1.000000 \\ -1.596400 \\ -1.322150 \\ 0.537091 \\ 1.913260 \end{bmatrix}, \begin{bmatrix} 0.100000 \\ -0.508579 \\ 0.684684 \\ 0.260771 \\ -0.625114 \end{bmatrix}, \text{ and } \begin{bmatrix} 0.100000 \\ -1.064080 \\ 4.497250 \\ -8.891040 \\ 5.818130 \end{bmatrix}$$

The flow chart of Fig. 5-3 demonstrates the implementation of the eigen-

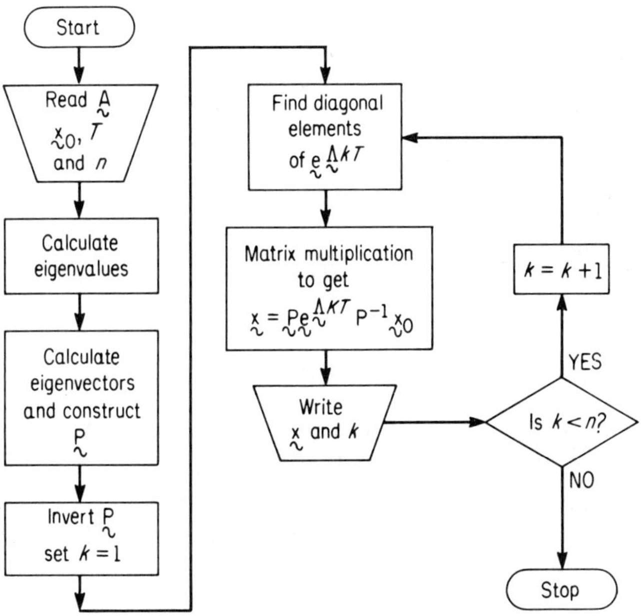

FIG. 5-3. Flow chart of eigenvector method.

vector method. The eigenvalues and eigenvectors are first found and the matrix **P** is constructed and inverted. These calculations need be done only once for the entire problem. The elements of the diagonal matrix $\mathbf{e}^{\Lambda t}$ are then calculated at each time step. It is important to note that the diagonal elements of this matrix are scalar exponentials which can be evaluated by means of library subroutines. After this diagonal matrix is found, a linear transformation involving only matrix multiplication is required to find $\mathbf{x}$ at each time step. The time step and the total number of time steps over which the response is to be calculated are given as input data. In the flow chart these parameters are T and n, respectively.

For this problem, the time step is chosen as 0.0225 sec, which is a factor of 10 less than the smallest eigenvalue. The solution was carried out at 2,000 time steps (far more than required). Table 5-1 lists the program which was written in FORTRAN 4 and executed on an IBM 1132 computer.

TABLE 5-1. Calculation of Response, Using Eigenvectors.

```
      REAL LAMDA (5)
      DIMENSION P(25),PI(25),X(25),M(5),Y(5),W(25),L(5)
      DIMENSION XO(5),OUT(5),A(25)
C     READ MATRIX P IN VECTOR FORM
      READ (8,5) (P(I),I=1,25)
C     READ EIGENVALUES
      READ(8,5) (LAMDA(I),I=1,5)
C     READ INITIAL STATE VECTOR
      READ (8,5) (XO(I),I=1,5)
C     READ MATRIX A IN VECTOR FORM
      READ(8,5) (A(I),I=1,25)
    5 FORMAT(5F10.6)
C     INVERT MATRIX P TO FORM PI
      DO 2 I=1,25
    2 PI(I)=P(I)
      CALL MINV(PI,5,D,L,M)
C     USE 2000 TIME STEPS IN SOLUTION
      DO 4 J=1,2000
      N=J
      Z=J
C     SELECT TIME INTERVAL OF 0.0225 SECONDS
      T=2.25E-2*Z
C     FORM THE NONZERO ELEMENTS OF DIAGONAL
C     EXPONENTIAL MATRIX Y
      DO 3 I=1,5
    3 Y(I)=EXP(T*LAMDA(I))
C     MATRIX MULTIPLICATION TO FORM STATE VECTOR OUT
      CALL MPRD(P,Y,X,5,5,0,2,5)
      CALL MPRD(X,PI,W,5,5,0,0,5)
      CALL MPRD(W,XO,OUT,5,5,0,0,1)
      WRITE(5,99) N,T,OUT
   99 FORMAT(' ','STEP',1X,I4,5X,E15.6,5(E15.6,' '))
    4 CONTINUE
      CALL EXIT
      END
```

Table 5-2 tabulates the calculated voltages on the capacitors at selected time intervals.

TABLE 5-2. Selected Results from Program of Table 5-1.

Time, Sec.	v_{c_1},Volts	v_{c_2},Volts	v_{c_3},Volts	v_{c_4},Volts	v_{c_5},Volts
0.0	0.0	4.0	0.0	6.0	10.0
0.225	0.523400	2.37775	3.05333	5.24771	6.94106
0.450	0.736436	2.15809	3.45018	4.78883	5.61487
0.675	0.857361	2.13079	3.35928	4.34832	4.87817
0.900	0.930082	2.10044	3.17621	3.96593	4.36413
1.125	0.968674	2.04685	2.98368	3.64020	3.96219
1.350	0.981797	1.97615	2.79983	3.36009	3.63056
1.575	0.976298	1.89545	2.62812	3.11567	3.34847
1.800	0.957632	1.80977	2.46851	2.89941	3.10343
2.250	0.896553	1.63524	2.18178	2.53053	2.69367
3.150	0.741987	1.31049	1.71276	1.96280	2.07824
4.500	0.530442	0.924359	1.19741	1.36501	1.44189
6.008	0.358096	0.622201	0.804422	0.915943	0.967019
9.000	0.162877	0.282833	0.365516	0.416088	0.439242

There is an interesting point that becomes prominent when applying the eigenvector method. The solution of the zero-input state equation is

$$\mathbf{x} = \mathbf{e}^{\mathbf{A}t}\mathbf{x}_0 \qquad [5\text{-}1]$$

where $\mathbf{e}^{\mathbf{A}t}$ can be expressed as

$$\mathbf{e}^{\mathbf{A}t} = \mathbf{P}\mathbf{e}^{\Lambda t}\,\mathbf{P}^{-1} \qquad [5\text{-}3]$$

Substituting the value for $\mathbf{e}^{\mathbf{A}t}$ given by Eq. 5-3 into Eq. 5-1 gives

$$\mathbf{x} = \mathbf{P}\mathbf{e}^{\Lambda t}\,\mathbf{P}^{-1}\,\mathbf{x}_0 \qquad (5\text{-}4)$$

If the initial vector $\mathbf{x}_0$ happens to correspond to one of the eigenvectors of $\mathbf{A}$ (for example the mth) we can write Eq. 5-4 in expanded form as

$$\mathbf{x} = \begin{bmatrix} u_{11} & u_{12} & \cdots & u_{1n} \\ u_{21} & u_{22} & \cdots & u_{2n} \\ \cdot & \cdot & & \cdot \\ \cdot & \cdot & & \cdot \\ \cdot & \cdot & & \cdot \\ u_{n1} & u_{n2} & & u_{nn} \end{bmatrix} \begin{bmatrix} e^{\lambda_1 t} & 0 & \cdots & 0 \\ 0 & e^{\lambda_2 t} & & 0 \\ \cdot & \cdot & \cdot\;\cdot & \cdot \\ \cdot & \cdot & & \cdot \\ \cdot & \cdot & \cdot\;\cdot & \cdot \\ 0 & 0 & \cdots & e^{\lambda_n t} \end{bmatrix}$$

$$\times \begin{bmatrix} v_{11} & v_{12} & \cdots & v_{1n} \\ v_{21} & v_{22} & & v_{2n} \\ \cdot & \cdot & & \cdot \\ \cdot & \cdot & & \cdot \\ \cdot & \cdot & & \cdot \\ v_{n1} & v_{n2} & \cdots & v_{nn} \end{bmatrix} \begin{bmatrix} u_{1m} \\ u_{2m} \\ \cdot \\ \cdot \\ \cdot \\ u_{nm} \end{bmatrix}$$

where

$$\begin{bmatrix} u_{1i} \\ u_{2i} \\ \cdot \\ \cdot \\ \cdot \\ u_{ni} \end{bmatrix}$$

is the ith eigenvector and the elements of the inverse matrix $\mathbf{P}^{-1}$ are given by v_{ij}. Since

$$\mathbf{P}^{-1}\,\mathbf{P} = \mathbf{U}$$

we know that the multiplication of a column of **P** by a row of $\mathbf{P}^{-1}$ always results in a zero element unless the column number of **P** corresponds to the row number of $\mathbf{P}^{-1}$ in which case the result is unity. That is, the elements of the product $\mathbf{P}^{-1}\,\mathbf{P}$ are given by

$$l_{ik} = \sum_{j=1}^{n} v_{ij}\, u_{jk} = \delta_{ik} \tag{5-5}$$

where the elements of the product matrix are represented by l_{ik} and δ_{ik} is the Kronecker delta function defined as

$$\delta_{ik} = 0, \quad i \neq k$$
$$\delta_{ik} = 1, \quad i = k$$

Since $\mathbf{x}_0$ corresponds to the mth eigenvector (or the mth column of **P**), the product $\mathbf{P}^{-1}\,\mathbf{x}_0$ is

$$\mathbf{P}^{-1}\,\mathbf{x}_0 = \begin{bmatrix} 0 \\ 0 \\ \cdot \\ \cdot \\ 1 \\ \cdot \\ \cdot \\ 0 \end{bmatrix}$$

where the nonzero element appears in the mth row. The product $\mathbf{P}\,e^{\Lambda t}$ is

$$\begin{bmatrix} u_{11}e^{\lambda_1 t} & u_{12}e^{\lambda_2 t} & \cdots & u_{1n}e^{\lambda_n t} \\ u_{21}e^{\lambda_1 t} & u_{22}e^{\lambda_2 t} & \cdots & u_{2n}e^{\lambda_n t} \\ \vdots & \vdots & & \vdots \\ u_{n1}e^{\lambda_1 t} & u_{2n}e^{\lambda_2 t} & \cdots & u_{nn}e^{\lambda_n t} \end{bmatrix}$$

Multiplying $\mathbf{P}^{-1}\,\mathbf{x}_0$ by $\mathbf{P}\,\mathbf{e}^{\Lambda t}$ gives

$$\mathbf{x} = \mathbf{P}\mathbf{e}^{\Lambda t}\,\mathbf{P}^{-1}\,\mathbf{x}_0 = \begin{bmatrix} u_{1m}\,e^{\lambda_m t} \\ u_{2m}\,e^{\lambda_m t} \\ \vdots \\ u_{nm}\,e^{\lambda_m t} \end{bmatrix} = e^{\lambda_m t}\,\mathbf{x}_0 \tag{5-6}$$

The output response when the initial conditions correspond to the mth eigenvector is proportional to that eigenvector and decays exponentially with a time constant determined by the mth eigenvalue.

When an eigenvector is complex, an initial vector cannot be chosen to correspond to this eigenvector. Complex eigenvalues and eigenvectors always occur in conjugate pairs for real systems. Just as poles or zeros of a Laplace function must occur in conjugate pairs to have physical meaning, both eigenvalues must be considered in the complete solution of the state equation. Thus it is impossible to excite a mode corresponding only to a single eigenvector and eigenvalue in the complex case.

EXAMPLE 5-3. Prove the validity of Eq. 5-6 by assuming an initial vector equal to the eigenvector corresponding to the eigenvalue $\lambda_1 = -0.263564$ for the network of Fig. 5-2. Calculate the overall response by applying Eq. 5-4 and compare to the result obtained by using Eq. 5-6.

The program of Table 5-1 is again executed with $\mathbf{x}_0$ taken as

$$\mathbf{x}_0 = \begin{bmatrix} 1.000000 \\ 1.736440 \\ 2.244040 \\ 2.554500 \\ 2.696640 \end{bmatrix}$$

The results are shown in Table 5-3. All voltages decay with the time constant $\tau = -1/\lambda_1$ in this case. Table 5-3 lists selected values.

TABLE 5-3. (a) Results of Execution of Program of Table 5-1. Initial Conditions Correspond to an Eigenvector. (b) Values Calculated from Eq. 5-6.

Time, Sec.	v_{c1}	v_{c2}	v_{c3}	v_{c4}	v_{c5}
			(a)		
0.0	1.000000	1.736440	2.244040	2.554500	2.696640
0.225	0.942421	1.63645	2.11483	2.40741	2.54137
0.450	0.888159	1.54223	1.99306	2.26880	2.39504
0.900	0.788826	1.36975	1.77015	2.01505	2.12718
1.800	0.622248	1.08049	1.39635	1.58953	1.67797
3.600	0.387193	0.672338	0.868877	0.989085	1.04412
7.200	0.149918	0.260325	0.336423	0.382967	0.404277
			(b)		
0.0	1.000	1.736	2.24	2.56	2.70
0.225	0.943	1.638	2.12	2.40	2.54
0.450	0.888	1.542	1.994	2.27	2.39
0.900	0.788	1.369	1.770	2.01	2.13
1.800	0.623	1.081	1.400	1.592	1.680
3.600	0.388	0.673	0.870	0.990	1.047
7.200	0.150	0.261	0.337	0.384	0.405

Comparison of the computer-generated results to hand calculations based on Eq. 5-6 indicates good agreement.

For machine calculation of the zero-input response a recursive method proposed by Liou [3] can be effective. If t is divided into fixed increments of time such that $t = nT$, the zero-input response is

$$\mathbf{x}(nT) = \mathbf{e}^{\mathbf{A}nT}\mathbf{x}_0 \tag{5-7}$$

At $t = (n+1)T$,

$$\mathbf{x}[(n+1)T] = \mathbf{e}^{\mathbf{A}(n+1)T}\mathbf{x}_0 \tag{5-8}$$

Solving for $\mathbf{x}_0$ in Eq. 5-7 and substituting this value into Eq. 5-8 results in an expression for $\mathbf{x}[(n+1)T]$ in terms of $\mathbf{x}(nT)$.

$$\mathbf{x}[(n+1)T] = \mathbf{e}^{\mathbf{A}t}\,\mathbf{x}(nT) \tag{5-9}$$

The function $\mathbf{e}^{\mathbf{A}t}$ is a constant matrix for $t = T$ and can be evaluated by means of a series expansion. Once $\mathbf{e}^{\mathbf{A}T}$ is evaluated, values of $\mathbf{x}(t)$ can be found at discrete points over the interval of interest by applying Eq. 5-9 recursively. Thus we calculate $\mathbf{x}(T)$, $\mathbf{x}(2T)$, $\mathbf{x}(3T)$, . . . $\mathbf{x}(bT)$, where bT is equal to the time interval of interest. Since

$$e^{\mathbf{A}t} = \sum_{n=0}^{\infty} \frac{1}{n!} \mathbf{A}^n t^n \qquad [2\text{-}42]$$

we can write

$$\begin{aligned} e^{\mathbf{A}T} &= \sum_{n=0}^{k} \frac{1}{n!} \mathbf{A}^n T^n + \sum_{n=k+1}^{\infty} \frac{1}{n!} \mathbf{A}^n T^n \qquad (5\text{-}10) \\ &= \mathbf{M} + \mathbf{R} \end{aligned}$$

where $\mathbf{M}$ is the matrix obtained by truncating the series expansion at $n = k$, and $\mathbf{R}$ is the remainder matrix. The number k can be chosen by an iterative scheme such that

$$|r_{ij}| << |m_{ij}|$$

or

$$|r_{ij}| \leqq 10^{-d} |m_{ij}| \qquad (5\text{-}11)$$

where d is the order of magnitude by which the elements of $\mathbf{R}$ are less than the corresponding elements of $\mathbf{M}$. An upper bound on the largest element of $\mathbf{R}$ can be established by considering the series expression for $\mathbf{R}$, that is

$$\mathbf{R} = \sum_{n=k+1}^{\infty} \frac{1}{n!} \mathbf{A}^n T^n \qquad (5\text{-}12)$$

We construct a matrix $\mathbf{A}_{\max}$ having the same order as $\mathbf{A}$, but all elements equal to the absolute value of the maximum element of $\mathbf{A}$, designated a_m. It is then obvious that all elements of $\mathbf{A}_{\max}^n$ will be greater than or equal to all elements of $\mathbf{A}^n$. If l is the dimension of $\mathbf{A}$, then each element of $\mathbf{A}_{\max}^n$ will be

$$b_{ij} = l^{n-1} a_m^n \qquad (5\text{-}13)$$

Since all elements of $\mathbf{A}^n$ are smaller than or equal to b_{ij}, then all elements of $\mathbf{R}$ are such that

$$|r_{ij}| \leqq \sum_{n=k+1}^{\infty} \frac{l^{n-1} a_m^n T^n}{n!} \qquad (5\text{-}14)$$

The infinite series of Eq. 5-14 can be expressed as

$$\sum_{n=k+1}^{\infty} \frac{l^{n-1} a_m^n T^n}{n!} = \frac{(la_m T)^{k+1}}{l(k+1)!} \left[1 + \frac{la_m T}{k+2} + \frac{(la_m T)^2}{(k+2)(k+3)} + \cdots \right] \qquad (5\text{-}15)$$

We note that

$$\frac{(la_m T)^{k+1}}{l(k+1)!}\left[1+\frac{la_m T}{(k+2)}+\left(\frac{la_m T}{k+2}\right)^2+\cdots\right]=\frac{(la_m T)^{k+1}}{l(k+1)!}\left[\frac{1}{1-la_m T/(k+2)}\right] \tag{5-16}$$

if $la_m T/(k+2)$ is less than unity, and that the series in Eq. 5-16 will be greater than that of Eq. 5-15. Therefore, we can pick T such that $la_m T = 1$ and write

$$|r_{ij}| < \frac{(la_m T)^{k+1}}{l(k+1)!}\left[\frac{1}{1-la_m T/(k+2)}\right]=\frac{1}{l(k+1)!}\left[\frac{1}{1-1/(k+2)}\right] \tag{5-17}$$

Equation 5-17 gives the upper bound on the magnitude of the largest element of r_{ij}. The matrix $\mathbf{e}^{\mathbf{A}T}$ can then be evaluated to the desired accuracy by

1. Choosing an initial value of k
2. Evaluating **M** as defined in Eq. 5-10
3. Evaluating the upper bound of $|r_{ij}|$ from Eq. 5-17
4. Comparing each element of **M** to the upper bound of $|r_{ij}|$ using Eq. 5-11. If Eq. 5-11 is satisfied the process stops, if not, k is increased and the process is repeated.

Although the series expression for $\mathbf{e}^{\mathbf{A}t}$ given by Eq. 2-42 is uniformly convergent [3], if the interval T is chosen to be too large, a great number of terms are required to satisfy Eq. 5-11. There is a practical upper limit on T which can be approximated by considering the diagonal form of $\mathbf{e}^{\mathbf{A}t}$. Equation 2-42 is expressible as

$$\mathbf{e}^{\mathbf{A}T} = \mathbf{P}\mathbf{e}^{\Lambda T}\mathbf{P}^{-1} = \mathbf{P}\left[\sum_{n=0}^{\infty}\frac{1}{n!}\Lambda^n T^n\right]\mathbf{P}^{-1} \tag{5-18}$$

Expanding the series in Eq. 5-18 gives

$$\mathbf{e}^{\mathbf{A}T} = \mathbf{P}\left[\mathbf{U}+\Lambda T+\frac{(\Lambda T)^2}{2!}+\cdots\right]\mathbf{P}^{-1} \tag{5-19}$$

The elements of $\mathbf{e}^{\Lambda T}$ will be smaller in absolute value than the corresponding elements of

$$[\mathbf{U}+\Lambda T+(\Lambda T)^2+(\Lambda T)^3+\cdots]$$

Collecting terms we find that this matrix is

$$\begin{bmatrix} (1+\lambda_1 T+(\lambda_1 T)^2+\cdots) & 0 & \cdots & 0 \\ 0 & (1+\lambda_2 T+(\lambda_2 T)^2+\cdots) & & 0 \\ \cdot & \cdot & & \cdot \\ \cdot & \cdot & & \cdot \\ \cdot & \cdot & & \cdot \\ 0 & 0 & \cdots & (1+\lambda_n T+(\lambda_n T)^2\cdots) \end{bmatrix}$$

$$= \begin{bmatrix} \frac{1}{1-\lambda_1 T} & 0 & \cdots & 0 \\ 0 & \frac{1}{1-\lambda_2 T} & & \cdot \\ \cdot & & & \cdot \\ \cdot & & & \cdot \\ \cdot & & & \\ 0 & \cdots & 0 & \frac{1}{1-\lambda_n T} \end{bmatrix}$$

All elements of the diagonal matrix will converge if

$$\lambda_{max} T < 1$$

The elements of $\mathbf{e}^{\mathbf{A}T}$ then have an upper bound if T is selected to be [1]

$$T < \frac{1}{\lambda_{max}}$$

This relationship is quite conservative, but shows that the number of points required to evaluate $\mathbf{x}$ over a given time interval is related to the maximum eigenvalue of $\mathbf{A}$. When $\mathbf{e}^{\mathbf{A}T}$ is evaluated, Eq. 5-9 is applied over the desired interval to find $\mathbf{x}$. It should be noted that truncation and round-off errors in the matrix $\mathbf{e}^{\mathbf{A}T}$ are cumulative with increasing time. The error can be checked [3] at any point in time by calculating the matrix $\mathbf{e}^{\mathbf{A}(n+1)T}$ and noting that

$$\mathbf{x}[(n+1)T] = \mathbf{e}^{\mathbf{A}(n+1)T}\mathbf{x}_0 \tag{5-20}$$

The value calculated from Eq. 5-20 is free from accumulated error, but will require several more iterations to evaluate than required to evaluate $\mathbf{e}^{\mathbf{A}T}$.

EXAMPLE 5-4. Work Example 5-2, using a finite power series expansion to evaluate $\mathbf{e}^{\mathbf{A}T}$.

Solution. There are two sections of the program for this method; first $\mathbf{M}$ must be found, then $\mathbf{x}[(n+1)T] = \mathbf{M}\mathbf{x}(nT)$ must be calculated for all values of n over the time interval of interest. The flow charts for this program are shown in Fig. 5-4.

Table 5-4 lists two separate programs for this example, however the two programs could easily be combined into one. These programs are written in Fortran IV. The first program was prepared for execution on the IBM 360/50 while the second was executed on the IBM 1132.

The program of Table 5-4a selects the appropriate time step to be used, the number of terms of the series expansion required, and evaluates the matrix $\mathbf{e}^{\mathbf{A}T}$. Table 5-4b lists the program that recursively applies Eq. 5-9 to evaluate the response of the network. The solution is carried out at 1,000 points of time in this example.

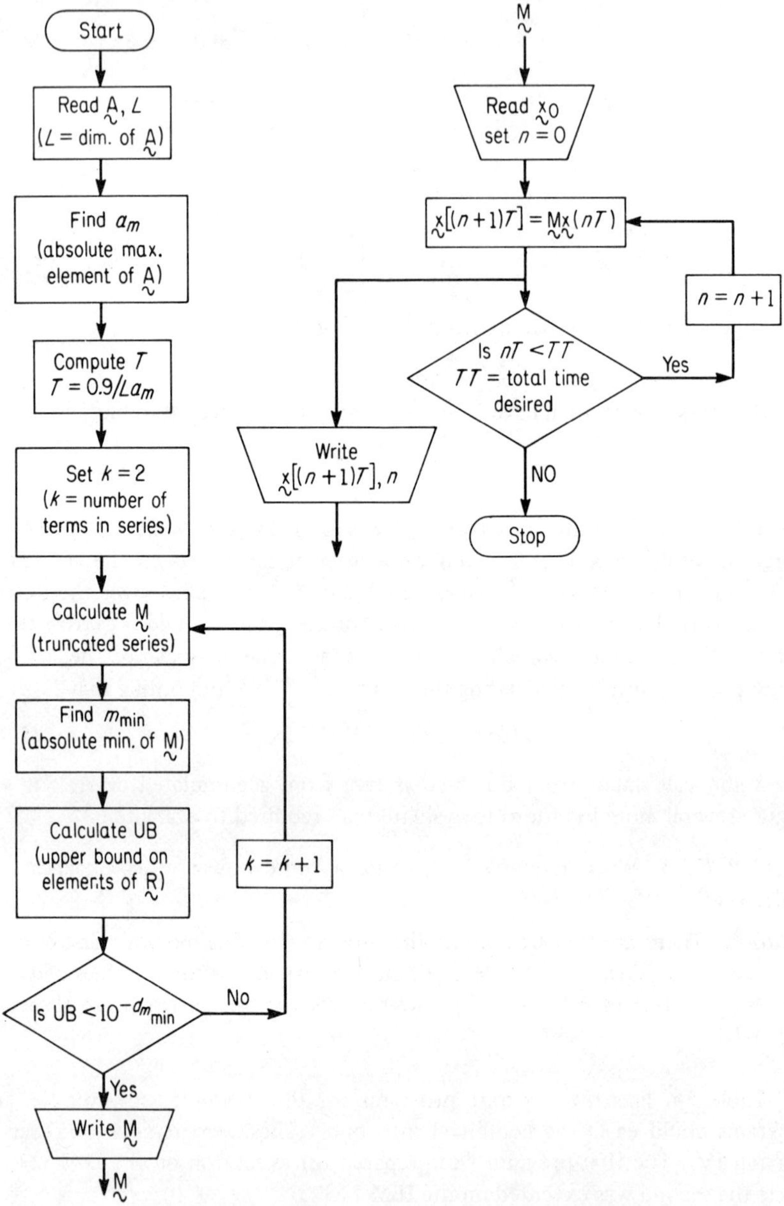

FIG. 5-4. Flow chart for series-expansion program.

The two subroutines of the first program, FACT and MPRD are to calculate factorials and matrix products respectively. The subroutine GMPRD of the second program is also to calculate matrix products. These subroutines are

TABLE 5-4. (a) Program to Find M. (b) Program to Find $\mathbf{x}[(n+1)T] = \mathbf{M}\mathbf{x}(nT)$

(a)

```
      DATA IN, I0/5,6/
C     SOLUTION FOR EXP(AT)
      REAL M,MMIN,NN
      DIMENSION A(5,5),B(5,5),C(5,5),D(5,5),E(5,5),M(5,5),R(5,5)
      L=5
C     MATRIX A IS READ
      DO 2 I=1,L
      READ(IN,3)(A(I,J),J=1,L)
    3 FORMAT(10F8.3)
    2 CONTINUE
C     MATRIX A IS PRINTED
      WRITE(I0,4)
    4 FORMAT(T30,'MATRIX A')
      DO 5 I=1,L
      WRITE(I0,6)(A(I,J),J=1,L)
    6 FORMAT('0',10(1PE14.6,'     '))
    5 CONTINUE
C     THE MAXIMUM ELEMENT OF A IS DETERMINED
      AM = A(1,1)
      DO 7 I=1,L
      DO 7 J=1,L
      IF(ABS(A(I,J)).GT.ABS(AM)) AM=A(I,J)
    7 CONTINUE
      WRITE(I0,8) AM
    8 FORMAT('0','AM = ',1PE14.6)
C     THE VALUE OF K IS DETERMINED
C     TIME INTERVAL NECESSARY FOR CONVERGENCE IS COMPUTED
      T = 0.9/(L*ABS(AM))
C     THE FIRST TWO TERMS OF THE EXPANSION ARE COMPUTED
      DO 11 I=1,L
      DO 11 J=1,L
      B(I,J) = A(I,J)*T
      C(I,J) = B(I,J)
      IF(I.EQ.J) C(I,J)=C(I,J)+1.0
   11 CONTINUE
C     THE TRUNCATED SERIES IS COMPUTED BY AN ITERATION
C     PROCESS
C     AND THE SMALLEST ELEMENT COMPARED AGAINST THE
C     REMAINDER
      CALL MPRD(L,L,L,B,B,D)
      DO 13 K=2,202
      CALL FACT(K,NN)
C     THE MINIMUM VALUE OF M IS COMPUTED
      MMIN = C(1,1)
      DO 12 I=1,L
      DO 12 J=1,L
      M(I,J) = C(I,J)+D(I,J)/NN
      IF(ABS(M(I,J)).LT.ABS(MMIN)) MMIN=M(I,J)
```

TABLE 5-4. (*continued*)

(a)

```
   12 CONTINUE
C     THE REMAINDER IS COMPUTED
      CALL FACT(K+1,FACK)
      RM = 0.9/(L*FACK*(1.0-1.0/(K+2)))
      X = 0.0001*MMIN
      WRITE(I0,19) T,K
   19 FORMAT('0',T10,'T=',1PE14.6,T40,'K=',I3)
C     CHECK ERROR
      IF(ABS(RM).LT.ABS(X)) GO TO 888
      CALL MPRD(L,L,L,B,D,E)
      DO 13 I=1,L
      DO 13 J=1,L
      C(I,J)=M(I,J)
      D(I,J)=E(I,J)
   13 CONTINUE
      WRITE(I0,15)
   15 FORMAT('   SOLUTION HAS NOT CONVERGED TO DESIRED
      1 TOLERANCE')
  888 WRITE(I0,16)
   16 FORMAT('0',T30,'MATRIX EXP(AT)')
C     WRITE FINAL MATRIX M
      DO 17 I=1,L
      WRITE(I0,18) (M(I,J),J=1,L)
   18 FORMAT('0',10(1PE14.6,'     '))
   17 CONTINUE
      CALL EXIT
      END
```

(b)

```
      REAL M(25)
      DIMENSION X(5), Y(5)
C     READ INITIAL STATE VECTOR
      READ(8,20)(X(I),I=1,5)
C     READ MATRIX M
      READ(8,20)(M(I),I=1,25)
   20 FORMAT(5E15.6)
      DO 50 J=1,1000
C     MULTIPLY M BY X(NT) TO FORM X(N+1)T
      CALL GMPRD(M,X,Y,5,5,1)
      N=J
      DO 49 I=1,5
      X(I)=Y(I)
   49 CONTINUE
      Z=N
C     CALCULATE TIME
      T=(2.25E-2)*Z
      WRITE(5,19) N,(X(I),I=1,5),T
   19 FORMAT('    ','MATRIX',1X,I4/'+',10X,5(E15.6,'    '),4X,E15.6)
   50 CONTINUE
      CALL EXIT
      END
```

listed in the Appendix. The time interval of 0.0225 sec calculated in this program was used in Example 5-2 to facilitate comparison of results. Table 5-5 shows selected results of the series-expansion program.

TABLE 5-5. Results of Series-Expansion Program.

MATRIX EXP(AT)

9.564744E-01	2.104710E-02	4.631386E-04	1.019141E-05	2.292106E-07
4.209422E-02	9.157695E-01	4.028243E-02	1.329414E-03	4.007794E-05
1.389416E-03	6.042359E-02	8.776827E-01	5.785477E-02	2.638559E-03
4.076560E-05	2.658827E-03	7.713962E-02	8.421810E-01	7.797909E-02
1.146052E-06	1.001949E-04	4.397593E-03	9.747380E-02	8.980269E-01

Time, Sec.	v_{c_1},Volts	v_{c_2},Volts	v_{c_3},Volts	v_{c_4},Volts	v_{c_5},Volts
0.0	0.0	4.0	0.0	6.0	10.0
0.225	0.523519	2.37749	3.05348	5.24739	6.94091
0.450	0.736568	2.15780	3.45020	4.78843	5.61458
0.675	0.857472	2.13051	3.35918	4.34785	4.87779
0.900	0.930160	2.10018	3.17601	3.96540	4.36366
1.125	0.968715	2.04659	2.98341	3.63961	3.96166
1.350	0.981803	1.97588	2.79949	3.35946	3.62999
1.575	0.976273	1.89518	2.62774	3.11501	3.34785
1.800	0.957579	1.80948	2.46809	2.89871	3.10277
2.250	0.896453	1.63493	2.18130	2.52980	2.69296
3.150	0.741826	1.31012	1.71221	1.96204	2.07748
4.500	0.530245	0.923969	1.19685	1.36428	1.44115
6.008	0.357905	0.621838	0.803912	0.915300	0.966364
9.000	0.162737	0.282575	0.365165	0.415660	0.438802

The zero-input response of the circuit of Fig. 5-2 calculated by the series expansion method agrees very well with the results of the eigenvector method. There are other important features of each program to be considered when selecting the approach to use. Some discussion has been presented regarding the number of required operations by both methods. The series-expansion method is often faster than the eigenvector method if the eigenvalues are not widely separated. Both methods suffer from accuracy problems. Finding the eigenvalues of a network or system correspond to finding pole locations of the system. Inaccuracies can arise in these calculations. The series-expansion method is not free of errors either. Since this method is recursive, roundoff error is accumulative in this case. However, as noted previously, an error criterion can be applied to the series-expansion method with results being corrected at preselected points of time at the expense of a fair amount of computer time.

Some caution must be used in applying the series-expansion program given. It is possible for certain active networks to possess an **A** matrix such that zero elements result for all powers of this matrix. In this instance the error criterion is never satisfied, thus some modification must be made in the program to account for this case.

While either of these methods may be used in the practical case [1,3], numerical integration schemes are more popular with the commercial programs. Numerical integration will be treated in the next section in connection with solution of the nonhomogeneous state equation.

B. THE NONHOMOGENEOUS STATE EQUATION

Because functions of matrices have properties very similar to scalar functions, the nonhomogeneous state equation can be solved much like a scalar differential equation. Starting with the nonhomogeneous state equation

$$\dot{\mathbf{x}} = \mathbf{A}\mathbf{x} + \mathbf{B}\mathbf{u}$$

we multiply both sides of the equation by the integrating factor $\mathbf{e}^{-\mathbf{A}t}$ giving

$$\mathbf{e}^{-\mathbf{A}t}\,\dot{\mathbf{x}} - \mathbf{e}^{-\mathbf{A}t}\,\mathbf{A}\mathbf{x} = \mathbf{e}^{-\mathbf{A}t}\,\mathbf{B}\mathbf{u} \qquad (5\text{-}21)$$

The left-hand side of the equation now equals

$$\frac{d}{dt}\,[\mathbf{e}^{-\mathbf{A}t}\,\mathbf{x}]$$

thus

$$\frac{d}{dt}\,[\mathbf{e}^{-\mathbf{A}t}\,\mathbf{x}] = \mathbf{e}^{-\mathbf{A}t}\,\mathbf{B}\mathbf{u} \qquad (5\text{-}22)$$

We can now write the integral equation

$$\mathbf{e}^{-\mathbf{A}t}\,\mathbf{x} = \int_0^t \mathbf{e}^{-\mathbf{A}\tau}\,\mathbf{B}\mathbf{u}\,d\tau + \mathbf{C}$$

Solving for $\mathbf{x}$ gives

$$\mathbf{x} = \int_0^t \mathbf{e}^{\mathbf{A}(t-\tau)}\,\mathbf{B}\mathbf{u}\,d\tau + \mathbf{e}^{\mathbf{A}t}\,\mathbf{C}$$

Evaluating $\mathbf{C}$ at $t = 0$ gives the final result

$$\mathbf{x} = \int_0^t \mathbf{e}^{\mathbf{A}(t-\tau)}\,\mathbf{B}\mathbf{u}\,d\tau + \mathbf{e}^{\mathbf{A}t}\,\mathbf{x}_0 \qquad (5\text{-}23)$$

The solution to the state equation when the initial vector is zero is called the zero-state response and is given by

$$\mathbf{x} = \int_0^t \mathbf{e}^{\mathbf{A}(t-\tau)}\,\mathbf{B}\mathbf{u}\,d\tau \qquad (5\text{-}24)$$

The general solution given by Eq. 5-23 is the sum of the zero-state and zero-input responses.

Writing the solution to the state equation as in Eq. 5-23 by no means completes the problem. For large systems or networks the evaluation of $\mathbf{e}^{\mathbf{A}t}$ and the integral term still presents a formidable problem. We will now consider various methods of solving the state equation by hand calculation and by the digital computer.

Laplace Transforms

This method avoids the evaluation of the integral term by transforming the state equation to the Laplace domain. The specific case of this method for ac signals has been considered in Chapter 2. The state equation is transformed to the Laplace domain giving

$$s\mathbf{X}(s) - \mathbf{x}_0 = \mathbf{A}\mathbf{X}(s) + \mathbf{B}\mathbf{u}(s) \tag{5-25}$$

A small **u** is used for the transform of the input matrix to distinguish this matrix from the unit matrix. The quantity $\mathbf{X}(s)$ can now be written as

$$\mathbf{X}(s) = [s\mathbf{U} - \mathbf{A}]^{-1}\,\mathbf{B}\mathbf{u}(s) + [s\mathbf{U} - \mathbf{A}]^{-1}\,\mathbf{x}_0 \tag{5-26}$$

The matrix $[s\mathbf{U} - \mathbf{A}]$ must be inverted to find $\mathbf{X}(s)$, then $\mathbf{X}(s)$ must be transformed back to the time domain to find $\mathbf{x}$.

EXAMPLE 5-5. Calculate the output voltage for the network of Fig. 5-5 if the switch is closed at $t = 0$. The initial capacitor voltage is 10 volts.

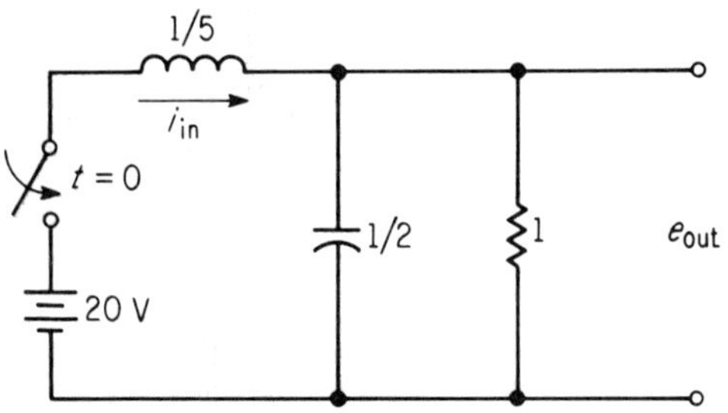

FIG. 5-5. *RLC* network.

Solution. From Example 5-1 the state equation is

$$\dot{\mathbf{x}} = \begin{bmatrix} 0 & -5 \\ 2 & -2 \end{bmatrix} \begin{bmatrix} i_{in} \\ e_{out} \end{bmatrix} + \begin{bmatrix} 5 \\ 0 \end{bmatrix} [e_{in}]$$

For this network Eq. 5-26 becomes

$$\mathbf{X}(s) = \begin{bmatrix} s & 5 \\ -2 & s+2 \end{bmatrix}^{-1} \begin{bmatrix} 5 \\ 0 \end{bmatrix} [20/s] + \begin{bmatrix} s & 5 \\ -2 & s+2 \end{bmatrix}^{-1} \begin{bmatrix} 0 \\ 10 \end{bmatrix}$$

Evaluating the inverse gives

$$\mathbf{X}(s) = \frac{1}{s^2 + 2s + 10} \begin{bmatrix} \dfrac{50s + 200}{s} \\ \dfrac{10s^2 + 200}{s} \end{bmatrix}$$

The Laplace state vector $\mathbf{X}(s)$ can be transformed to the time domain to give

$$\mathbf{x}(t) = \begin{bmatrix} 20\,u(t) - 20\,\mathrm{e}^{-t}\cos 3t + 10\,\mathrm{e}^{-t}\sin 3t \\ 20\,u(t) - 10\,\mathrm{e}^{-t}\cos 3t - 10\,\mathrm{e}^{-t}\sin 3t \end{bmatrix}$$

The two most difficult steps in this method, especially for higher-order systems or networks, are the evaluation of the inverse matrix and the partial fraction expansions required to transform $\mathbf{X}(s)$ back to the time domain. Inverting a matrix requires several multiplications for large matrices. Furthermore, the roots of an nth-order polynomial in s must be found before a partial fraction expansion is effected.

It is worth noting that this method can be treated as an eigenvector problem also. We have noted before that the roots of the determinant of $[s\mathbf{U} - \mathbf{A}]$ are the eigenvalues of the matrix $\mathbf{A}$. Consequently an evaluation of the eigenvalues provides the necessary information to start the partial fraction expansion. If a set of eigenvectors of $\mathbf{A}$ are found, the inverse of $[s\mathbf{U} - \mathbf{A}]$ is more readily calculated. We can write

$$\begin{aligned} [s\mathbf{U} - \mathbf{A}]^{-1} &= [\mathbf{P}(s\mathbf{U} - \Lambda)\,\mathbf{P}^{-1}]^{-1} \\ &= \mathbf{P}\,(s\mathbf{U} - \Lambda)^{-1}\,\mathbf{P}^{-1} \end{aligned}$$

The matrix $(s\mathbf{U} - \Lambda)$ is diagonal and can be inverted with only n multiplications.

Using the eigenvector approach requires the same computational effort as the direct approach in general. However, as pointed out in Chapter 2, for frequency response problems where $s = j\omega$, this method can be most effective.

EXAMPLE 5-6. Calculate the magnitude of the output voltage for the network of Fig. 5-6 as a function of frequency.

Solution. Returning to Eq. 2-55 we can express the output voltage (which is the second component of the state vector) as

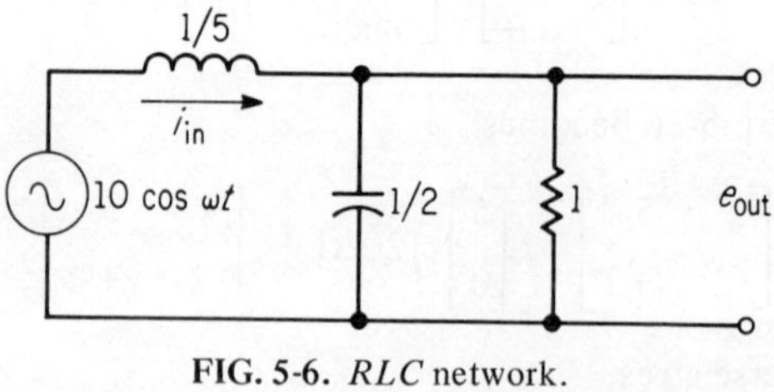

FIG. 5-6. *RLC* network.

$$x_2(\omega) = \sum_{j=1}^{2} \frac{p_{2j}\, d_j}{j\omega - \lambda_j} \qquad [2\text{-}55]$$

where the symbols are defined in connection with Eq. 2-55. We must express the input in terms of rotating phasors, since the derivation of Eq. 2-55 is the result of an input signal of $g_0 e^{j\omega t}$. In the present example the input signal is

$$e_{\text{in}} = 5e^{j\omega t} + 5e^{-j\omega t}$$

We can consider this signal to be the superposition of two rotating phasors, one of which is the complex conjugate of the other. This will lead to two output signals of the form of Eq. 2-55, but one will be the complex conjugate of the other. Thus, we can find the response due to the signal $5e^{j\omega t}$ and the complex conjugate of this signal will be the response due to $5e^{-j\omega t}$. The matrix **P** is found in Example 5-1 to be

$$\mathbf{P} = \begin{bmatrix} 1 & 1 \\ 1/5 + j3/5 & 1/5 - j3/5 \end{bmatrix}$$

and

$$\mathbf{P}^{-1} = -j\frac{5}{6}\begin{bmatrix} -1/5 + j3/5 & 1 \\ 1/5 + j3/5 & -1 \end{bmatrix}$$

The state equation is

$$\dot{\mathbf{x}} = \begin{bmatrix} 0 & -5 \\ 2 & -2 \end{bmatrix} \mathbf{x} + \begin{bmatrix} 5 \\ 0 \end{bmatrix} [e_{\text{in}}]$$

The matrix **g** is

$$\mathbf{g} = \begin{bmatrix} 25\, e^{j\omega t} + 25\, e^{-j\omega t} \\ 0 \end{bmatrix}$$

If we consider the term associated with $e^{j\omega t}$, then

$$\mathbf{g}(j\omega) = \begin{bmatrix} 25\, e^{j\omega t} \\ 0 \end{bmatrix}$$

For this case the matrix **D** is

$$\mathbf{D} = \mathbf{P}^{-1}\, \mathbf{g}_0(j\omega) = -j\frac{5}{6}\begin{bmatrix} -1/5 + j3/5 & 1 \\ 1/5 + j3/5 & -1 \end{bmatrix}\begin{bmatrix} 25 \\ 0 \end{bmatrix}$$

$$= -j\frac{25}{6}\begin{bmatrix} -1 + j3 \\ 1 + j3 \end{bmatrix}$$

Applying Eq. 2-55 gives for $x_2(\omega)$ (the output voltage for $e_{in} = 5e^{j\omega t}$)

$$x_2(\omega) = \frac{50(10 - \omega^2 + j2\omega)}{\omega^4 - 16\omega^2 + 100}$$

The output signal for an input of 10 cos ωt or $5e^{j\omega t} + 5e^{-j\omega t}$ is

$$\frac{50(10 - \omega^2 + j2\omega)}{\omega^4 - 16\omega^2 + 100}[e^{j\omega t} + e^{-j\omega t}]$$

$$= \frac{100(10 - \omega^2 + j2\omega)}{\omega^4 - 16\omega^2 + 100} \cos \omega t$$

The magnitude of this signal is

$$|e_{out}| = \frac{100}{\sqrt{\omega^4 - 16\omega^2 + 100}}$$

Of course, for this simple example the result could be obtained easily by conventional circuit analysis, but as the number of elements increases the above method becomes attractive for machine calculation of frequency responses.

A flow chart of the method of Example 5-6 is shown in Fig. 5-7.

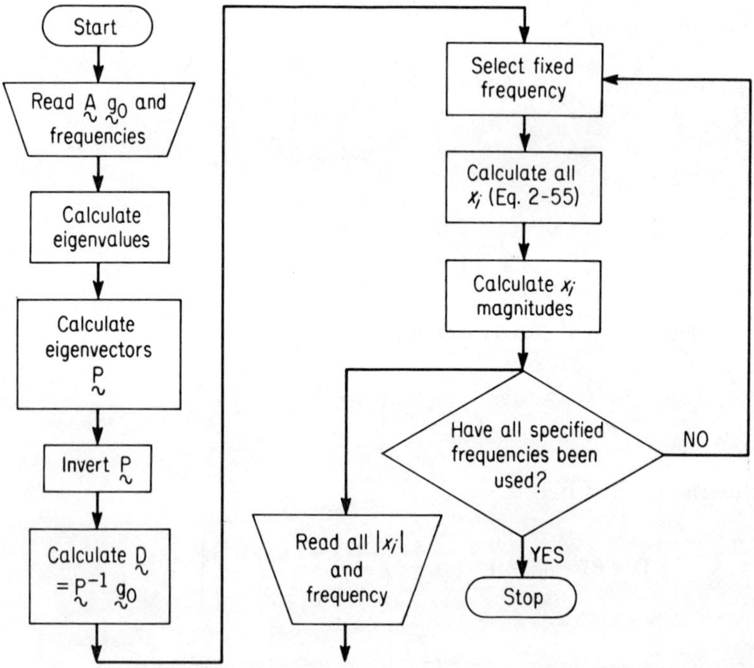

FIG. 5-7. Flow chart for frequency-response calculations.

It was pointed out in Chapter 2 that frequency response of networks can be calculated by other methods such as the nodal method of Chapter 3. This method requires inversion of a matrix at each frequency of interest whereas the eigenvector approach requires matrix inversion only one time. Of course, the eigenvector problem must be initially solved and this may require a fair amount of computer time. If many frequency points are of interest the eigenvector method can be much more efficient than the nodal method [1].

We might also note that the program can be written for the eigenvector method such that only a single variable (say the output signal) need be calculated. In problems where it is unnecessary to know the response of all variables, further time savings are possible. The nodal method solves for all variables at the same time and this amounts to extra computer time when the responses of only one or two variables are of interest.

Power Series Expansion for Step or Pulse Response [3]

The solution to the nonhomogeneous state equation was found to be

$$\mathbf{x} = \int_0^t e^{\mathbf{A}(t-\tau)}\, \mathbf{Bu}\, d\tau + e^{\mathbf{A}t}\, \mathbf{x}_0 \qquad [5\text{-}23]$$

If $t = nT$, we can write

$$\mathbf{x}(nT) = e^{\mathbf{A}nT} \int_0^{nT} e^{-\mathbf{A}\tau}\, \mathbf{Bu}\, d\tau + e^{\mathbf{A}nt}\, \mathbf{x}_0 \qquad (5\text{-}27)$$

We can also write

$$\mathbf{x}[(n+1)T] = e^{\mathbf{A}(n+1)T} \int_0^{(n+1)T} e^{-\mathbf{A}\tau}\, \mathbf{Bu}\, d\tau + e^{\mathbf{A}(n+1)T}\, \mathbf{x}_0 \qquad (5\text{-}28)$$

Expressing $\mathbf{x}[(n+1)T]$ in terms of $\mathbf{x}(nT)$ results in

$$\mathbf{x}[(n+1)T] = e^{\mathbf{A}t}\, \mathbf{x}(nT) + e^{\mathbf{A}(n+1)T} \int_{nT}^{(n+1)T} e^{-\mathbf{A}\tau}\, \mathbf{Bu}\, d\tau \qquad (5\text{-}29)$$

When all inputs are constant in the interval $nT \leqslant t \leqslant (n+1)T$, the input vector $\mathbf{u}$ is a constant. The integral term becomes

$$e^{\mathbf{A}(n+1)T} \left(-\mathbf{A}^{-1}\, e^{-\mathbf{A}\tau} \Big|_{nT}^{(n+1)T} \right) \mathbf{Bu} = \mathbf{A}^{-1}\, [e^{\mathbf{A}T} - \mathbf{U}]\, \mathbf{Bu}(nT)$$

Equation 5-29 can now be expressed as

$$\mathbf{x}[(n+1)T] = e^{\mathbf{A}T}\, \mathbf{x}(nT) + \mathbf{A}^{-1}\, [e^{\mathbf{A}t} - \mathbf{U}]\, \mathbf{Bu}(nT) \qquad (5\text{-}30)$$

In this case, if $e^{\mathbf{A}T}$ and $\mathbf{A}^{-1}\, [e^{\mathbf{A}T} - \mathbf{U}]\, \mathbf{B}$ are evaluated, the discrete values of

$\mathbf{x}(nT)$ can be found starting with $\mathbf{x}(T)$ and working up to $\mathbf{x}(kT)$, where k is chosen to include the response interval of interest.

The quantity $\mathbf{e}^{\mathbf{A}T}$ is found exactly as in the zero-input case. Although it appears necessary to invert the matrix $\mathbf{A}$ to find the quantity $\mathbf{A}^{-1}[\mathbf{e}^{\mathbf{A}T} - \mathbf{U}]\mathbf{B}$, such is not the case. Using the series expression for $\mathbf{e}^{\mathbf{A}T}$ given by Eq. 2-42 we can write

$$\mathbf{A}^{-1}[\mathbf{e}^{\mathbf{A}T} - \mathbf{U}] = \mathbf{A}^{-1}\left[\sum_{n=0}^{\infty} \frac{1}{n!} \mathbf{A}^n T^n - \mathbf{U}\right]$$

$$= T \sum_{n=0}^{\infty} \frac{1}{(n+1)!} \mathbf{A}^n T^n \tag{5-31}$$

thus

$$\mathbf{A}^{-1}[\mathbf{e}^{\mathbf{A}T} - \mathbf{U}]\mathbf{B} = \left[T \sum_{n=0}^{\infty} \frac{1}{(n+1)!} \mathbf{A}^n T^n\right] \mathbf{B} \tag{5-32}$$

The series expression in Eq. 5-32 converges more rapidly than the expression for $\mathbf{e}^{\mathbf{A}T}$, since each term of the above expression has a larger denominator than the corresponding term in the exponential series. A suggested procedure [3] for evaluating the quantities necessary to apply Eq. 5-30 starts by evaluating $\mathbf{e}^{\mathbf{A}T}$ using the iterative procedure discussed in connection with the zero-input case. Once the number of terms necessary to obtain the prescribed accuracy is found, the same number of terms can be used to evaluate Eq. 5-32.

If repetitive pulse inputs are applied to the network, Eq. 5-30 is still applicable provided that when an input transition occurs just prior to $t = nT$, the new value of $\mathbf{u}(nT)$ is used to calculate $\mathbf{x}(n+1)T$. This allows the response due to rectangular-wave inputs to be treated assuming the transition times of the input pulses are small compared to T.

EXAMPLE 5-7. Evaluate the state-vector response for the circuit of Fig. 5-8.

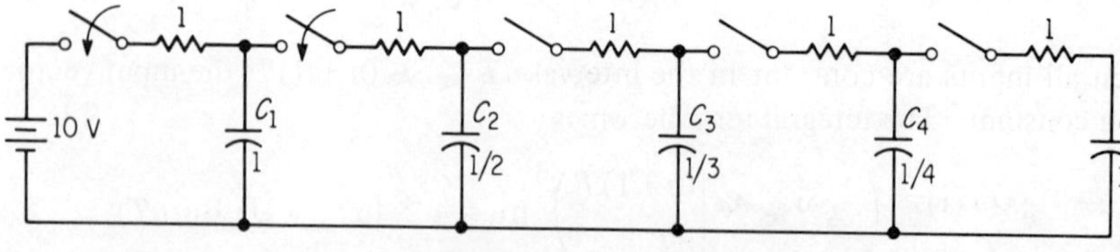

FIG. 5-8. *RC* network.

The initial voltages are $v_{c1}(0) = 0$, $v_{c2}(0) = 4$, $v_{c3}(0) = 0$, $v_{c4}(0) = 6$, $v_{c5}(0) = 10$. All switches are closed at $t = 0$.

Solution. The series expansion method is programmed in accordance with the flow chart of Fig. 5-9.

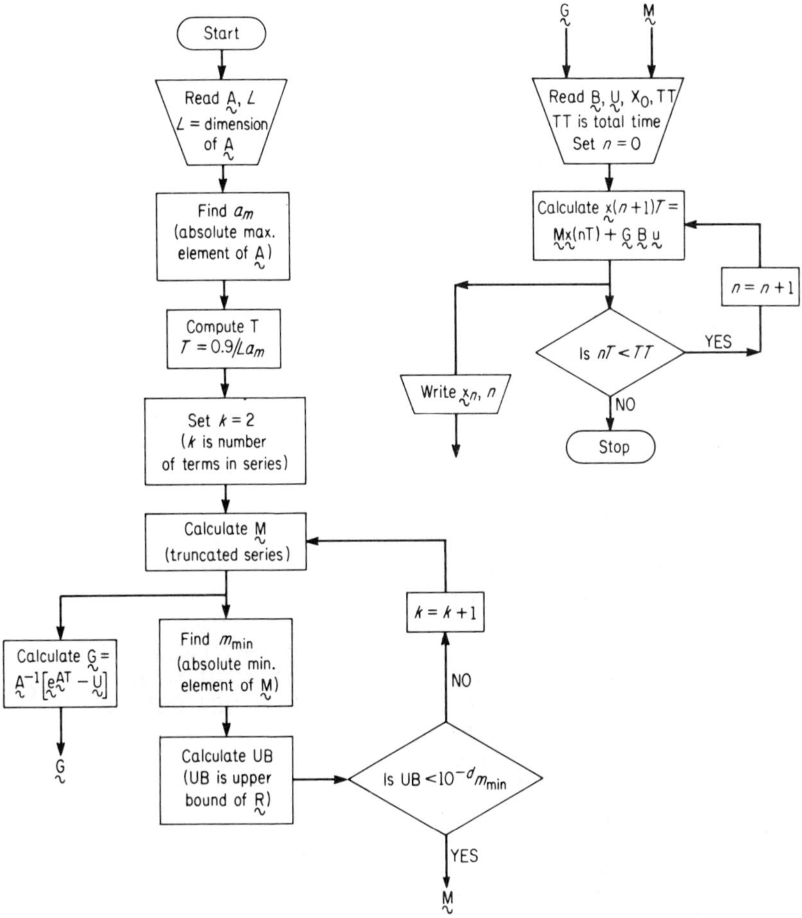

FIG. 5-9. Flow chart for series-expansion method.

The method of solution for the zero-input case is similar to that of the general case. The number of terms required is found by comparing the truncated series to the upper bound of the remainder. However, at each iteration the value of $\mathbf{A}^{-1}$ $[\mathbf{e}^{\mathbf{A}T} - \mathbf{U}]$ is also calculated or updated. Equation 5-30 is implemented once the values of $\mathbf{e}^{\mathbf{A}T}$ and $\mathbf{A}^{-1}$ $[\mathbf{e}^{\mathbf{A}T} - \mathbf{U}]$ are evaluated to sufficient accuracy.

The program for calculating $\mathbf{M} = \mathbf{e}^{\mathbf{A}T}$ and $\mathbf{G} = \mathbf{A}^{-1}$ $[\mathbf{e}^{\mathbf{A}T} - \mathbf{U}]$ is given as a subroutine in Table 5-6. This program is written in Fortran IV to be executed by the IBM 360/50.

Once **M** and **G** are calculated the recursion relation is applied to find $\mathbf{x}(nT)$ as n varies over the values of interest. Table 5-7 shows a program to calculate the response assuming **M** and **G** must be read into storage. The programs would normally be combined to avoid the necessity of this read step. In this program,

TABLE 5-6. Subroutine to Calculate M and G.

```
      SUBROUTINE EXPAT(L,A,B,C,D,E,R,F,MAX,I0,M,G,T)
      REAL M,MMIN,NN
      DIMENSION A(L,L),B(L,L),C(L,L),D(L,L),E(L,L),M(L,L),R(L,L)
      DIMENSION F(L,L),G(L,L)
C     THE MAXIMUM ELEMENT OF A IS DETERMINED
      AM = A(1,1)
      DO 7 I=1,L
      DO 7 J=1,L
      IF(ABS(A(I,J)).GT.ABS(AM)) AM=A(I,J)
    7 CONTINUE
C     TIME INTERVAL NECESSARY FOR CONVERGENCE IS
C        COMPUTED
      T = 0.9/(L*ABS(AM))
C     THE FIRST TWO TERMS OF BOTH SERIES ARE COMPUTED
      DO 11 I = 1,L
      DO 11 J = 1,L
      B(I,J) = A(I,J)*T
      C(I,J) = B(I,J)
      IF(I.EQ.J) C(I,J) = C(I,J) + 1.0
      F(I,J) = 0.5*B(I,J)*T
      IF(I.EQ.J) F(I,J) = F(I,J) + T
   11 CONTINUE
C     THE MINIMUM ELEMENT OF M AND N + 2 TERMS OF
C        BOTH SERIES ARE COMPUTED AND THE MINIMUM
C        ELEMENT OF M IS COMPARED AGAINST THE UPPER
C        BOUND OF THE REMAINDER
      CALL MPRD(L,L,L,B,B,D)
      DO 13 K = 2,MAX
C     THE VALUE OF K IS DETERMINED
      KK = K
      CALL FACT(K,NN)
      CALL FACT(K + 1,FACK)
      MMIN = C(1,1)
      DO 12 I = 1,L
      DO 12 J = 1,L
      M(I,J) = C(I,J) + D(I,J)/NN
      IF(ABS(M(I,J)).LT.ABS(MMIN)) MMIN = M(I,J)
      G(I,J) = F(I,J) + T*D(I,J)/FACK
   12 CONTINUE
C     THE UPPER BOUND OF THE REMAINDER IS COMPUTED
      RM = 0.9/(L*FACK*(1.0-1.0/(K + 2)))
      X = 0.0001*MMIN
      IF(ABS(RM).LT.ABS(X)) GO TO 888
      CALL MPRD(L,L,L,B,D,E)
      DO 13 I = 1,L
      DO 13 J = 1,L
      C(I,J) = M(I,J)
      D(I,J) = E(I,J)
      F(I,J) = G(I,J)
   13 CONTINUE
```

TABLE 5-6 (*continued*)

```
      WRITE(I0,15)
   15 FORMAT(' SOLUTION HAS NOT CONVERGED TO DESIRED
     1  TOLERANCE')
C     THE OUTPUT IS PRINTED
  888 WRITE(I0,16)
   16 FORMAT('0',T30,'MATRIX EXP(AT)')
      DO 17 I = 1,L
      WRITE(I0,18)(M(I,J),J = 1,L)
   18 FORMAT('0',5(1PE16.6,'  '))
   17 CONTINUE
      WRITE(I0,19) T,KK
   19 FORMAT('0',T10,'T = ',1PE14.6,T40,'K = ',I3)
      WRITE(I0,20)
   20 FORMAT('0',T30,'MATRIX A**-1*(EXPAT-I)')
      DO 21 I = 1,L
      WRITE(I0,18) (G(I,J),J = 1,L)
   21 CONTINUE
      RETURN
      END
```

the matrix **M** is denoted by **E**. This program was executed on the IBM 1132. Selected results are listed in Table 5-8. Appendix 5-1 lists a program that combines those of Table 5-6 and 5-7 to be run on the IBM 360/50.

TABLE 5-7. Program for Calculation of Response Using Series Expansion.

```
      DIMENSION X(5),BU(5),G(25),E(25),Y(5),U(5),XN(5)
      READ(8,20) (X(I),I = 1,5)
      READ(8,20) (BU(I),I = 1,5)
      READ(8,20) (G(I),I = 1,25)
      READ(8,20) (E(I),I = 1,25)
   20 FORMAT(5E15.6)
C     FORM MATRIX PRODUCT G*B*U
      CALL GMPRD(G,BU,U,5,5,1)
      DO 50 J = 1,2000
C     FORM MATRIX PRODUCT M*X (E = M)
      CALL GMPRD(E,X,Y,5,5,1)
C     ADD RESULTS
      CALL GMADD(U,Y,XN,5,1)
      N = J
      DO 49 I = 1,5
      X(I) = XN(I)
   49 CONTINUE
      Z = N
C     T IS TIME INTERVAL
      T = 2.25E - 2*Z
      WRITE(5,19) N,(X(I),I = 1,5),T
   19 FORMAT('  ','MATRIX',1X,I4/' + ',10X,5(E15.6,' '),2X,E15.6)
   50 CONTINUE
      CALL EXIT
      END
```

The time step calculated by the program of Table 5-6 is $T = 0.0225$ seconds. The entire response of the state vector is completed after approximately 800 steps or 18 seconds. The matrix $\mathbf{e}^{\mathbf{A}T}$ was calculated using the first 13 terms of the series expansion. It must be pointed out that the extreme eigenvalues in this case differed only by a factor of 50. If this factor had been much greater, for example 1,000 or 10,000, the number of steps required to complete the solution would be much greater.

TABLE 5-8. Selected Results of Series-Expansion Method Applied to Example 5-7.

Time Sec.	v_{c1} Volts	v_{c2} Volts	v_{c3} Volts	v_{c4} Volts	v_{c5} Volts
0.0	0.0	4.0	0.0	6.0	10.0
0.09	1.10694	3.05295	1.90572	5.55629	8.45118
0.18	1.98347	2.74058	2.83803	5.34412	7.35813
0.36	3.31455	2.89687	3.58392	5.02345	6.05130
0.72	5.06131	3.87905	4.06238	4.58760	4.93630
1.44	6.88651	5.44700	4.84980	4.63409	4.57109
2.88	8.27839	7.12114	6.37553	5.93925	5.74454
4.00	8.77170	7.88975	7.29231	6.93073	6.76648
5.00	9.06376	8.37989	7.91098	7.62491	7.49453
6.00	9.28526	8.76025	8.39883	8.17773	8.07694
7.00	9.44969	9.04473	8.76557	8.59458	8.51672
8.00	9.57840	9.26794	9.05378	8.92248	8.86281
9.00	9.67492	9.43547	9.27021	9.16880	9.12282
10.00	9.75070	9.56704	9.44022	9.36231	9.32710

All final values = 10 V

Solving the State Equation Using Eigenvectors

Another approach to this problem that is especially effective for step or pulse inputs has been suggested by Branin [1]. Again, the eigenvalues and eigenvectors must be found allowing a diagonal representation of **A**. In the general case, this implies nonrepeated eigenvalues. We can now express Eq. 5-29 as

$$\mathbf{x}[(n+1)T] = \mathbf{P}\,\mathbf{e}^{\Lambda T}\,\mathbf{P}^{-1}\,\mathbf{x}(nT) + \mathbf{P}\left[\int_{nT}^{(n+1)T} \mathbf{e}^{\Lambda[(n+1)T-\tau]}\,d\tau\right]\mathbf{P}^{-1}\,\mathbf{B}\mathbf{u}(nT) \tag{5-33}$$

The limits of integration can be changed such that $\mathbf{x}[(n+1)T]$ becomes

$$\mathbf{x}[(n+1)T] = \mathbf{P}\,\mathbf{e}^{\Lambda T}\,\mathbf{P}^{-1}\,\mathbf{x}(nT) + \mathbf{P}\left[\int_{0}^{T} \mathbf{e}^{\Lambda(T-\tau)}\,d\tau\right]\mathbf{P}^{-1}\,\mathbf{B}\mathbf{u}(nT) \tag{5-34}$$

Integrating the diagonal matrix gives

$$\int_{0}^{T} \mathbf{e}^{\Lambda(T-\tau)}\,d\tau = \Lambda^{-1}\,[\mathbf{e}^{\Lambda T} - \mathbf{U}] = \mathbf{G} \tag{5-35}$$

The matrix **G** is easily evaluated by the computer, since it appears as

$$\mathbf{G} = \begin{bmatrix} (e^{\lambda_1 T} - 1)/\lambda_1 & 0 & 0 & \cdots & 0 \\ 0 & (e^{\lambda_2 T} - 1)/\lambda_2 & 0 & \cdots & 0 \\ 0 & 0 & (e^{\lambda_3 T} - 1)/\lambda_3 & \cdots & 0 \\ \vdots & \vdots & \vdots & & \vdots \\ 0 & 0 & 0 & \cdots & (e^{\lambda_n T} - 1)/\lambda_n \end{bmatrix} \tag{5-36}$$

Equation 5-34 is the basis for a recursion formula giving the value of $\mathbf{x}(t)$ in terms of $\mathbf{x}(t - T)$. We can write

$$\mathbf{x}[(n+1)T] = \mathbf{P}\,\mathbf{e}^{\Lambda T}\,\mathbf{P}^{-1}\,\mathbf{x}(nT) + \mathbf{P}\,\mathbf{G}\,\mathbf{P}^{-1}\,\mathbf{Bu}(nT) \tag{5-37}$$

In addition to finding the eigenvalues and eigenvectors of **A**, we note that the inverse of matrix **P** must be evaluated to apply Eq. 5-37.

If the input vector consists of rectangular pulses, Eq. 5-37 can be used to calculate the pulse response. This requires that the appropriate $\mathbf{u}(nT)$ be used when a transition of the input signal occurs between $(n - 1)T$ and nT.

While this approach is quite effective in computing the response to step or pulse inputs, Branin [1] reports that it can be useful for arbitrary inputs if these input signals can be approximated by polynomials in t. This approach can be used when widely separated eigenvalues occur to minimize computer time. However, widely separated eigenvalues can lead to inaccuracies in evaluating the eigenvectors.

Numerical Integration

The most obvious method of solving the state equation is to integrate. There are several subroutines available for numerical integration with the digital computer. These subroutines can treat matrix equations and generally iterate at each step of time until the error is reduced to a negligible value. Unfortunately, this process can be slow resulting in excessive computer time. For many problems numerical integration provides a satisfactory method of solution. On the other hand, certain types of problems that occur often in the field of electrical engineering are very inefficient to solve by numerical integration. In solving for the response of systems with widely varying time constants or eigenvalues of the **A** matrix, the digital computer becomes very inefficient. The smallest time constant which corresponds to the reciprocal of the largest eigenvalue determines the maximum value of the integration step size The largest time constant, equal to the reciprocal of the smallest eigenvalue determines the total length of time over which the integration must be carried out [1]. If a large ratio of extreme time constants occurs, the step size is chosen to be quite small while the total number of steps is very large. It would appear that the obvious solution to this problem would be to use a small step size up to a time equal to five or ten times the smallest time constant. A larger step size could then be selected, compatible with the next smallest time constant. Repetition of this procedure while integrating should result in a minimum number of steps and sufficient accuracy over the total response time. While this method appears on the surface to be quite logical, it cannot be used due to a well-known peculiarity called numerical instability. The effects of numerical instability can be visualized in terms of a circuit having two widely separated time constants. Consider the circuit of Fig. 5-10 where A is an ideal voltage amplifier with infinite input and zero

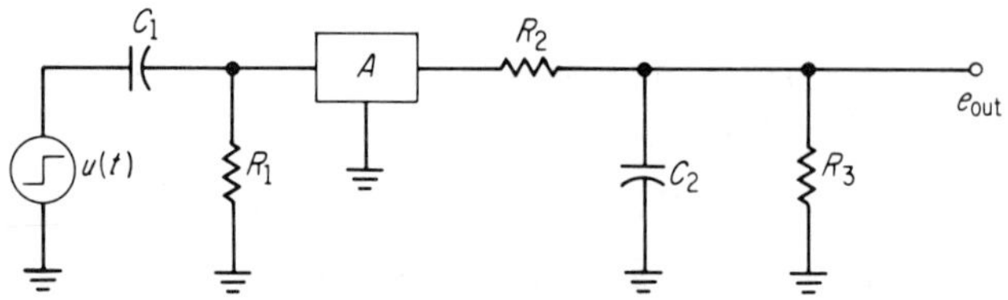

FIG. 5-10. *RC* amplifying circuit.

output impedances. The input capacitor C_1 might represent a coupling capacitor having a large value, whereas C_2 could correspond to stray capacitance or input capacitance of a following stage with a relatively small value. The response to a unit step function can be found to be

$$e_{out} = \frac{AR_3\tau_1}{(R_2 + R_3)(\tau_1 - \tau_2)} [e^{-t/\tau_1} - e^{-t/\tau_2}]$$

where $\tau_1 = C_1R_1$, $\tau_2 = C_2R$, and $R = R_2R_3/(R_2 + R_3)$. The output voltage as a

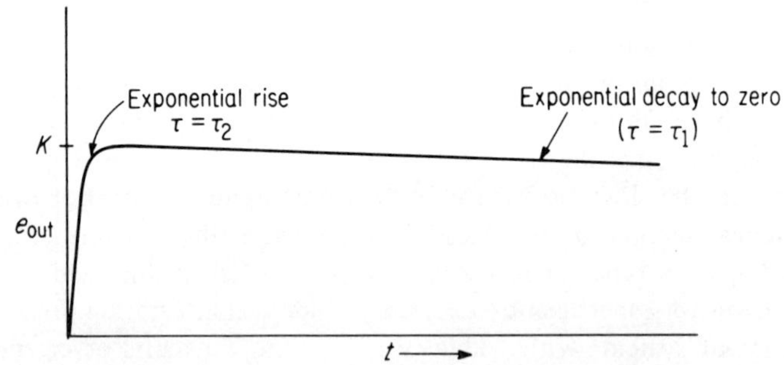

FIG. 5-11. Transient response of the circuit of Fig. 5-6.

function of time is shown in Fig. 5-11, where K is defined as

$$K = \frac{AR_3\tau_1}{(R_2 + R_3)(\tau_1 - \tau_2)}$$

The output voltage has a rise time governed by τ_2 and a fall time governed by τ_1. If $C_1 >> C_2$, then τ_2 will be much smaller than τ_1. The state equation for this circuit is

$$\frac{d}{dt}\begin{pmatrix} v_{c1} \\ v_{c2} \end{pmatrix} = \begin{bmatrix} -1/R_1C_1 & 0 \\ -A/R_2C_2 & -1/RC_2 \end{bmatrix} \begin{bmatrix} v_{c1} \\ v_{c2} \end{bmatrix} + \begin{bmatrix} 1/R_1C_1 \\ A/R_2C_2 \end{bmatrix} [e_{in}]$$

$$= \mathbf{A}\mathbf{x} + \mathbf{B}\,\mathbf{e}_{in}$$

If e_{in} is taken as a unit step, the output voltage is found using numerical integration.

A fourth-order Runge-Kutta numerical integration scheme, given in Table 5-9, was used to solve for v_{c1} and v_{c2}. Element values used were $R_1 = 10\ k\Omega$, $R_2 = 1\ k\Omega$, $R_3 = 9\ k\Omega$, $A = 20$, $C_1 = 1\ \mu f$ and $C_2 = 6{,}000$ pf. The time constant governing rise time is 5.4 μsec, while that governing fall time is 10 msec.

The Runge-Kutta scheme for numerical integration was chosen because it is a well-known method that is available in subroutine form for many different types of computer. Development of the theory behind this method will not be attempted here, since it is available in most standard texts on numerical methods.

TABLE 5-9. Fourth-Order Runge-Kutta Integration Program.

```
      REAL K11,K12,K21,K22,K31,K32,K41,K42
      WRITE(5,1)
    1 FORMAT(///'NUMERICAL INTEGRATION'///)
      T = 0.
      V0 = 0.
      V1 = 0.
      V2 = 0.
      READ(8,40)H,A11,A12,A21,A22,B1,B2,L,M,C1,C2,HL
C     L IS MAXIMUM NO. OF INCREMENTS, M IS NO. BEFORE
C        STEP SIZE CHANGE
C     H IS SMALL INCREMENT SIZE
C     HL IS LARGE INCREMENT SIZE
   40 FORMAT(7F10.0/2I5,3F10.0)
      DO5 I = 1,L
      IF(I - M)15,11,15
C     EVALUATE RUNGE-KUTTA CONSTANTS
   15 K11 = H*((A11*V1) + (A12*V2) + B1)
      K12 = H*((A21*V1) + (A22*V2) + B2)
      K21 = H*( A11* (V1 + K11/2.) + A12*(V2 + K12/2.) + B1)
      K22 = H*( A21* (V1 + K11/2.) + A22*(V2 + K12/2.) + B2)
      K31 = H*( A11* (V1 + K21/2.) + A12*(V2 + K22/2.) + B1)
      K32 = H*( A21* (V1 + K21/2.) + A22*(V2 + K22/2.) + B2)
      K41 = H*( A11* (V1 + K31) + A12* (V2 + K32) + B1)
      K42 = H*( A21* (V1 + K31) + A22* (V2 + K32) + B2)
C     EVALUATE VOLTAGES
      V1 = V1 + (K11 + 2. *K21 + 2. *K31 + K41)/6
      V2 = V2 + (K12 + 2. *K22 + 2. *K32 + K42)/6.
      V0 = C1*V1 + C2*V2
```

TABLE 5-9 (*continued*)

```
      T = T + H
    5 WRITE(5,2)I,T,V1,V2,V0
    2 FORMAT(1HK,10X,2HI = ,I4,10X,2HT = ,F8.6,15X,3HV1 = ,
     1    F10.6,15X,3HV2 = ,F10.6,15X,3HVO = ,F10.6)
      GO TO 33
   11 WRITE(5,13)
   13 FORMAT(1H1,'INCREMENT CHANGES',///)
      H = HL
      GO TO 15
   33 CONTINUE
      CALL EXIT
      END
```

The initial step size was chosen to be 1 μsec to obtain resolution near the start of the pulse. After 54 steps (10 time constants) the effects of the capacitor C_2 should be negligible. Certainly, after 500 steps (amost 100 time constants) one would expect no problems in now increasing the step size to one-tenth of the time constant τ_1.

Figure 5-12 shows the results of increasing the step size from 1 μsec after

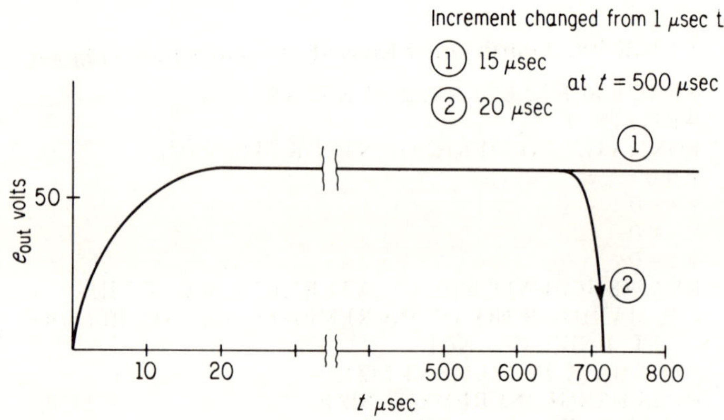

FIG. 5-12. Results of numerical integration using program of Table 5-9.

500 μsec. If the step size is changed to 15 μsec, the integration process continues to completion. However, if the step size is increased to 20 μsec at the same point in time, numerical instability immediately results. We can note that the fall time constant is 500 times larger than the step size causing instability. There are two simple (though very inefficient) methods of guaranteeing completion of the integration process. The first is to continue the process with a small step size. For most integration formulas the step size must be such that [1]

$$h \leqslant c / |\lambda_{\max}(\mathbf{A})| \tag{5-38}$$

where $\lambda_{\max}(\mathbf{A})$ is the largest eigenvalue of **A** and c is a constant. This constant

TABLE 5-10. Integration Program for Example 5-7.

```
      REAL K11,K12,K13,K14,K15,K21,K22,K23,K24,K25,K31,
     1     K32,K33,K34,K35,K41,K42,K43,K44,K45
      WRITE(5,1)
    1 FORMAT(///'NUMERICAL INTEGRATION'///)
      T  =  0.
      V1 =  0.
      V2 =  4.
      V3 =  0.
      V4 =  6.
      V5 = 10.
      H  =  0.01
      DO5 I = 1,1200
   15 K11 = H*((-2.*V1) + V2 + 10.)
      K12 = H*((2.*V1)  - (4.*V2) + (2.*V3))
      K13 = H*((3.*V2)  - (6.*V3) + (3.*V4))
      K14 = H*((4.*V3)  - (8.*V4) + (4.*V5))
      K15 = H*((5.*V4)  - (5.*V5))
      K21 = H*(-2.* (V1 + K11/2.) + V2 + K12/2. + 10.)
      K22 = H*( 2.* (V1 + K11/2.) - 4.*(V2 + K12/2.) +
     1     2.*(V3 + K13/2.))
      K23 = H*( 3.* (V2 + K12/2.) - 6.*(V3 + K13/2.) +
     1     3.*(V4 + K14/2.))
      K24 = H*( 4.* (V3 + K13/2.) - 8.*(V4 + K14/2.) +
     1     4.*(V5 + K15/2.))
      K25 = H*( 5.* (V4 + K14/2.) - 5.*(V5 + K15/2.))
      K31 = H*(-2.* (V1 + K21/2.) + V2 + K22/2. + 10.)
      K32 = H*( 2.* (V1 + K21/2.) - 4.*(V2 + K22/2.) +
     1     2.*(V3 + K23/2.))
      K33 = H*( 3.* (V2 + K22/2.) - 6.*(V3 + K23/2.) +
     1     3.*(V4 + K24/2.))
      K34 = H*( 4.* (V3 + K23/2.) - 8.*(V4 + K24/2.) +
     1     4.*(V5 + K25/2.))
      K35 = H*( 5.* (V4 + K24/2.) - 5.*(V5 + K25/2.))
      K41 = H*(-2.* (V1 + K31/2.) + V2 + K32/2. + 10.)
      K42 = H*( 2.* (V1 + K31/2.) - 4.*(V2 + K32/2.) +
     1     2.*(V3 + K33/2.))
      K43 = H*( 3.* (V2 + K32/2.) - 6.*(V3 + K33/2.) +
     1     3.*(V4 + K34/2.))
      K44 = H*( 4.* (V3 + K33/2.) - 8.*(V4 + K34/2.) +
     1     4.*(V5 + K35/2.))
      K45 = H*( 5.* (V4 + K34/2.) - 5.*(V5 + K35/2.))
      V1  = V1 + (K11 + 2.*K21 + 2.*K31 + K41)/6.
      V2  = V2 + (K12 + 2.*K22 + 2.*K32 + K42)/6.
      V3  = V3 + (K13 + 2.*K23 + 2.*K33 + K43)/6.
      V4  = V4 + (K14 + 2.*K24 + 2.*K34 + K44)/6.
      V5  = V5 + (K15 + 2.*K25 + 2.*K35 + K45)/6.
      T = T + H
    5 WRITE(5,2)T,V1,V2,V3,V4,V5
    2 FORMAT(1HO,2HT = ,F8.4,4X,3HV1 = ,F10.6,5X,3HV2 = ,F10.6,
           5X,3HV3 = ,F10.6,5X,3HV4 = ,F10.6,5X,3HV5 = ,F10.6)
      CONTINUE
      CALL EXIT
      END
```

may have an approximate value of 1 or 2, depending on the integration method used. The second means of avoiding instability is to include a convergence criterion along with a variable step size in the integration process. In this scheme the step size is decreased as necessary to satisfy an appropriate convergence criterion before advancing to the next time interval. Both methods require far more computing time than should be required to arrive at the solution.

More practical approaches to the solution of the instability problem are (1) separation of a network into high frequency and low frequency equivalents or (2) modification of the integration process to remove the causes of instability when the step size is changed. While these methods can be applied in the case of a simple circuit such as that of Fig. 5-10, satisfactory application of these methods to complex networks or systems has yet to be accomplished. Some progress in this direction has recently been made [2], but more progress is necessary to apply effective numerical integration procedures to systems with widely separated eigenvalues.

As an example of numerical integration, the problem of Example 5-7 is solved using a step time of 10 ms in the fourth order Runge-Kutta process. The program is indicated in Table 5-10 while pertinent results are shown in Table 5-11. Note in this case that all eigenvalues are of the same order of magnitude, thus instability problems are avoided. This program was written for execution on the IBM 1132.

The program of Table 5-10 is written specifically to solve the problem of Example 5-7. A general program accepting different **A** and **B** matrices can easily be written.

TABLE 5-11. Solution to Example 5-7.

T Sec	V_1 Volts	V_2 Volts	V_3 Volts	V_4 Volts	V_5 Volts
0.00	0.000	4.000	0.000	6.000	10.000
0.01	0.138	3.846	0.290	5.924	9.803
0.02	0.272	3.707	0.556	5.857	9.612
0.04	0.528	3.466	1.027	5.746	9.247
0.08	0.999	3.113	1.765	5.586	8.595
0.10	1.216	2.989	2.053	5.499	8.305
0.20	2.158	2.716	2.980	5.305	7.158
0.40	3.560	2.989	3.669	4.959	5.854
0.80	5.350	4.094	4.145	4.539	4.810
1.00	5.954	4.584	4.356	4.489	4.615
2.00	7.618	6.247	5.482	5.083	4.917
4.00	8.771	7.888	7.291	6.929	6.765
8.00	9.578	9.268	9.054	8.923	8.864
10.00	9.751	9.568	9.442	9.364	9.329

All final values = 10V

A comparison of these results to those of the series expansion method shows little disagreement.

C. DISCUSSION OF NUMERICAL METHODS

This chapter has discussed some of the various methods that are in use or have been proposed for solving the state equation by machine. Hopefully, none of these methods has emerged in the mind of the reader as clearly superior to the others. Such is not the case. Each method possesses peculiarities making it more applicable to certain types of problem.

Several of the commercial programs use numerical integration techniques at present. This method would be expected to be the obvious choice since numerical methods for solving first-order differential equations in matrix form have been developed over several years. Alternative methods have been appearing in recent years and newer programs will apply more efficient methods.

ECAP applies a numerical integration routine that avoids the possibility of numerical instability at the expense of computer time [5]. A very small, non-variable step time is necessary to obtain an accurate solution for transient analysis problems with this program. SCEPTRE allows the user to specify a choice of one of three different numerical integration methods [4], however widely separated time constants still lead to excessive computer time.

It is clear that more development is required in the area of solving transient problems. Optimization schemes, requiring a new solution each time a parameter is varied, consume a tremendous amount of computer time. Of course, if an optimization scheme is developed to the point that only a few iterations are required to optimize all parameter values, computer time is decreased. At the present time it appears more likely that the largest savings in computer time will result from more efficient methods of numerical solution.

We should now return to the general design procedure indicated by Fig. I-1. Chapters 3 and 4 were primarily concerned with equation formulation requiring network topology and element value information as input data. It was assumed that the designer specified this information. The latter part of Chapter 3 and all of Chapter 5 considered machine solution methods requiring the appropriate matrix equations as input data. The equation formulation and solution programs can easily be combined. Commercial programs generally require only topology and element values to evaluate the response of the network. Formulation and solution is thus accomplished by the computer freeing the designer from these tedious, error-prone tasks.

APPENDIX 5-1

This appendix contains a program written for the IBM 360/50 computer to solve the state equation by the series expansion method. The various subroutines are also listed. Both zero-input response and step response can be calculated.

```
C     SOLUTION OF THE STATE EQUATION USING SERIES
C     EXPANSION
```

```
C     ************************************************
C     DIMENSIONING:
C         DIMENSION A(L,L),B(L,1),U(M,1),XO(L,1),XNT(L,1),
C         XNT1(L,1)
C         XNT2(L,1),X(L,1),EXP1(L,L),EXP2(L,L),OUT1(L,FTIME/
C         (T*STEP)),
C         OUT2(L,FTIME/(T*STEP)),TT(FTIME/(T*STEP)),
C         Y(FTIME/(T*STEP)),
C         WV1-6(L,L),XX(L,M)
C         WHERE FTIME IS THE FINAL TIME DESIRED FOR
C         SOLUTION,
C         T IS THE INCREMENT OF TIME BETWEEN STEPS GIVEN
C         BY THE EQUATION T = 0.9/(L* ABS. VALUE OF THE
C         LARGEST ELEMENT IN A).  STEP IS THE NUMBER OF
C         TIME INTERVALS BETWEEN OUTPUT POINTS GIVEN
C         BY THE EQUATION STEP = FTIME/(T*#OF OUTPUT
C         POINTS DESIRED).
C     DATA CARDS:
C         THE FOLLOWING IS THE ORDER OF THE DATA CARDS
C            CARD 1 L-CC.1-3,M-CC.4-6,STEP-CC.7-9,K-CC.10-12
C            CARD 2-FTIME IN THE FIRST FIELD OF 10
C            CARD 3-FORMAT, INCLUDING PARENTHESES,
C            ANYWHERE ON CARD
C            CARDS FOR MATRIX A
C            CARDS FOR MATRIX B
C            CARDS FOR MATRIX U
C            CARDS FOR MATRIX XO
C            CARD FOR TITLE
C         USAGE
C            THE FORMAT ON DATA CARD 3 DETERMINES THE
C            FORMAT FOR READING ALL MATRICES
C            FOR THE VALUE K = 1 THE PROGRAM COMPUTES
C            ONLY THE INITIAL CONDITION RESPONSE, THE
C            VALUE K = 2 COMPUTES ONLY THE STEP RESPONSE,
C            ANY OTHER VALUE FOR K COMPUTES BOTH
      INTEGER STEP
      DIMENSION A(5,5),B(5,1),U(1,1),XO(5,1),XNT(5,1),XNT1(5,1)
      DIMENSION XNT2(5,1),X(5,1),XX(5,1)
      DIMENSION FMT(20), EXP1(5,5),EXP2(5,5),OUT1(5,30),
     1 OUT2(5,30)
      DIMENSION TT(30),Y(30),WV1(5,5),WV2(5,5),WV3(5,5),
     1 WV4(5,5)
      DIMENSION WV5(5,5),WV6(5,5),TITLE(20)
      DATA IN,IO/5,6/
      READ(IN,1) L,M,STEP,K
    1 FORMAT(4I3)
      READ(IN,2) FTIME
    2 FORMAT(F10.5)
      READ(IN,3) FMT
    3 FORMAT(20A4)
```

```
      READ(IN,FMT) ((A(I,J),J = 1,L),I = 1,L)
      READ(IN,FMT) ((B(I,J),J = 1,M),I = 1,L)
      READ(IN,FMT) (U(I,1),I = 1,M)
      READ(IN,FMT) (XO(I,1),I = 1,L)
      READ(IN,33) TITLE
   33 FORMAT(20A4)
      WRITE(IO,4) ((A(I,J),J = 1,L),I = 1,L)
    4 FORMAT('1',T30,'MATRIX A'/('0',5(1PE15.6,'   ')))
      WRITE(IO,5) ((B(I,J),J = 1,M),I = 1,L)
    5 FORMAT('0',T30,'MATRIX B'/('0',5(1PE15.6,'   ')))
      WRITE(IO,6) (U(I,1),I = 1,M)
    6 FORMAT('1',T34,'MATRIX U'/('0',T30,1PE15.6))
      WRITE(IO,66) (XO(I,1),I = 1,L)
   66 FORMAT('+',T64,'MATRIX XO'/('0',T60,1PE15.6))
      CALL EXPAT(L,A,WV1,WV2,WV3,WV4, WV5,WV6,200,IO,
      1 EXP1,EXP2,T)
      IF(K.EQ.2) GO TO 111
C     INITIAL CONDITION RESPONSE IS COMPUTED HERE
      N = FTIME/T
      CALL MPRD(L,L,1,EXP1,XO,XNT1)
      KK = 1
      NN = 0
      DO 777 J = 1,N
      NN = NN + 1
      DO 7 I = 1,L
      IF(NN.EQ.STEP) OUT1(I,KK) = XNT1(I,1)
    7 XNT(I,1) = XNT1(I,1)
      IF(NN - STEP) 8,9,9
    9 TT(KK) = J*T
      KK = KK + 1
      NN = 0
    8 CALL MPRD(L,L,1,EXP1,XNT,XNT1)
  777 CONTINUE
      WRITE(IO,10)
   10 FORMAT('1','INITIAL CONDITION RESPONSE')
      KK = KK - 1
      DO 11 I = 1,L
      DO 12 J = 1,KK
   12 Y(J) = OUT1(I,J)
      WRITE(IO,333) TITLE
  333 FORMAT('1',T25,20A4)
      WRITE(IO,13) I
   13 FORMAT('0',T55,'NODE',1X,12)
      CALL PRTPLT(KK,IO,TT,Y)
   11 CONTINUE
  111 CONTINUE
      IF(K.EQ.1) GO TO 200
C     STEP RESPONSE IS COMPUTED HERE
      CALL MPRD(L,L,1,EXP1,XO,XNT1)
      CALL MPRD(L,L,M,EXP2,B,XX)
```

```
      CALL MPRD(L,M,1,XX,U,X)
      KK = 1
      NN = 0
      DO 888 J = 1,N
      NN = NN + 1
      DO 14 I = 1,L
      XNT2(I,1) = XNT1(I,1) + X(I,1)
      IF(NN.EQ.STEP) OUT2(I,KK) = XNT2(I,1)
   14 XNT(I,1) = XNT2(I,1)
      IF(NN - STEP) 15,16,16
   16 TT(KK) = J*T
      KK = KK + 1
      NN = 0
   15 CALL MPRD(L,L,1,EXP1,XNT,XNT1)
  888 CONTINUE
      WRITE(IO,18)
   18 FORMAT('1','STEP RESPONSE')
      KK = KK - 1
      DO 20 I = 1,L
      DO 19 J = 1,KK
   19 Y(J) = OUT2(I,J)
      WRITE(IO,333) TITLE
      WRITE(IO,13) I
      CALL PRTPLT(KK,IO,TT,Y)
   20 CONTINUE
  200 CONTINUE
      CALL EXIT
      END

      SUBROUTINE EXPAT(L,A,B,C,D,E,R,F,MAX,IO,M,G,T)
      REAL M,MMIN,NN
      DIMENSION A(L,L),B(L,L),C(L,L),D(L,L),E(L,L),M(L,L),R(L,L)
      DIMENSION F(L,L),G(L,L)
C     THE MAXIMUM ELEMENT OF A IS DETERMINED
      AM = A(1,1)
      DO 7 I = 1,L
      DO 7 J = 1,L
      IF(ABS(A(I,J)).GT.ABS(AM)) AM = A(I,J)
    7 CONTINUE
C     TIME INTERVAL NECESSARY FOR CONVERGENCE IS
C     COMPUTED
      T = 0.9/(L*ABS(AM))
C     THE FIRST TWO TERMS OF BOTH SERIES ARE COMPUTED
      DO 11 I = 1,L
      DO 11 J = 1,L
      B(I,J) = A(I,J)*T
      C(I,J) = B(I,J)
      IF(I.EQ.J) C(I,J) = C(I,J) + 1.0
      F(I,J) = 0.5*B(I,J)*T
      IF(I.EQ.J) F(I,J) = F(I,J) + T
   11 CONTINUE
```

```
C     THE MINIMUM VALUE OF M AND N + 2 TERMS OF BOTH
C     SERIES ARE COMPUTED AND THE MINIMUM VALUE OF
C     M IS COMPARED AGAINST THE REMAINDER
      CALL MPRD(L,L,L,B,B,D)
      DO 13 K = 2,MAX
C     THE VALUE OF K IS DETERMINED
      KK = K
      CALL FACT(K,NN)
      CALL FACT(K + 1,FACK)
      MMIN = C(1,1)
      DO 12 I = 1,L
      DO 12 J = 1,L
      M(I,J) = C(I,J) + D(I,J)/NN
      IF(ABS(M(I,J)).LT.ABS(MMIN)) MMIN = M(I,J)
      G(I,J) = F(I,J) + T*D(I,J)/FACK
   12 CONTINUE
C     THE REMAINDER IS COMPUTED
      RM = 0.9/(L*FACK*(1.0 - 1.0/(K + 2)))
      X = 0.0001*MMIN
      IF(ABS(RM).LT.ABS(X)) GO TO 888
      CALL MPRD(L,L,L,B,D,E)
      DO 13 I = 1,L
      DO 13 J = 1,L
      C(I,J) = M(I,J)
      D(I,J) = E(I,J)
      F(I,J) = G(I,J)
   13 CONTINUE
      WRITE(IO,15)
   15 FORMAT('1','MATRIX ITERATION HAS NOT CONVERGED
     1 TO TOLERANCE')
C     THE OUTPUT IS PRINTED
  888 WRITE(IO,16)
   16 FORMAT('0',T30,'MATRIX EXP(AT)')
      DO 17 I = 1,L
      WRITE(IO,18) (M(I,J),J = 1,L)
   18 FORMAT('0',5(1PE16.6,'  '))
   17 CONTINUE
      WRITE(IO,19) T,KK
   19 FORMAT('0',T10,'T = ',1PE14.6,T40,'K = ',I3)
      WRITE(IO,20)
   20 FORMAT('0',T30,'MATRIX A** - 1*(EXPAT-I)')
      DO 21 I = 1,L
      WRITE(IO,18) (G(I,J),J = 1,L)
   21 CONTINUE
      RETURN
      END

      SUBROUTINE FACT(I,ANS)
C     SUBROUTINE FACT IS USED FOR COMPUTING
C     FACTORIALS
      ANS = 1.0
```

```
      DO 1 J = 1,I
      F = J
      ANS = ANS*F
    1 CONTINUE
      RETURN
      END

      SUBROUTINE MPRD(N,K,M,A,B,R)
C     THIS SUBROUTINE FORMS THE MATRIX PRODUCT AXB
C     WHERE A IS AN NXK MATRIX AND B IS A KXM MATRIX
      DIMENSION A(N,K),B(K,M),R(N,M)
      DO 1 I = 1,N
      DO 1 J = 1,M
      R(I,J) = 0.0
      DO 1 L = 1,K
    1 R(I,J) = R(I,J) + A(I,L)*B(L,J)
      RETURN
      END

      SUBROUTINE PRTPLT(N,IO,X,Y)
C     SUBROUTINE PRTPLT IS A GENERAL SUBROUTINE
C     GRAPHICALLY DISPLAYING TWO VARIABLES, X AND Y,
C     WHERE X IS THE INDEPENDENT VARIABLE AND Y IS THE
C     SINGLE VALUED DEPENDENT VARIABLE.  VARIABLE N IS THE
C     THE NUMBER OF POINTS TO BE PLOTTED.  VARIABLE IO
C     IS THE DEVICE NUMBER OF THE PRINTER.
      DIMENSION X(N),Y(N)
      INTEGER SCAN(91), BLANK,STAR
      DATA BLANK,STAR/'    ','****'/
      YMAX = Y(1)
      YMIN = Y(1)
      DO 1 I = 1,N
      IF(YMAX.LT.Y(I)) YMAX = Y(I)
    1 IF(YMIN.GT.Y(I)) YMIN = Y(I)
      WRITE(IO,2)
    2 FORMAT('0',T5,'MIN',T25,'MAX')
      WRITE(IO,3) YMIN,YMAX
    3 FORMAT(1PE15.6,5X,1PE15.6)
      WRITE(IO,5)
    5 FORMAT('0',T2,'IND.-VARIABLE',T22,'DEP.VARIABLE')
      SCFACT = 90.0/(YMAX-YMIN)
      DO 7 I = 1,N
      YSCALE = (Y(I) - YMIN)*SFACT + 1.001
      K = YSCALE
      DO 4 J = 1,91
      IF(J.LE.K) SCAN(J) = STAR
    4 IF(J.GT.K) SCAN(J) = BLANK
      WRITE(IO,6) X(I),Y(I),SCAN
    6 FORMAT(1PE15.6,5X,1PE15.6,5X,91A1)
    7 CONTINUE
```

```
      RETURN
      END
$ENTRY
```

BIBLIOGRAPHY

1. Branin, F.H., Jr., "Computer Methods of Network Analysis," *Proc. IEEE*, Vol. 55, No. 11 (November 1967).
2. Lee, H.B., "Matrix Filtering as an Aid to Numerical Integration," *Proc. IEEE*, Vol. 55, No. 11 (November 1967).
3. Liou, M. L., "Time and Frequency Domain Analysis of Linear Time-Invariant Systems," in F. F. Kuo and J. F. Kaiser (eds.), *System Analysis by Digital Computer*, John Wiley & Sons, Inc., New York, 1966.
4. Sedore, S. R., "SCEPTRE: A Program for Automatic Network Analysis," *IBM Journal*, Vol. 11 (November 1967), pp. 627–637.
5. 1620 Electronic Circuit Analysis Program (ECAP), (1620-EE-02x) User's Manual, IBM Publication H20-0170-1, IBM Corporation, White Plains, N.Y.

PROBLEMS

5-1 to 5-3 Evaluate the zero-input responses of the circuits shown with the given initial conditions.

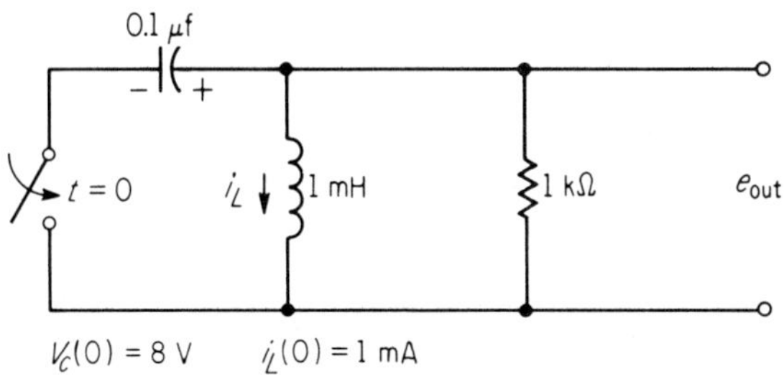

PROB. 5-1

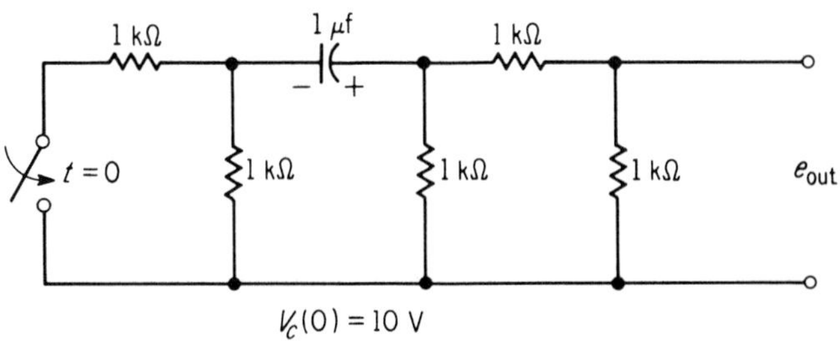

PROB. 5-2

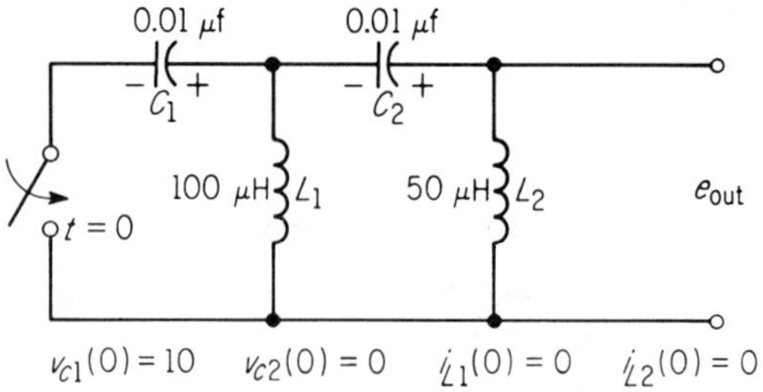

PROB. 5-3

5-4 Use a computer to solve the zero-input response of the circuit of Prob. 4-1. Assume that a switch is closed across the input terminals at $t = 0$ and that $i_{L1}(0) = 0$, $i_{L2}(0) = 0$, $v_{c1}(0) = 20$ V, and $v_{c2}(0) = 0$.

5-5 Repeat Prob. 5-4 for the circuit of Prob. 4-4. Assume that $i_{L1}(0) = 0$, $i_{L2}(0) = 2$ ma, $v_{c1}(0) = 8$ V, and $v_{c2}(0) = 4$ V.

5-6 If the switch in Prob. 5-1 closes on a +10-V source, evaluate the output signal using Laplace transforms.

5-7 Solve Prob. 5-6, using the series expansion method.

5-8 If the switch in Prob. 5-3 closes on a +10-V source, evaluate the output signal using Laplace transforms.

5-9 Solve Prob. 5-8 using the series-expansion method.

5-10 Write a computer program to generate a frequency response of a general network. The **A** matrix will be limited to sixth order. Base your program on Eq. 2-55. Assume that subroutines are available for finding eigenvalues, and eigenvectors, and for matrix inversion.

5-11 Write a fourth-order Runge-Kutta program to solve Prob. 5-1.

5-12 Write a fourth-order Runge-Kutta program to solve Prob. 5-3.

5-13 Write a fourth-order Runge-Kutta program to solve Prob. 5-6.

5-14 Write a fourth-order Runge-Kutta program to solve Prob. 5-8.

chapter 6

DEVICE MODELING

A. INTRODUCTION TO MODELING

The underlying physical principles governing the behavior of most electronic devices are fairly complex, although the actual electrical behavior of the device may be quite straightforward. Rather than attempt to relate physical effects of the device directly to network analysis, an intermediate step can be undertaken. This intermediate step generally takes one of two forms: (1) representation of the behavior of the device by voltage-current or other appropriate characteristics (these characteristics may be specified graphically or analytically), or (2) representation of the device by an equivalent circuit made up of basic components such as resistors or passive elements, voltage or current sources, or other ideal elements. In the latter case, the device is easily analyzed in terms of standard circuit theory.

For manual analysis the standard approach has traditionally been to apply an equivalent circuit to linear (small-signal) problems. Either a graphical analysis or piecewise linear analysis is applied to the solution of large signal circuits. Large-signal or nonlinear networks are often too difficult to analyze, and it is not uncommon to resort to several simplifying assumptions to obtain a very approximate solution. Even small-signal problems might require undue calculations if a high degree of accuracy is to be maintained. Thus with manual analysis it is almost always necessary to simplify the device model to reduce the complexity of the overall circuit. In many practical applications these approximations are quite acceptable, but for critical design work more accurate modeling is required, generally leading to very tedious calculations.

At first thought one might think that computer-aided design programs eliminate all necessity of approximations. Certainly the computer should be able to carry out the additional calculations required by a precise model of the device. While there are several instances wherein this assumption is valid, there are many others to which this reasoning does not apply. If there are many transistors in a given network and the equivalent circuit for each requires several elements, storage capacity of the computer can be exceeded. Very accurate modeling can also lead to problems of excessive computer time under certain conditions. While additional circuit complexity may cause more computer time, this increased solution time may not always be significant. There are two specific cases, however, that should be avoided. These are (1) inclusion of nonlinear elements that do not greatly affect the overall circuit performance, and (2) inclusion of high-frequency elements when only midband or low-frequency performance is desired. The reverse of the latter situation should also be avoided—that is, low-frequency elements should not be included in a circuit if midband or high-frequency performance is of interest. Nonlinear elements in a circuit require iterative solutions, and these elements should not be considered if their deletion obviously has little effect on circuit behavior. The numerical instability problem is very significant for networks with widely separated time constants leading to excessive solution time. Here, too, elements can often be deleted from the equivalent circuit to minimize the spread of time constants. It becomes obvious that use of a general model or typical model for each electronic device does not always lead to the most efficient solution. Selection of an appropriate device model might then depend on the specific application, the frequency range of interest, the required accuracy, or the storage capability of the computer being applied.

When utilizing a specific program, various constraints on device models will be imposed. For example, ECAP recognizes only resistors, capacitors, inductors, independent voltage and current sources, and dependent current sources for a-c analysis. Dependent voltage sources are not recognized by the program and therefore cannot be used in modeling. Furthermore, this program does not have the capability of modeling by equation or characteristics that other programs have.

Since the same types of devices are used repeatedly throughout networks, modeling can become quite redundant. If six transistors are used in an amplifier circuit, the six models may be the same or very similar. Some of the newer commercial programs such as SCEPTRE, NET-1, or NET-2 offer the capability of stored models. That is, a model for the transistor can be stored on tape and called by the program each time the model is required in the circuit. The SCEPTRE stored model will be discussed in more detail later in this chapter.

Other programs that do not allow stored models require a model for each device even if the devices are similar. ECAP is such a program. Typically, the network is prepared for analysis by including appropriate equivalent circuits for each electronic device. The entire circuit then consists of basic elements recognizable by the program; passive elements, independent voltage and current sources, and dependent current sources. Entry of the model consists of entering single elements of the equivalent circuit into the program in the same way that other elements are entered. This approach has the obvious disadvantage of being very time-consuming.

In summary, modeling for computer-aided design programs is generally accomplished by using either an equivalent circuit or an expression of the characteristics of the device. This information may be stored in some programs, to be used as many times as appropriate. Other programs require entry of the complete model data for each electronic device. This chapter will consider the development of equivalent circuits for various electronic devices in terms of elements that are accepted by the bulk of commercial computer-aided design programs.

B. LINEAR CIRCUITS

Since resistors, capacitors, inductors, switches, and ideal sources can by analyzed in an orderly manner, an attempt is made to relate devices such as active circuits to these elements. The basic elements all have known voltage-current characteristics that can be characterized by constant, time-invariant parameters. The characteristics of devices such as the diode, pentode, or transistor are certainly nonlinear, and one might predict that these devices cannot be modeled in terms of constant-parameter elements. Many important applications require the device to operate over only a small area of the possible operating region within which the characteristics will be approximately linear. This is referred to as *small-signal operation*, and the device can be modeled by constant-parameter elements and analyzed as a linear circuit. One disadvantage of this method is that once the device is modeled by constant-parameter elements, the area over which linear operation takes place is not apparent. Thus a given input signal applied to the actual circuit may be large enough to cause a highly distorted output, but the small-signal circuit model will predict a nondistorted output. Some attention is then required in prescribing the limits over which a given small-signal model is valid, especially when automatic analysis programs are utilized.

Although it is possible to propose an equivalent circuit that is valid in more than one region, the a-c portion of ECAP is an example of a program utilizing a single-region circuit. A manual inspection of results is then necessary to determine whether the active region of each device has been exceeded. If so, the results are invalid.

Small-Signal Model of the Diode

For d-c and low-frequency analysis, the diode model can be derived from the *V-I* characteristics. A typical characteristic curve in the forward-biased

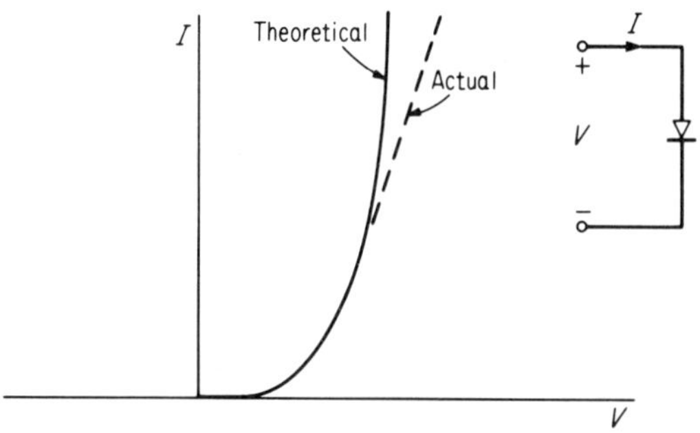

FIG. 6-1. Voltage-current characteristic of a semiconductor diode.

region is shown in Fig. 6-1. The characteristic of the intrinsic diode is given by [2]

$$I = I_s(e^{V/\delta} - 1) \tag{6-1}$$

where

$$\delta = kT/q$$

and k is Boltzmann's constant, T the absolute temperature, and q the electronic charge. In practice, the current for higher voltages does not increase as rapidly as predicted by Eq. 6-1. This is due to the slight ohmic voltage drop across the contacts to the semiconductor material and to the bulk resistance of the material. Thus, not all of the applied voltage is dropped across the junction.

If a diode is forward-biased by a d-c current, a small change of current will cause a corresponding change in voltage. This voltage change can be obtained graphically from the diode characteristic curve. It can also be obtained by assuming that the curve is linear over the small range of operation, and thus

$$\Delta V = \frac{\Delta I}{m} = r_d \, \Delta I \tag{6-2}$$

where m is the slope of the curve and r_d, called the *small-signal diode resistance*, is equal to the reciprocal of the slope. The diode resistance r_d can be calculated at a given bias current from Eq. 6-1. If ΔI and ΔV are approximated as differential quantities, Eq. 6-1 can be differentiated to give

$$\frac{dI}{dV} = \frac{I_s}{\delta} e^{V/\delta} \approx \frac{I_s}{\delta} (e^{V/\delta} - 1) = \frac{I}{\delta} \tag{6-3}$$

When the diode is biased to a current I, the slope of the characteristic curve is equal to I/δ and the small-signal resistance is

$$r_d = \delta / I \tag{6-4}$$

At room temperature (300°K), δ is equal to 0.026 V. For small signal, low frequency analysis we can represent the diode as a resistor of value r_d to account for the behavior indicated by Eq. 6-1 and a series resistance r_0 to account for ohmic resistance as shown in Fig. 6-2. Typically r_0 will be 1 to 2 Ω, while r_d will be 26 Ω when the bias current is 1 ma.

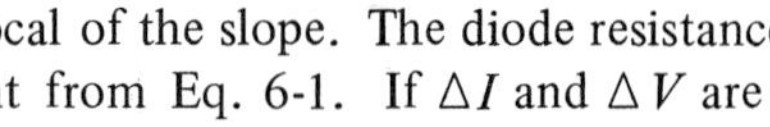

$r_d = \delta/I$
$= 0.026/I$ (300° K)

FIG. 6-2. Small-signal equivalent circuit of the diode.

Of course we should consider the region over which the diode equivalent circuit is valid. The region of validity unfortunately cannot be specified without regard to the application. In many cases a current variation equal to 10 percent of the d-c bias current is the maximum allowable variation. In other cases a greater swing can be tolerated. Since this problem can only be answered in conjunction with specific applications it will not be discussed further. We now have a model for the diode. which can be analyzed according to well-developed methods.

For high-frequency small-signal analysis we must include the reactive effects of the diode [6]. One such model is shown in Fig. 6-3 along with an approximate model that is often acceptable.

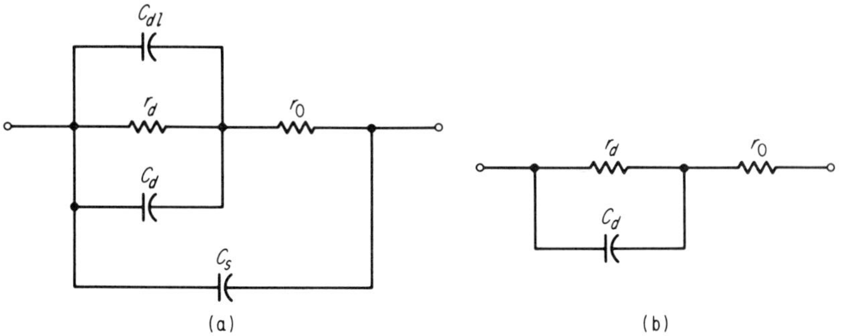

FIG. 6-3. (a) Accurate small-signal equivalent circuit of forward biased diode. (b) Approximate equivalent circuit.

The capacitance C_d is the diffusion capacitance which varies with the diode current while C_{dl} is the depletion layer capacitance and C_s is stray capacitance. The value of C_d is generally much greater than the other capacitances and the circuit model of Fig. 6-3b is reasonably accurate for the forward-biased diode.

Any computer-aided design program will accept all elements of this simple equivalent circuit. The parameters must be determined or estimated with sufficient accuracy. The problems of parameter measurement will be considered later in this chapter.

Small-Signal Model of the Transistor

There are many models that have been proposed to represent the transistor at low frequencies. Any of these small-signal equivalent circuits is valid only when the transistor operates in the active region and produces negligible distortion in the output signal. Fortunately the output characteristics of transistors are reasonably linear over a fairly large range, often allowing output swings of several volts with little distortion.

Three possible useful circuit configurations further complicate the modeling problem. The common-emitter configuration is used in practical circuits far more than the common-base or common-collector configurations, hence only this configuration will be discussed here. Two popular models for the common-emitter configuration are the t equivalent circuit [3] shown in Fig. 6-4 and the hybrid π circuit [2] of Fig. 6-5.

The parameter β is the short-circuit current gain from base to collector. The resistance r_b accounts for the ohmic base resistance plus base-width modula-

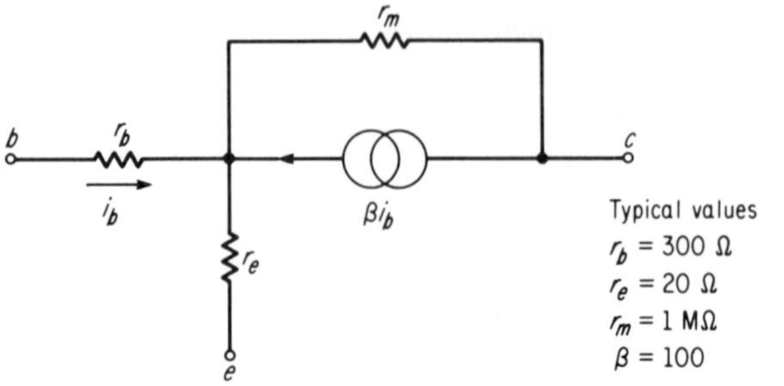

FIG. 6-4. Common-emitter *t* equivalent circuit.

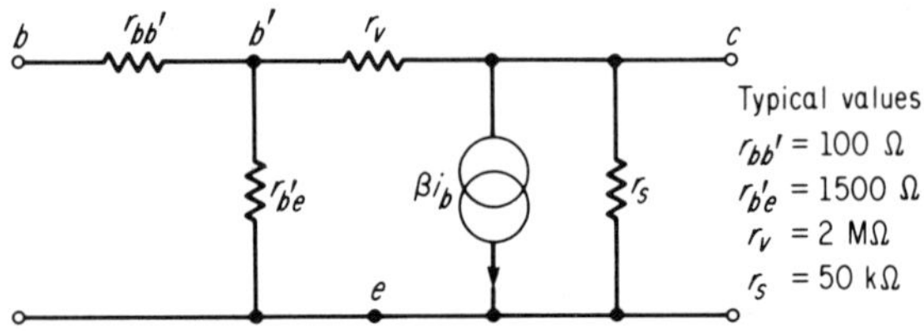

FIG. 6-5. Hybrid π equivalent circuit.

tion effects; r_e differs from the small-signal diode resistance due to base-width modulation. Since all resistances of this circuit relate to a second-order effect it is difficult to approximate element values. This is a problem common to all equivalent circuits, although the elements of the hybrid π model can be estimated somewhat more accurately. In critical design applications, such as high-frequency amplifiers, the transistor parameters must be known with sufficient accuracy to maximize circuit performance. It is necessary in such cases to measure the required values. For production circuits inaccurate estimates of element values rarely affect the design due to the required dependence of performance on element values external to the device. This point will be discussed more fully later in the present section.

In the hybrid π circuit, $r_{bb'}$ is the ohmic base resistance, $r_{b'e}$ is equal to $(\beta + 1)\, r_d$, where r_d is the dynamic emitter-base diode resistance given by

$$r_d = \frac{kT}{qI_E} \tag{6-4a}$$

$$r_d = \frac{0.026}{I_E} \quad (300^\circ\text{K}) \tag{6-4b}$$

where I_E is the d-c emitter current. The resistances r_v and r_s again account for base-width modulation effects and are difficult to determine without resorting to measurement. Production work allows r_s and r_v to be neglected as second-order effects and approximate values of $r_{bb'}$, $r_{b'e}$, and β to be used with surprisingly good results. Critical work would require accurate measurements.

High-frequency models for the transistor are more complex than low-frequency models, since additional reactive elements must be included. A high-frequency hybrid π model is shown in Fig. 6-6a [6].

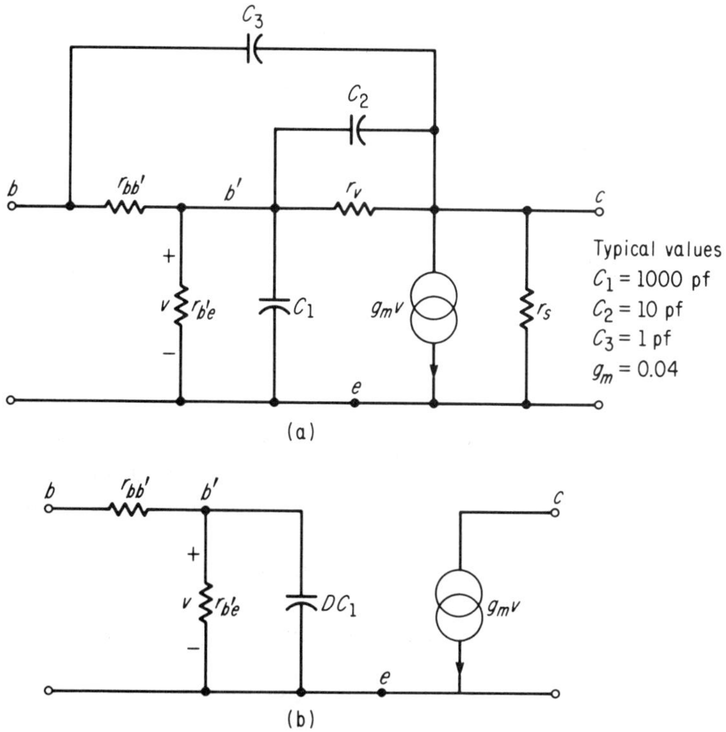

FIG. 6-6. (a) Complete hybrid π model. (b) Simplified model.

Figure 6-6a includes header capacitance C_3, collector-base depletion capacitance C_2, and emitter-base diffusion capacitance C_1 (plus depletion capacitance if significant). The capacitance C_2 depends on base to collector bias voltage, while C_1 depends on emitter current. The complete circuit can be reduced to the simplified model of Fig. 6-6b for the majority of high frequency applications. The capacitance C_3 is neglected as also are resistances r_v and r_s. These resistances are usually negligible; r_s because the shunt-load resistance is necessarily small for high-frequency work and r_v because the feedback through C_2 is greater than that through r_v at higher frequencies. For narrow band applications involving a high resonant collector load, these simplifications are not allowable. The effects of C_2 are reflected to the input circuit by means of the Miller effect [2] and combined with C_1. This model is referred to as the unilateral equivalent circuit. The D factor is given by

$$D = 1 + R_L \omega_t C_2 \tag{6-5}$$

where ω_t is the short-circuit-current gain-bandwidth and R_L is the load resistance. One might question the motivation behind the proposed circuit simplification. For manual calculations, benefits of the simplification are obvious. Circuit response is easier to determine and the influence of each element is more apparent. In choosing between the models for computer-aided design however, one might expect the more accurate circuit to be chosen, since the problem of more complex computation is minimized by the computer. For multistage design it sometimes becomes necessary to use the unilateral model due simply to the fact that the computer program cannot treat problems with a large number of elements. The unilateral circuit requires four elements whereas the accurate model requires eight. When the associated circuitry is added to a wideband multistage amplifier, the maximum number of elements allowable by the computer program may be exceeded. Thus the unilateral model would be expedient in this case.

Both models suffer from the ever-present problem of parameter measurement. Some of the elements of the models are not directly measurable, and it can be difficult to apply these models to practical design work. A certain amount of "worst-case design" can be done, based on a knowledge of the extremes of the element values. In addition to this very practical aspect of production design, important trends in performance can also be predicted by approximating the element values.

It is obvious from the preceding discussions that the determination of element values for a given model is an important area. This is probably the most troublesome area in computer-aided circuit design. There are several possible solutions among which are (1) control of the device fabrication processes to produce element values that fall in a narrow range, and (2) automatic methods of parameter measurement and possibly automatic entry of the model parameters into the computer program. Of course fabrication methods have been improving for years and should continue to do so. Selection of transistors having element values in a prescribed range is also presently possible, but often too costly to justify. Measurement and program entry techniques are being developed, and at least on a limited basis appear to offer a possible solution to the problem. A typical design must meet specifications within some tolerable range. When a design, based on specific devices, is completed by the computer, new device parameters could be measured with the results automatically entered into the analysis program. The computer would then reanalyze the circuit to determine if performance specifications are again met. The obvious disadvantages of this scheme are the cost of the extra equipment and the time required for measurement.

C. NONLINEAR CIRCUIT MODELS

The small-signal equivalent circuits discussed up to this point can represent the transistor only when the device operating region is small. For large signals the nonlinearities inherent in the device characteristics begin to introduce distortion into the response of the actual circuit. In fact, the nonlinearities of the device are often used to create the desired waveform. Switching circuits produce rectangular waveforms by extending the region of operation into a highly nonlinear portion of the characteristics. We must be able to describe the nonlinearities by circuit models or analytical methods in order to design large-signal circuits.

One method used to analyze nonlinear circuits is that of piecewise linear approximation. The nonlinear characteristics are averaged over the swing of interest and represented approximately by linearized characteristics. A linear circuit model yielding the linearized characteristics can then be proposed. For devices passing from one operating region to another, such as a transistor moving from the active region to saturation, a different, linearized equivalent circuit can be proposed for each region. For computer-aided design models some type of measurement circuit must be used to determine which equivalent circuit is appropriate at a given point in time. For example, a switching circuit output may switch through the active region to saturation, requiring that the active region model be replaced by the saturated model as the saturation region is entered. Generally ideal diodes included in the model can automatically affect the necessary changes. This section will discuss the general theory of piecewise linear models, while Chapter 7 will discuss the implementation of some models by the Electronic Circuit Analysis Program.

A second method of nonlinear analysis describes the nonlinearities analytically. By segmenting a network or system into a linear section and a nonlinear section the computer can apply iterative schemes to find a solution which satisfies both linear and nonlinear sections of the network. This approach will also be discussed in Chapter 7.

Piecewise Linear Approximation

The diode characteristic. The static voltage-current characteristic of a diode can be approximated to any degree of accuracy by straight-line segments as shown in Fig. 6-7. The necessary accuracy, of course, determines the number of straight line segments required. Some possible approximants to the forward diode characteristic are shown in Fig. 6-8.

The piecewise linear characteristics given in Fig. 6-8 can be obtained with the circuits of Fig. 6-9. The diodes shown in Fig. 6-9 are ideal diodes or switches that are controlled by the direction of current flow. For forward current the diode represents a short circuit or a closed switch. A tendency for reverse current results in no current flow as the diode represents an open switch in this case. The resistors are chosen properly to result in the desired slope of the V-I characteristic. The resistor R in Fig. 6-9 has a value given by $R = 1/S$ where S is the slope of the linear segment of Fig. 6-8a. The circuit of Fig. 6-9b will allow no current flow until the input voltage exceeds V_0. For voltages higher than V_0 the V-I curve will be a linear one with a slope equal to the reciprocal of

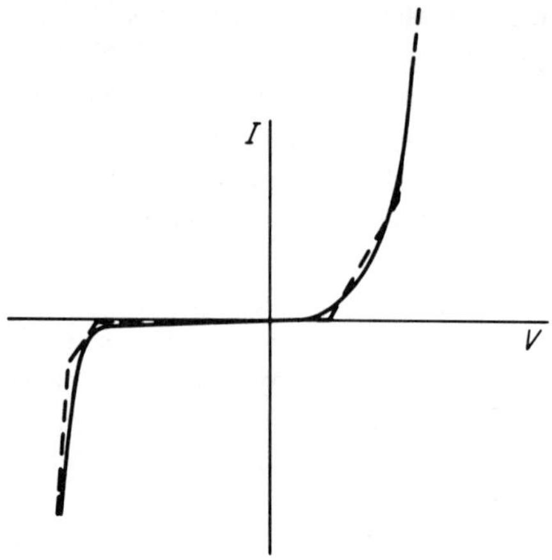

FIG. 6-7. Approximation by piecewise linear characteristics.

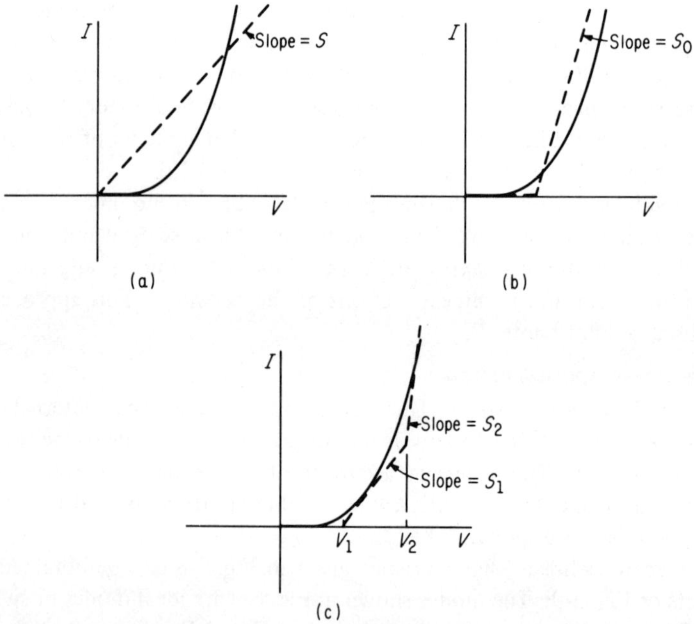

FIG. 6-8. Curves approximating forward-diode characteristic.

R_0 producing the characterisitc of Fig. 6-8b. To produce the characteristic of Fig. 6-8c requires two branches, each consisting of a source, a resistor, and an ideal diode. Diode D_1 conducts when the input voltage exceeds V_1. The slope of this portion of the curve is given by $S_1 = 1/R_a$. When the input voltage exceeds V_2, both diodes conduct and the slope is determined by the reciprocal

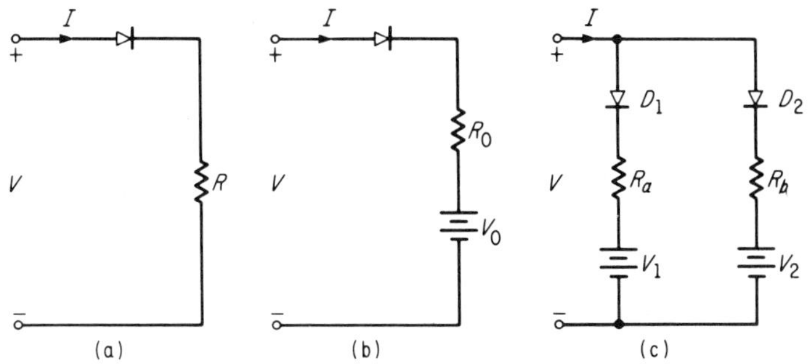

FIG. 6-9. Circuits used to approximate diode characteristics.

of the parallel combination of R_a and R_b, that is $S_2 = 1/R_{eq}$ where $R_{eq} = R_aR_b/(R_a + R_b)$. More breaks in the piecewise approximation can be effected by additional branches, however this is generally not required. In fact, the characteristic given by the circuit of Fig. 6-9b is often sufficient to represent a diode. It is to be noted that all equivalent circuits possess elements with constant parameters to which linear analysis applies, avoiding difficulties that arise in analyzing nonlinear behavior.

All elements of the proposed circuits are recognizable by most commercial computer-aided design programs. Ideal diodes are represented in ECAP by current controlled switches. The number of line segments required to approximate the actual diode characteristic depends on the desired accuracy. The designer often can only make engineering estimates when selecting the model for the program.

Synthesis of General Characteristics

Any static *V-I* characteristic can be synthesized using constant-parameter elements. We have discussed circuits which cause a break point in the slope of a *V-I* curve. In the case of parallel branches each consisting of a diode, a resistor, and a voltage source, the slope of the curve increases at each succeeding break point. The slope of the curve can be caused to decrease at a break point with the circuit of Fig. 6-10.

When the input current I is less than the source current, a forward diode current equal in magnitude to $I_0 - I$ must flow to satisfy Kirchhoff's current law at the input node. Since forward current is flowing the diode must be forward biased with zero voltage drop across it. Thus there is no voltage across the resistor and no resistor current. As I increases above I_0 a current equal to $I - I_0$ must leave the input node through either R or the diode. Since no reverse current can flow through the diode, there will be a resistive current causing a voltage drop across the circuit. The slope of the characteristic below the break point is infinite while the slope above the break point is determined

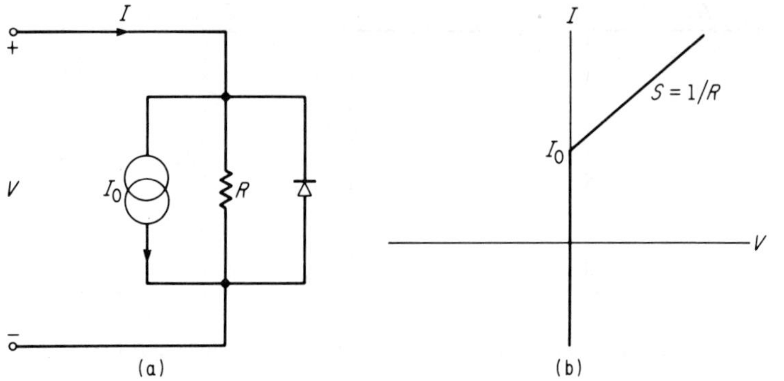

FIG. 6-10. (a) Circuit having a decreased slope above the break point. (b) *V-I* characteristic.

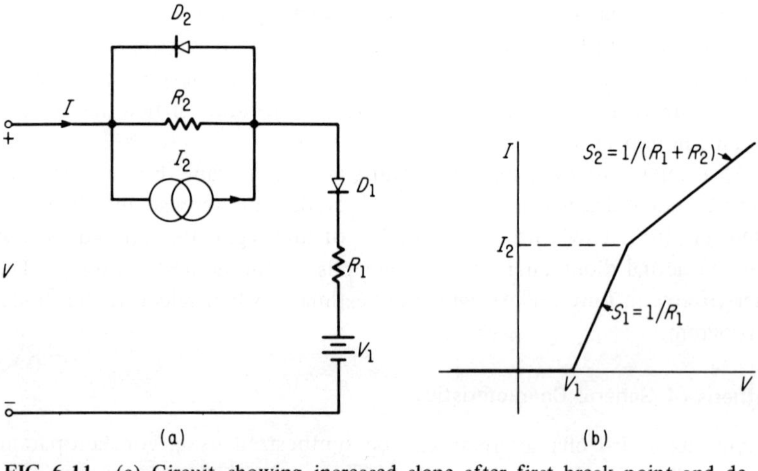

FIG. 6-11. (a) Circuit showing increased slope after first break point and decreased slope after second break point. (b) *V-I* characteristic.

by the value of R. The current source branch can be combined with a voltage source branch as shown in Fig. 6-11.

Diode D_1 will not allow current to flow until the input voltage exceeds V_1. The *V-I* relationship for values of I less than I_2 is determined by R_1, since D_2 is forward-biased and appears as a short circuit. As V increases to the point that I equals I_2 the second break point occurs. Diode D_2 becomes reverse biased and the resistance of the shunt combination equals R_2. The slope of the curve above this point is determined by the sum of R_1 and R_2. It is to be noted that the same characteristics can be realized using current sources only, or voltage sources only rather than one of each. These possibilities are shown in Fig. 6-12.

With the basic building blocks networks with very complex piecewise linear characteristics can be synthesized. The network of Fig. 6-13a possesses a *V-I* relationship as shown.

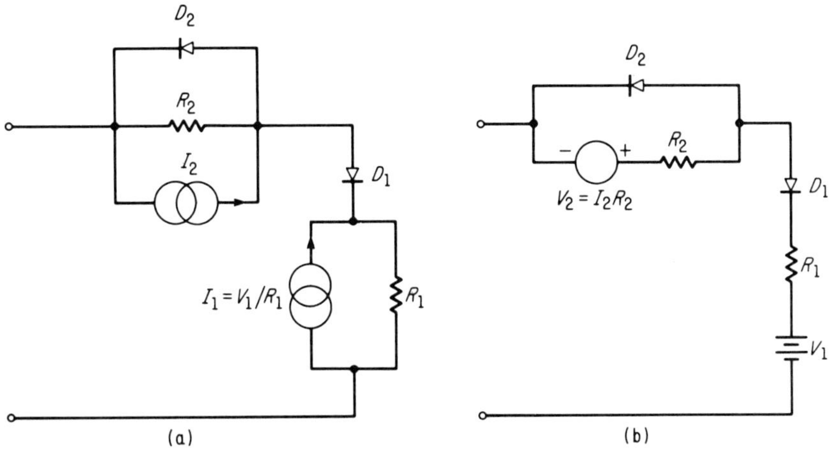

FIG. 6-12. Alternative circuits having the *V-I* characteristic of Fig. 6-11b.

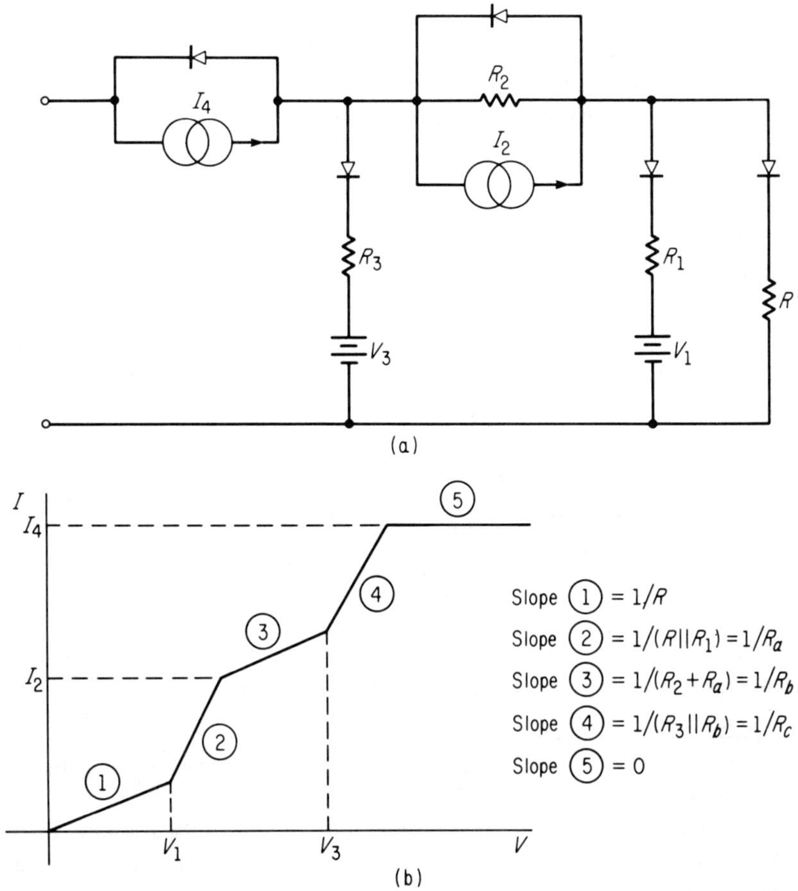

FIG. 6-13. (a) Piecewise linear network. (b) *V-I* characteristics.

The entire V-I curve can be shifted in any direction by inserting either a series voltage source or a shunt current source or both. Consider the situation depicted in Fig. 6-14. Increasing E shifts the original curve horizontally to the

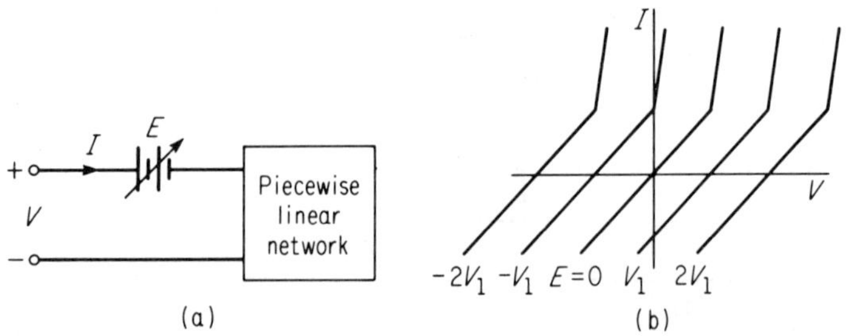

FIG. 6-14. (a) Piecewise linear network with series voltage. (b) V-I characteristics.

right, while decreasing E shifts the curve to the left. The curve can be shifted up or down using the scheme of Fig. 6-15. When J is positive the original characteristics are shifted upward while a negative value for J causes a downward shift.

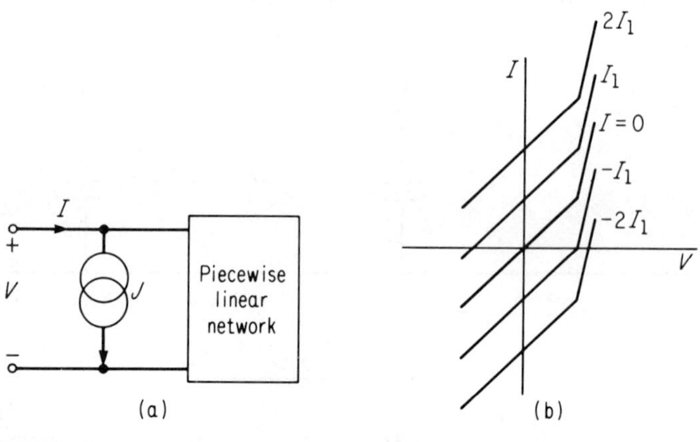

FIG. 6-15. (a) Piecewise linear network with shunt current source. (b) V-I characteristic.

Transistor Characteristics (Common-Emitter Configuration)

A typical set of characteristic curves for a common-emitter *n-p-n* transistor is given in Fig. 6-16. One possible approximation to these curves is the set of equally spaced curves of Fig. 6-17.

The piecewise linear characteristics are certainly less accurate than the actual characteristics for two obvious reasons; (1) the actual curves are not equally spaced, and (2) the slopes of the actual curves become higher for

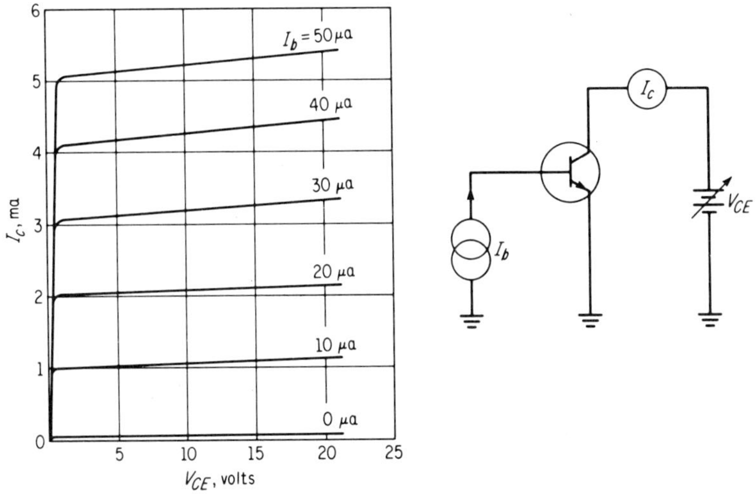

FIG. 6-16. Common-emitter characteristics.

larger base currents. In spite of these differences, the piecewise linear circuits lead to usable results in a surprisingly large number of problems.

Only the output characteristics are represented in Fig. 6-17. It is often

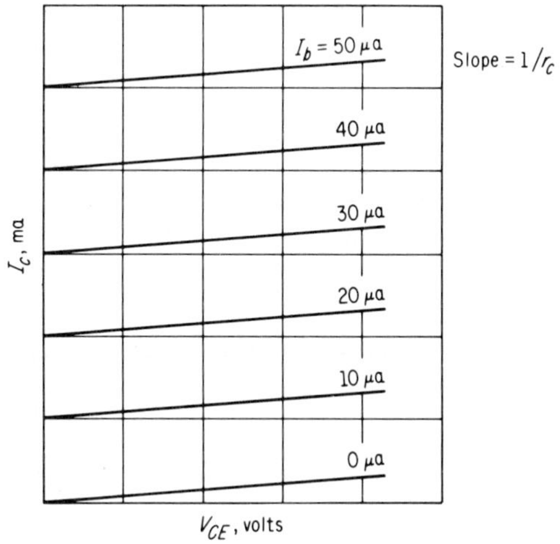

FIG. 6-17. Piecewise linear approximation to transistor characteristics.

useful to know the input characteristics which are given in Fig. 6-18. Constant-parameter equivalent circuits can now be proposed for both the input and output which satisfy the appropriate piecewise linear characteristics. These networks are shown in Fig. 6-19.

The current generator in the collector-emitter circuit is proportional to the

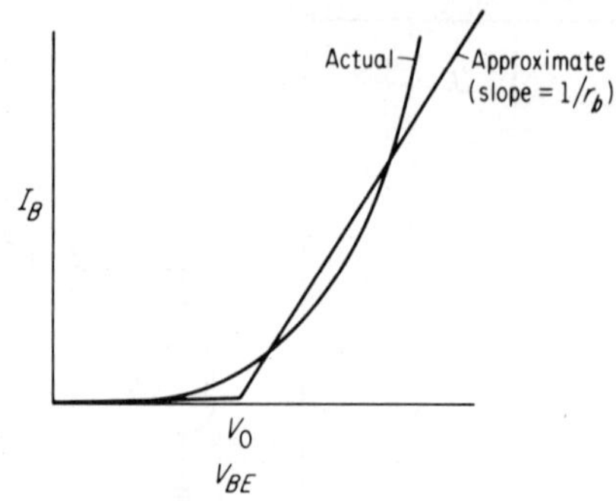

FIG. 6-18. Input characteristics for the common-emitter configuration.

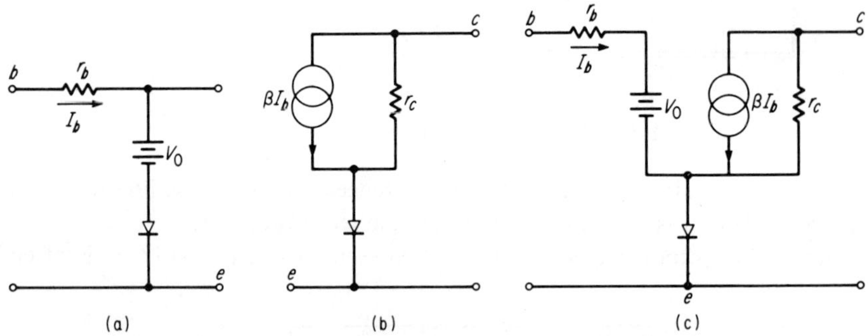

FIG. 6-19. (a) Input circuit of an *n-p-n* transistor. (b) Output circuit. (c) Combined circuit.

total instantaneous base current that flows. It is sometimes convenient to split the current source into two parallel sources, one associated with the d-c component of collector current and one proportional to the incremental base current [3]. If the output swing is limited to a particular region, the value of incre-

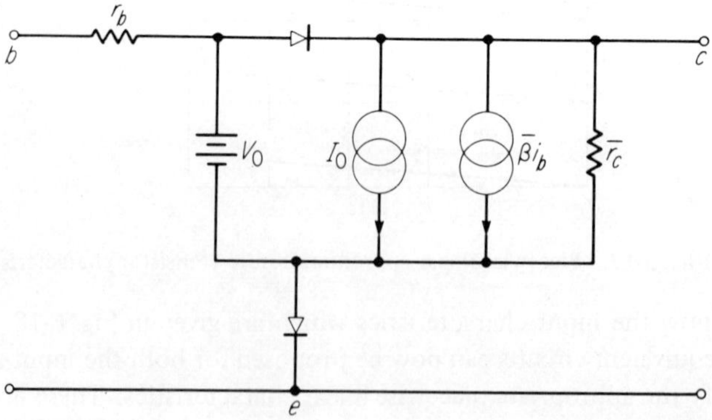

FIG. 6-20. Alternate piecewise linear equivalent circuit for an *n-p-n* transistor.

mental current gain β can be evaluated over the region of interest. The resulting equivalent circuit is shown in Fig. 6-20.

The diodes are present to create the boundaries of the saturation and cutoff regions. If only the active region is to be used, no diodes need appear. The current source I_0 depends on the d-c base current that flows. Returning to Fig. 6-17, if the d-c base current equals 30 μa, then I_0 must be 3 ma. The circuit of Fig. 6-20 is more complex than that of Fig. 6-19 and requires the evaluation of I_0. However, the accuracy can be greater, since $\bar{\beta}$ and $\bar{r}_c$ are averaged over the actual output swing and are defined by

$$\bar{\beta} = \left.\frac{\Delta I_c}{\Delta I_b}\right|_{V_{CE}=\text{const.}} \qquad (\Delta I_c \text{ taken over the output swing})$$

and

$$1/\bar{r}_c = \left.\frac{\Delta I_c}{\Delta V_{CE}}\right|_{I_b=\text{const.}} \qquad (\Delta I_c \text{ taken over the output swing})$$

For the circuit of Fig. 6-19, β and r_c are averaged over the entire operating region of interest, even though the actual output swing may be smaller giving

$$\beta = \text{average of } \left.\frac{I_c}{I_b}\right|_{V_{CE}=\text{const.}}$$

and

$$1/r_c = \text{average of } \left.\frac{\Delta I_c}{\Delta V_{CE}}\right|_{I_b=\text{const.}}$$

The latter two values are subject to larger errors.

The depletion region capacitances of a transistor lead to some problems when large-signal operation takes place. These capacitances depend on the voltages appearing across the depletion regions in a complex fashion. The small signal capacitance for a uniformly doped *p-n* junction is [6]

$$C = \frac{k_1}{(\phi - V)^{1/2}} \tag{6-6a}$$

where ϕ is the built-in barrier voltage, k_1 is a constant dependent on doping and geometry, and V is the voltage applied to the junction. The applied voltage has a negative value for reverse bias and a positive value for forward bias. When the doping is nonuniform or graded, the small-signal capacitance is approximated by

$$C = \frac{k_2}{(\phi - V)^{1/3}} \tag{6-6b}$$

Most high-frequency transistors such as planar or mesa transistors exhibit a capacitance which can be approximated by Eq. 6-6b.

As a transistor switches from one state to another, the voltages from base to collector and base to emitter vary. These variations cause the capacitances to change and some method of relating the effect of this variation to circuit performance must be found. One method that has been used [3] is to define an average value of capacitance as

$$\overline{C} = \frac{\Delta Q}{\Delta V} = \frac{Q_2 - Q_1}{V_2 - V_1} \tag{6-7}$$

where ΔQ is the change in charge corresponding to a change in voltage ΔV. The small-signal capacitances given by Eq. 6-6 are based on the definition

$$C \equiv \frac{dQ}{dV} \tag{6-8}$$

Calculation of the total charge can be based on Eq. 6-8, since

$$\Delta Q = \int_{Q_1}^{Q_2} dQ = \int_{V_1}^{V_2} \frac{dQ}{dV}\, dV = \int_{V_1}^{V_2} C\, dV \tag{6-9}$$

The linearized capacitance $\overline{C}$ is then

$$\overline{C} = \frac{\int_{V_1}^{V_2} C\, dV}{V_2 - V_1} \tag{6-10}$$

For the uniformly doped junction, evaluation of the integral yields

$$\overline{C} = \frac{2k_1}{V_2 - V_1} \left[(\phi - V_1)^{1/2} - (\phi - V_2)^{1/2}\right] \tag{6-11}$$

whereas the graded junction exhibits a linearized capacitance of

$$\overline{C} = \frac{3k_2}{2(V_2 - V_1)} \left[(\phi - V_1)^{2/3} - (\phi - V_2)^{2/3}\right] \tag{6-12}$$

For many switching circuits the voltage levels before and after switching are known or easily calculated and thus the linearized value of capacitance can be used to determine switching times.

The depletion-region capacitance is normally required to calculate switching times when the corresponding junction is reverse-biased. Thus both depletion-layer capacitances may be required to calculate delay time [3], since the transistor operates in the cutoff region during this time. In the active region the collector-base depletion layer capacitance is still significant, but the emitter-base depletion-layer capacitance becomes less important due to the rapid increase in diffusion capacitance as the base-emitter junction becomes forward-biased. The small-signal diffusion capacitance of a junction is proportional to the bias current of the junction. When the junction is reverse-biased, the current is small and the depletion-layer capacitance is dominant; for the forward-biased case, the current is such to cause the diffusion capacitance to be dominant.

The small-signal hybrid π model is often used in large-signal, active region switching analysis with certain modifications. Figure 6-21 shows this model.

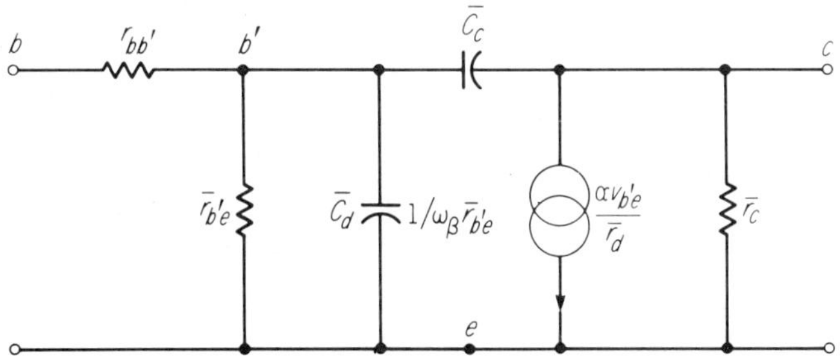

FIG. 6-21. Linearized hybrid π model for switching-circuit analysis.

The quantity $r_{bb'}$ is the ohmic base resistance in the hybrid π circuit. A linearized value of $\overline{r}_{b'e}$ is selected, for example

$$\overline{r}_{b'e} = (\overline{\beta} + 1)\frac{kT}{qI_{E_{\text{ave}}}} \tag{6-13}$$

where the emitter current is averaged over the total swing. The diffusion capacitance is calculated so that the time constant of $\overline{r}_{b'e}$ and $\overline{C}_d$ is equal to $1/\omega_\beta$ where ω_β is the β-cutoff frequency (radian) of the transistor. A three-region equivalent circuit is considered in Chapter 7.

There are certain shortcomings in this model as in all proposed models. To account for all physical effects of a device generally requires such a complex model that calculations are difficult and inefficient. For some purposes a very complex model can be used. In developing a model to predict voltage-current terminal relationships, a model containing several nodes might be used [5]. However, if several transistors are to be used in a network, the model must be simplified to limit the problem to a practical size.

Another serious problem with this type of equivalent circuit is that certain output variables must be known before parameter values can be entered. For example, $I_{E_{\text{ave}}}$ must be estimated to find $\overline{r}_{b'e}$ and the extreme collector voltage levels must be known to calculate $\overline{C}_c$. Again, engineering estimates are often required in cases such as this.

We have previously touched on some of the difficulties of accurate device modeling. Perhaps it would be worthwhile at this point to expand on these problems. The measurement problem is quite important. Although we can propose device models that can represent actual devices with almost arbitrary precision, a very accurate model may be almost impossible to characterize in terms of actual measured parameters. In general, a complex model maximizes the measurement problem. These models are limited in application to critical design work rather than production. In production design, measurement or selection of devices adds considerably to design cost, thus most production circuits are designed to minimize effects on performance by device parameters. Since device characteristics are so variable, the overall circuit performance must not be affected greatly by these parameters. These networks are so designed that external elements with accurately known values, such as resistors and capacitors, determine the behavior of the network. Production circuits must exhibit this dependence on external elements in order that similar units will behave uniformly and allow devices to be replaced as required. For many industrial circuits, then, neither accurate modeling nor measurement is required. A model reflecting the fundamental device behavior and possibly the more important second-order effects is sufficient.

D. NONIDEAL TRANSFORMER [7]

An equivalent circuit for the ideal transformer was introduced in Chapter 3. The absence of frequency dependent elements from the simple model limits its application to some range of frequencies wherein transformer reactive effects are negligible. A more accurate model is necessary to use in transient response or frequency-response calculations. The transformer exhibits losses that can be represented by resistances in series with both primary and secondary windings. Each winding exhibits self-inductance and there is a mutual inductance coupling the primary and secondary. The coefficient of coupling k can be related to the inductances by

$$k = M/\sqrt{L_p L_s} \tag{6-14}$$

where M is the mutual inductance, L_p the primary inductance, and L_s the secondary inductance. In developing a model we can initially neglect the resistive losses, since these elements can be added in series with the windings after we have developed a circuit to represent the inductive effects. The winding voltages can be expressed as

$$e_p = L_p \frac{di_p}{dt} + M \frac{di_s}{dt} \tag{6-15}$$

$$e_s = M \frac{di_p}{dt} + L_s \frac{di_s}{dt} \tag{6-16}$$

where the p subscript refers to the primary and the s to the secondary. Solving these equations for i_p and i_s results in

$$i_p = [1/L_p(1 - k^2)] \int e_p\, dt - \frac{M}{L_p} [1/L_s(1 - k^2)] \int e_s\, dt \tag{6-17}$$

$$i_s = \frac{-M}{L_s}[1/L_p(1-k^2)]\int e_p\,dt + [1/L_s(1-k^2)]\int e_s\,dt \qquad (6\text{-}18)$$

Since the voltage and current of an inductor are related by

$$i = (1/L)\int v\,dt$$

the first term of Eq. 6-17 and the second term of Eq. 6-18 can be represented by self-inductances of value

$$L_1 = L_p(1-k^2) \qquad (6\text{-}19)$$

and

$$L_2 = L_s(1-k^2) \qquad (6\text{-}20)$$

respectively. The negative terms in the two equations can be modeled by current-controlled current sources. The complete model is shown in Fig. 6-22, including resistive losses.

The use of the equivalent circuit of Fig. 6-22 requires that the transformer

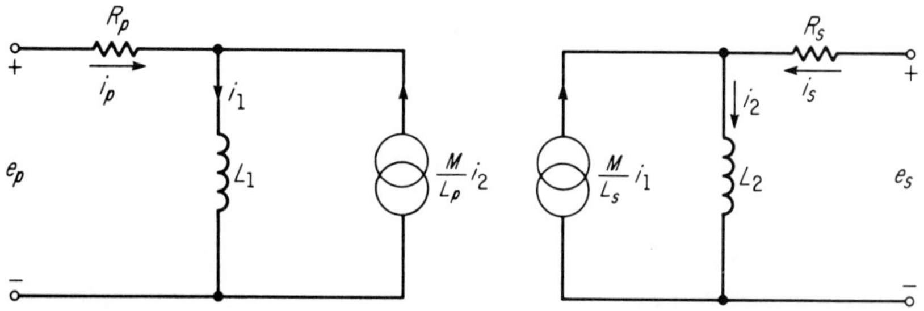

FIG. 6-22. Transformer equivalent circuit.

parameters R_p, R_s, L_p, L_s, and M be known. For critical work these parameters can be measured; for noncritical work the ideal circuit of Chapter 3 can be used. Transformer parameters for transformers of the same type are quite uniform in comparison to transistor parameters. In production work, the characterization of a transformer type is effected by measuring the parameters of a typical transformer. In ECAP the model of Fig. 6-22 is used only for transient analysis, since a-c analysis can be done by specifying a mutual inductance.

There are other important networks that can be modeled and used in computer-aided design such as the gyrator and negative-impedance converter [1]. Although these circuits are important in network synthesis, they will not be covered here.

E. SCEPTRE STORED MODELS [8]

Programs such as NET-1 and SCEPTRE provide a very significant modeling feature. These programs allow the use of stored models for transistors and other devices. NET-1 contains a fixed transistor model requiring that the necessary parameters be supplied by the user. SCEPTRE is more flexible in this regard allowing the user to construct a particular model for any device that can be represented by passive elements, voltage or current sources, and mutual inductors. Since SCEPTRE is used mainly for transient analysis, large-signal circuits are generally stored in this program.

The equivalent circuit most often used for the transistor is the basic Ebers-Moll equivalent circuit of Fig. 6-23. This circuit is valid for all possible operating

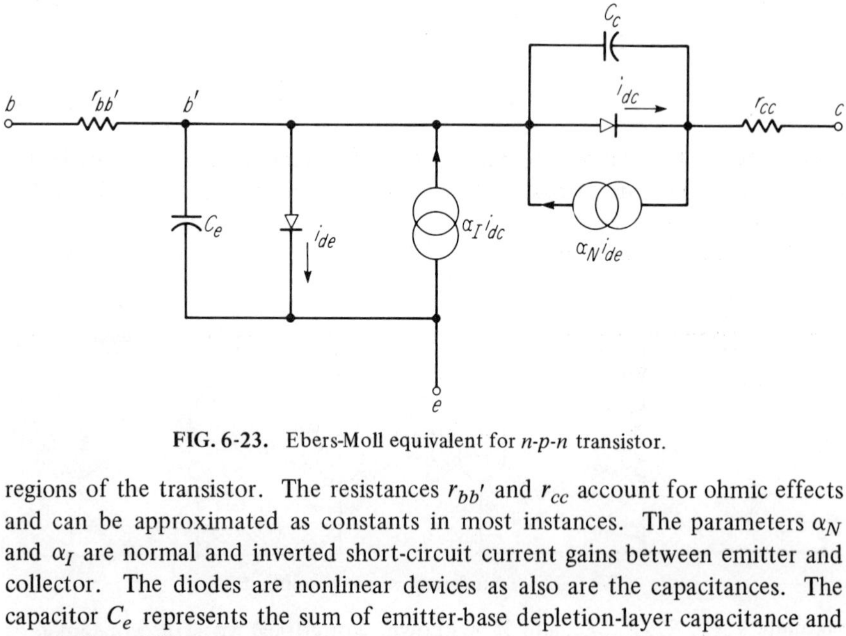

FIG. 6-23. Ebers-Moll equivalent for *n-p-n* transistor.

regions of the transistor. The resistances $r_{bb'}$ and r_{cc} account for ohmic effects and can be approximated as constants in most instances. The parameters α_N and α_I are normal and inverted short-circuit current gains between emitter and collector. The diodes are nonlinear devices as also are the capacitances. The capacitor C_e represents the sum of emitter-base depletion-layer capacitance and emitter-base diffusion capacitance, while C_c represents the sum of collector-base depletion and diffusion capacitances.

SCEPTRE allows nonlinear functions to be represented analytically. The currents through the diodes depend on the diode voltages in the usual way—that is,

$$I_{de} = I_{e1}\left(e^{V_e/\delta} - 1\right) \tag{6-21}$$

and

$$I_{dc} = I_{c1}\left(e^{V_c/\delta} - 1\right) \tag{6-22}$$

where I_{e1} and I_{c1} are constants for a given transistor. The value of δ is given by

$$\delta = \frac{kT}{q} \tag{6-23}$$

where k is Boltzmann's constant, q the electronic charge, and T the absolute temperature. These diodes can be represented by voltage-dependent current sources obeying the functional relations given by Eqs. 6-21 and 6-22. The parameters I_{e_1} and I_{c_1} must be known or assumed in order to enter the current-source equations into the program. SCEPTRE also offers the capability of representing the current sources with tabular data rather than by analytic expressions.

Since the current gains α_N and α_I vary with operating point, an equation or table can be used for these values if constant values are not sufficiently accurate.

The capacitances are complex functions of the junction voltages. Diffusion capacitance varies roughly as the diode current or exponentially with junction voltage. Depletion-layer capacitance varies inversely as a fractional power of junction voltage [6] depending on geometry and construction process of the transistor. Again the values for these parameters (C_e and C_c) can be expressed in terms of appropriate equations or in tabular form. When operation of the transistor is limited to certain regions, approximations to the nonlinear elements can be made.

Once the parameters of the model are determined the model can be stored in the program and used to represent any or all transistors in a circuit. In fact, the model can be called with modified parameter values if the situation warrants. This flexibility of the SCEPTRE stored model is extremely useful.

Of course, other models can be used for the transistor besides the Ebers-Moll model and this choice is left to the designer. Obviously other devices can be represented by an appropriate model and stored by this program. An interesting extension of the stored model approach is to store complete circuits that are often used by a particular designer. For example, in logic design perhaps simplified models of gates or multivibrators could be stored and used with modified element values as often as needed.

While the flexibility of this concept is tremendous we must not forget that nonlinear equation solution is difficult and time-consuming with a digital computer. The Newton-Raphson iterative procedure is applied in SCEPTRE, and this scheme will be discussed in Chapter 7. Nevertheless, the stored-model capability represents a significant improvement in computer-aided design techniques.

In summarizing the equivalent-circuit approach to device modeling for computer-aided design we note that:

1. Equivalent circuits of devices must consist of circuit elements that are recognizable by the computer program being used.
2. Entry of device models into programs with no stored-model capability is done by entering the individual elements of the model in the same way all other circuit elements are entered. Generally this information is specified on punched cards. Of course, the possibility of creating a model deck exists; that is, a deck that characterizes a given device could be used each time the device is called for in the program deck.
3. For stored-model programs, a model is entered in the normal way

(punched cards) and stored on tape to be called by the program when necessary.

4. A general model for a device does not always lead to the most efficient analysis since certain troublesome elements of the model may not actually be required by the problem at hand. For example, nonlinear or high frequency elements may not affect the circuit response greatly, but could easily affect the solution time significantly.

BIBLIOGRAPHY

1. Calahan, D. A., *Computer-Aided Network Design*, preliminary edition, McGraw-Hill Book Company, New York, 1968.
2. Comer, D. J., *Introduction to Semiconductor Circuit Design*, Addison-Wesley Publishing Company, Reading, Mass., 1968.
3. Comer, D. T., *Large-Signal Transistor Circuits*, Prentice-Hall, Inc., Englewood Cliffs, N.J., 1967.
4. Ghausi, M. S., *Principles and Design of Linear Active Circuits*, McGraw-Hill Book Company, New York, 1965.
5. Ghosh, H. N., F. H. De La Moneda, and N. R. Dono, "Computer-aided Transistor Design, Characterization, and Optimization," *Proc. IEEE*, Vol. 55, No. 11 (November 1967), pp. 1897-1912.
6. Gray, P. E., D. DeWitt, A. R. Boothroyd, and J. F. Gibbons, *Physical Electronics and Circuit Models of Transistors*, SEEC Vol. 2, John Wiley & Sons, Inc., New York, 1964.
7. Jensen, R. W., and M. D. Lieberman, *IBM Electronic Circuit Analysis Program*, Prentice-Hall, Inc., Englewood Cliffs, N.J., 1968.
8. Sedore, S. R., "SCEPTRE: A Program for Automatic Network Analysis," *IBM Journal*, Vol. 11 (November 1967), pp. 627-637.

PROBLEMS

6-1 Calculate the small-signal diode resistance at room temperature when the bias current is (1) 0.5 ma, (b) 2 ma, and (c) 6 ma.

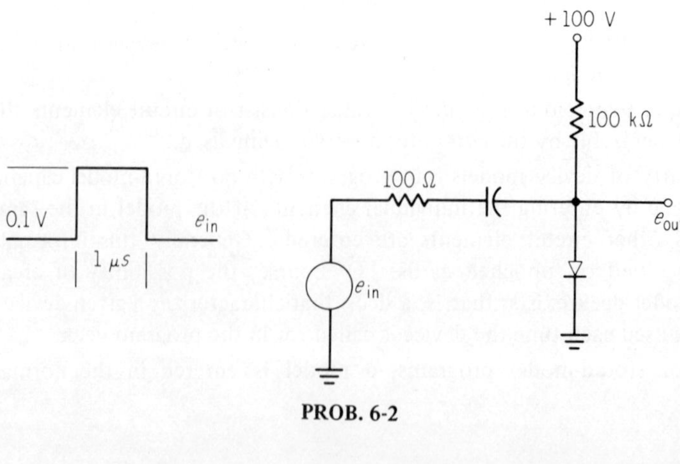

PROB. 6-2

6-2 Neglecting ohmic resistance and assuming that $C_d = 4{,}000\,\text{pf}$, use the diode model of Fig. 6-3b to calculate the time response of the output voltage. Assume that the coupling capacitor appears as a short circuit for the input pulse.

6-3 Repeat Prob. 6-2 when $R = 20\,\text{k}\Omega$. Explain why the amplitude of the output voltage changes when R is changed.

6-4 Calculate the a-c voltage gain of the circuit if $\beta = 100$, $r_{bb'} = 200\,\Omega$, $I_E = 2\,\text{ma}$, $r_s = 100\,\text{k}\Omega$, and $r_v = 2\,\text{m}\Omega$. Assume that C is a short circuit for a-c signals.

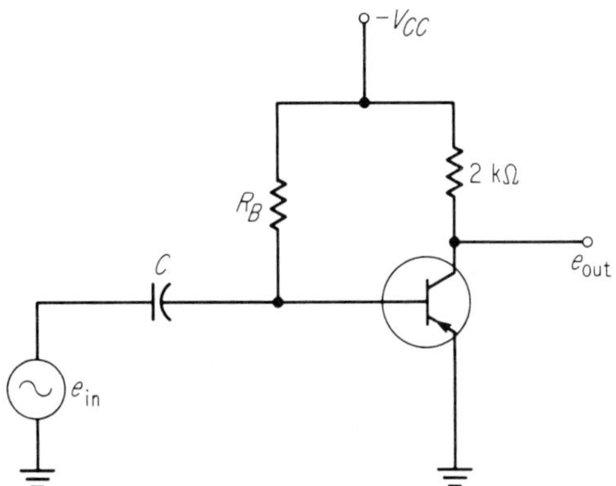

PROB. 6-4

6-5 Repeat Prob. 6-4 assuming r_s and r_v are infinite. Compare the voltage gains calculated for the two cases.

6-6 The transistor is an abrupt junction transistor with $\omega_t = 6.28 \times 10^6$ rad/-sec, $g_m = 40$ millimhos, $C_2 = 20\,\text{pf}$ at $V_{CB} = 6\,\text{V}$. Assuming that $\phi = 0.8\,\text{V}$,

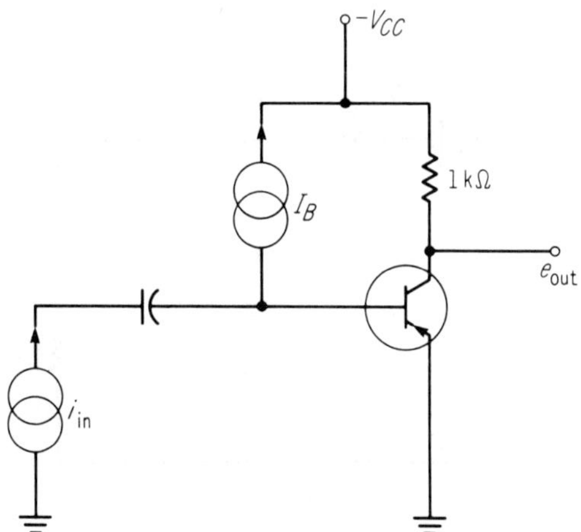

PROB. 6-6

find the upper 3-db point of e_{out}/i_{in} when $I_E = 1$ ma. Use the unilateral equivalent circuit.

6-7 Repeat Prob. 6-6 if I_E is changed to 5 ma.

6-8 When emitter current decreases to the point that diffusion capacitance is not several factors larger than depletion layer capacitances, the formula for f_t is

$$f_t = \frac{1}{2\pi r_d(C_1 + C_2 + C_3)}$$

Explain why this quantity decreases at low values of emitter bias current.

6-9 to 6-12 Sketch the V-I characteristics for the circuits shown, assuming ideal diodes.

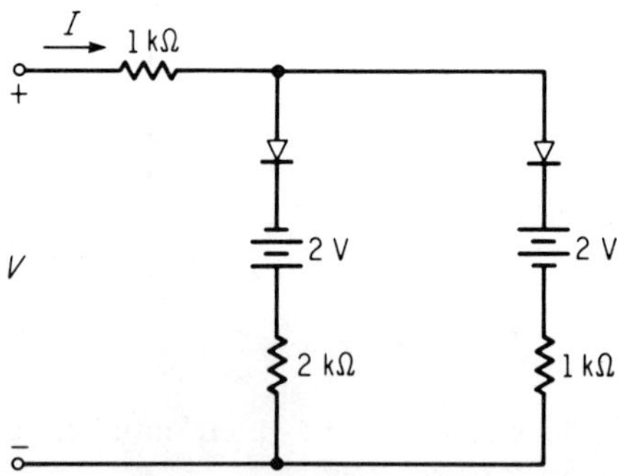

PROB. 6-9

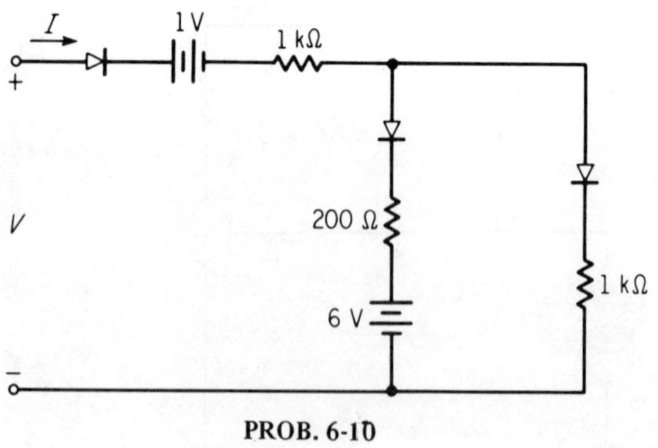

PROB. 6-10

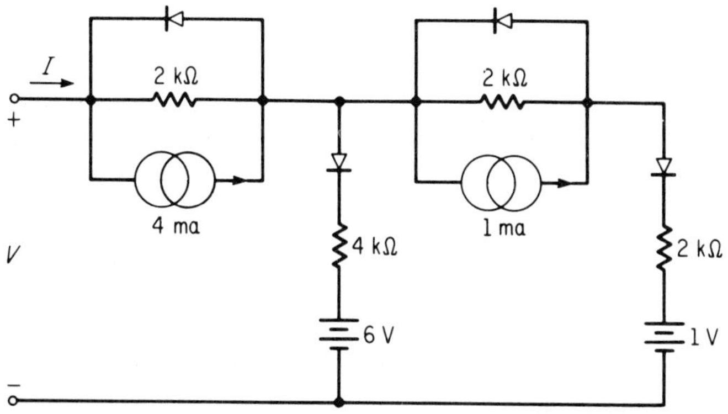

PROB. 6-11

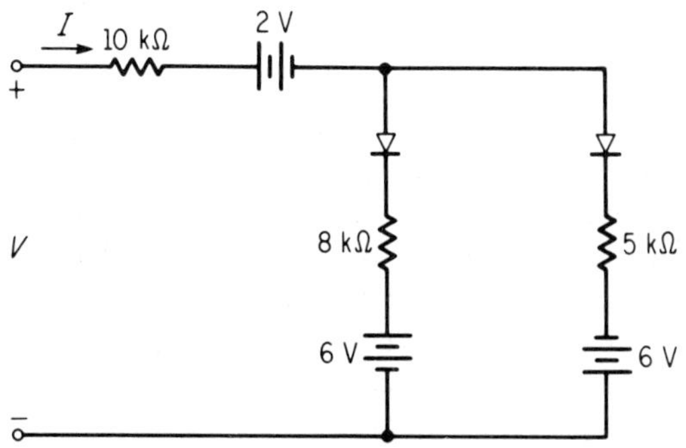

PROB. 6-12

6-13 Plot the V-I characteristics of the circuit of Prob. 6-9 if a 10-V source is inserted in the input lead. The positive terminal of the source is connected to the 1-k Ω resistance.

6-14 Calculate the output voltage as a function of frequency for the circuit

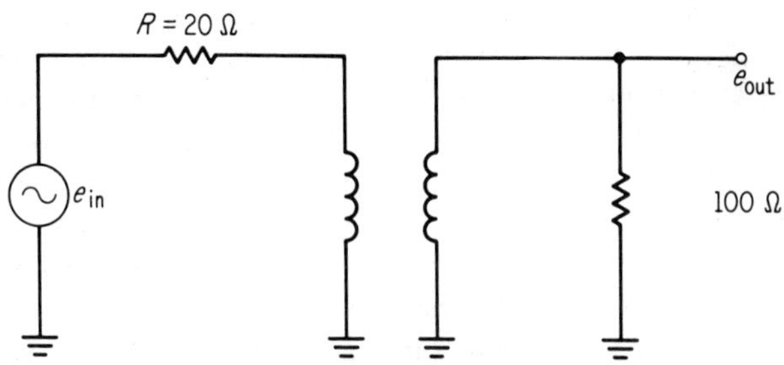

PROB. 6-14

shown. The transformer parameters are $R_p = 10\Omega$, $R_s = 5\Omega$, $L_p = 0.4\,\text{H}$, $L_s = 0.1\,\text{H}$, and $k = 0.999$.

chapter 7

NONLINEAR ANALYSIS

In the field of engineering there are few problems that are completely linear over the entire range of possible input or output variables. It may seem strange that, in light of the vast number of nonlinear circuits and systems, most design procedures are based on linear analysis. Upon closer consideration one finds several justifications for applying linear techniques to many areas of design. Certainly most passive networks lead to linear-response equations since resistors, capacitors and inductors generally show negligible dependence on currents or voltages. Active devices such as vacuum tubes and transistors, while possessing pronounced nonlinearities, can be analyzed as linear circuits when the operating region is limited to specified areas. Thus we speak of small-signal circuits to describe a circuit containing a nonlinear element with a limited operating region within which the element behaves as a linear device. Linear amplifiers are examples of small-signal circuits using nonlinear devices. These two classes of circuits, linear amplifiers and passive networks, are quite important in electrical engineering, but there is another major reason for studying linear analysis—because nonlinear analysis is most efficiently accomplished using linear methods. The two popular methods of nonlinear analysis are

1. The use of models having constant, linearized parameters over given regions of operation allowing linear analysis to be valid for each region.
2. Iterative schemes wherein linear equations guide successive iterations to updated solutions.

The latter of these methods is inherently more accurate than the first. For a very large class of important circuits, namely digital circuits, accuracy is often not a prime requisite in analysis. For example, in logic circuits a logical level may be defined as a voltage greater then or equal to 8 volts, while the complementary level may be defined as any voltage less than or equal to 2 volts. Thus fairly large inaccuracies in the calculation of voltage levels would not lead to incorrect results for these logic circuits and the first method of analysis would be appropriate. Of course, very critical design might require more accuracy, but even in several of these cases the nonlinear elements can be replaced with suitably linearized elements to effect a design procedure.

Modeling of nonlinear elements can take several forms all of which cannot be treated in a single chapter. Thus the first section of this chapter is devoted to the development of models that can be used in ECAP. While the underlying principles discussed will apply to modeling for other computer programs, they are specifically directed toward use with ECAP. This limits the coverage to an appropriate amount of material and also provides background for the discussion of ECAP which follows in Chapter 8.

A consideration of iterative techniques would also require several pages if all aspects of this subject are to be treated. Consequently, to limit the topic to a viable discussion, only one technique is considered. Treatment of Newton-Raphson techniques should not imply that other methods are less important. In fact, the Fletcher-Powell method generally offers advantages over Newton-Raphson methods, but space limitations prevent coverage of the Fletcher-Powell method. The second section of the chapter is then devoted to a development of Newton-Raphson methods and discusses the implementation of these methods on a digital computer.

A. PIECEWISE LINEAR ANALYSIS

The nonlinear circuit easiest to analyze is one that can be represented by linearized single-valued, constant parameter elements. Linear techniques can then be applied to calculate the network response. An example of this is a transistor switching circuit with a large output-signal swing that never exceeds the limits of the active region.

A more difficult problem, but one that arises often, consists of analyzing a circuit that exceeds the bounds of the active region and requires a second linearized equivalent circuit to be used. Again, a transistor switching circuit can exhibit this type of behavior as the operating point moves from the active region to either saturation or cutoff. A different equivalent circuit must be used for operation in each of the three regions. In computer-aided design the circuit to be used must be selected automatically as dictated by the instantaneous value of output quantities. One practical approach to the solution of this type of nonlinear problem is demonstrated in the Electronic Circuit Analysis Program. The ECAP method of analysis will now be considered, noting that much of the information included in this section is derived from the ECAP user's manual [7].

Piecewise linear models have been discussed in detail in Chapter 6. The ideal diode plays a very important part in modeling nonlinear circuits and thus must be included in any program that bases nonlinear analysis on piecewise linear models. The ideal diode can be modeled by a current-controlled switch. The computer program can monitor the current in the branch containing the switch to determine which state the switch should assume. Since the ECAP program uses the concept of a standard branch as defined in Chapter 3, the branch impedance must always be finite. Neither short circuits nor open circuits are allowed in the program, thereby guaranteeing that certain matrices involving branch impedances or admittances possess an inverse. The current-controlled switch cannot represent an ideal diode, but if a suitably large resistor represents an open circuit and a suitably small resistor represents a short circuit, the ideal diode is very closely approximated. The switch is actually more versatile in this program than an ideal diode could be. A switch can control any or all parameters in its branch. Furthermore, when a switch changes states, circuit elements in other branches can be changed also (passive elements, independent sources, and controlled current sources). When the value of a dependent source is to be changed, the switch must control the branch upon which the source is dependent (called the "from" branch).

Diode Analysis

Figure 6-8 demonstrates three piecewise linear characteristics that can be used to approximate the diode characteristics. Two of these curves are reproduced in Fig. 7-1. In ECAP the switch is assumed to be ON when branch current is greater than zero and OFF when current is less than or equal to zero.

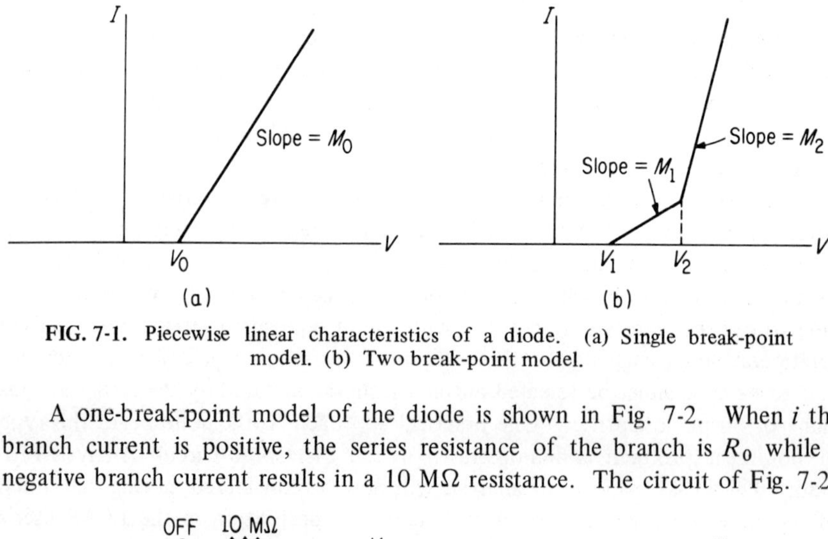

FIG. 7-1. Piecewise linear characteristics of a diode. (a) Single break-point model. (b) Two break-point model.

A one-break-point model of the diode is shown in Fig. 7-2. When i the branch current is positive, the series resistance of the branch is R_0 while a negative branch current results in a 10 MΩ resistance. The circuit of Fig. 7-2b

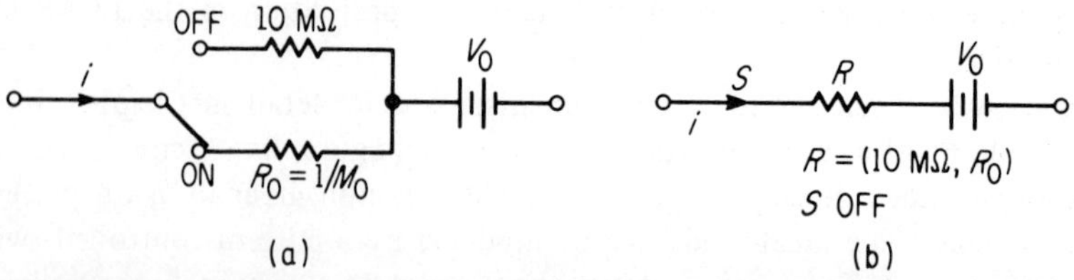

FIG. 7-2. (a) One break-point model of the diode. (b) ECAP equivalent.

corresponds more closely with the format used in ECAP than does the circuit of Fig. 7-2a. The resistance values are indicated in parentheses with the first value corresponding to the state of the switch indicated. Equivalent information could be expressed as R = (R_0, 10 MΩ), S ON in this example. The two break point model is shown in Fig. 7-3 assuming that M_1 = 0.01 ℧ and M_2 = 0.1 ℧ As voltage becomes more positive at node 1 with respect to node 2 by an amount V_1, current through branch 1 becomes positive and the branch pre-

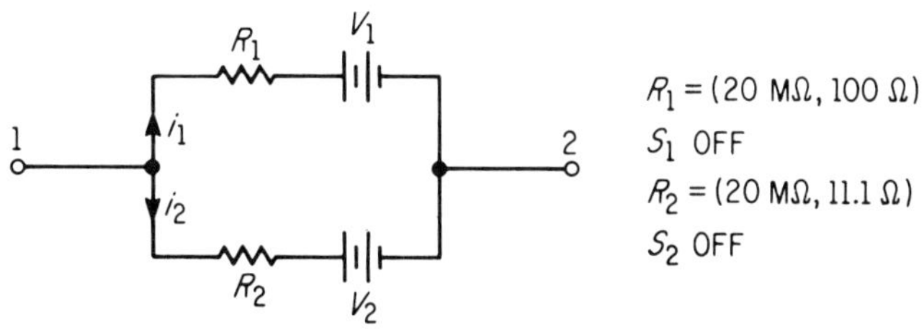

FIG. 7-3. Two break-point model of the diode.

sents a 100-Ω resistance to the current flow. Increase of this voltage to a value greater than V_2 causes current to become positive in branch 2 also. This branch now presents a resistance of 11.1 Ω resulting in an equivalent resistance of 10 Ω between nodes 1 and 2. If the input voltage is less than V_1, both branches contain an impedance of 20 MΩ, giving a 10-MΩ resistance between nodes.

EXAMPLE 7-1. Sketch the voltage across the resistance assuming the diode characteristics are given by Fig. 7-lb with $V_1 = 0.2$ V, $V_2 = 0.3$ V, $M_1 = 0.01$ ℧ and $M_2 = 0.1$ ℧.

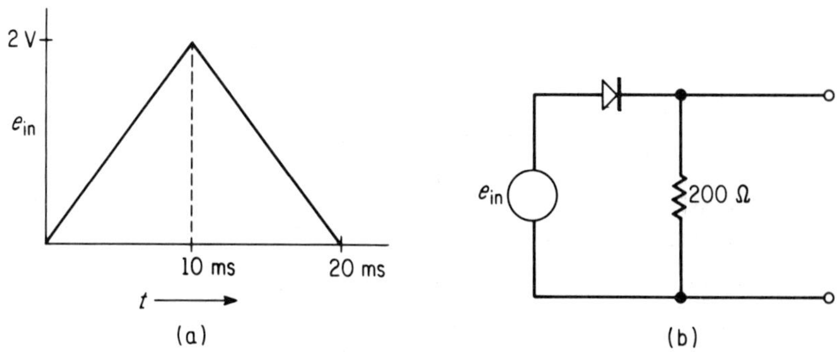

FIG. 7-4. Data for Example 7-1. (a) Input signal. (b) Diode circuit.

Solution. The model of Fig. 7-3 can replace the diode and conducts negligible current until e_{in} exceeds 0.2 V. The current i_1 becomes positive at this point resulting in the equivalent circuit of Fig. 7-5a. As e_{in} increases, the voltage across the output is given by $(e_{in} - 0.2) \times 2/3$. When the voltage across the diode reaches 0.3 V, the equivalent circuit of Fig. 7-5b applies. The second break point occurs at a value of e_{in} satisfying the equation

$$e_{in} - (e_{in} - 0.2) \times 2/3 = 0.3 \text{ V}$$

or

$$e_{in} = 0.5 \text{ V}$$

Above this value, the output voltage varies as

$$e_{out} = 0.952\, e_{in} - 0.276$$

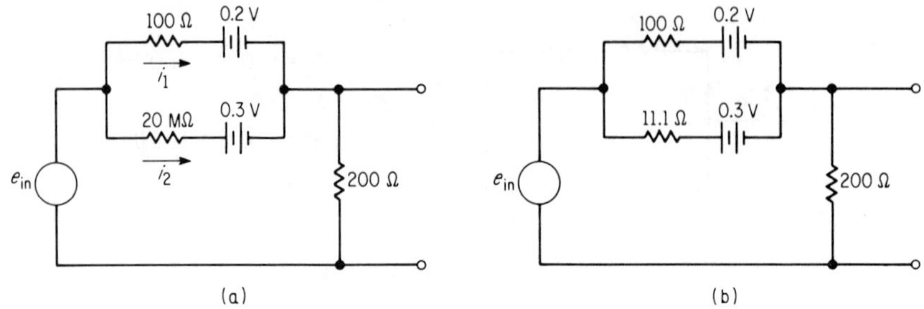

FIG. 7-5. Equivalent circuits. (a) For $0.2\text{ V} < e_{in} < 0.5\text{ V}$. (b) For $e_{in} > 0.5\text{ V}$.

A sketch of the output waveform is shown in Fig. 7-6.

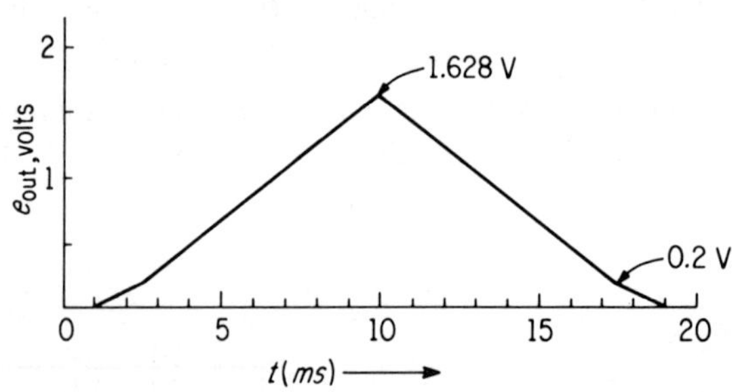

FIG. 7-6. Output for Example 7-1.

More accuracy in modeling the diode can be obtained by using additional branches, but they are usually not required.

Transistor Analysis

Switching circuits are often designed such that some or all of the transistors in a given circuit enter the three possible regions of operation during the switching cycle. We will discuss a model that has been proposed for use with ECAP to account for operation in the three regions, but first models for the separate regions will be considered.

In the cutoff region when base current tends to some very small negative value, the resistances between all terminals are quite large. In a small number of cases, mainly when germanium transistors are used, leakage current must be considered. Because of the wide application of silicon stages in lumped and integrated circuits, we will neglect leakage effects in the succeeding analysis. A suitable cutoff equivalent circuit is shown in Fig. 7-7. The source V_o is not necessary, but is included to simplify the transition from the cutoff to the active region equivalent circuit. This source represents the forward-bias voltage of the base emitter junction, which will be approximately 0.2 V for germanium and 0.6 V for silicon.

The active-region equivalent circuit shown in Fig. 7-8 (from Fig. 6-19 not

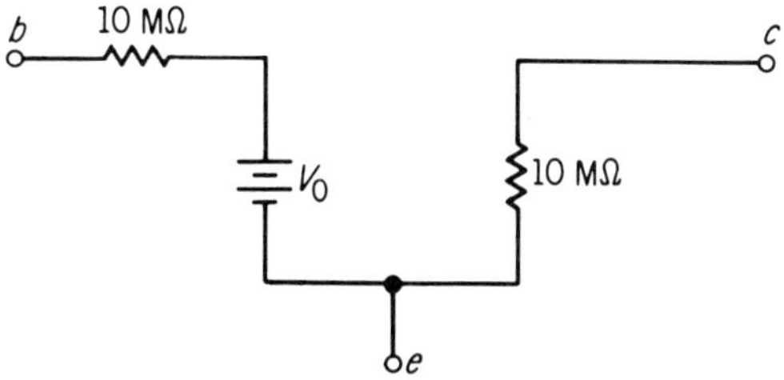

FIG. 7-7. Cutoff equivalent circuit (*n-p-n* transistor).

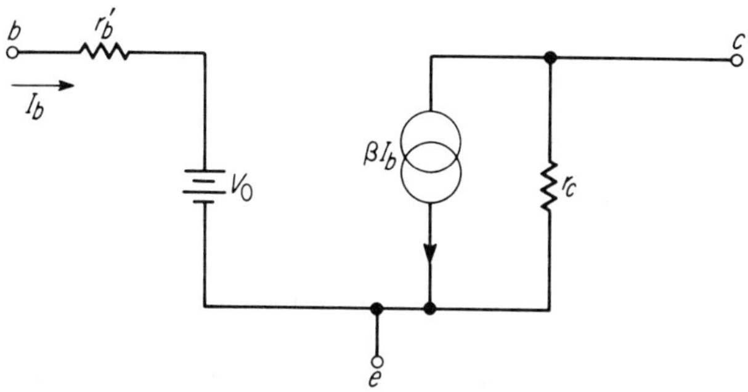

FIG. 7-8. Active region equivalent (*n-p-n* transistor).

including diodes) can be applied to the present discussion. The parameters r_b', r_c, and β must be known for the particular transistor being modeled. These values will vary over the active region and must be suitably linearized. The boundary between the cutoff region and the active region is defined by the base current. If I_b is zero or negative, the transistor is cutoff. The circuit of Fig. 7-9 with a single switch can represent the transistor in both of these re-

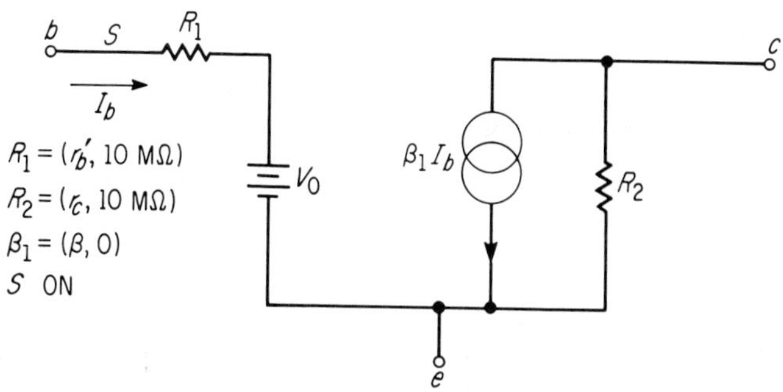

FIG. 7-9. Equivalent circuit valid for cutoff and active regions (*n-p-n* transistor).

gions. When I_b is positive the switch is ON resulting in the appropriate values for R_1, R_2, and β_1. As I_b becomes zero or negative, R_1, and R_2 assume a value of 10 MΩ, and no current gain from base to collector is exhibited.

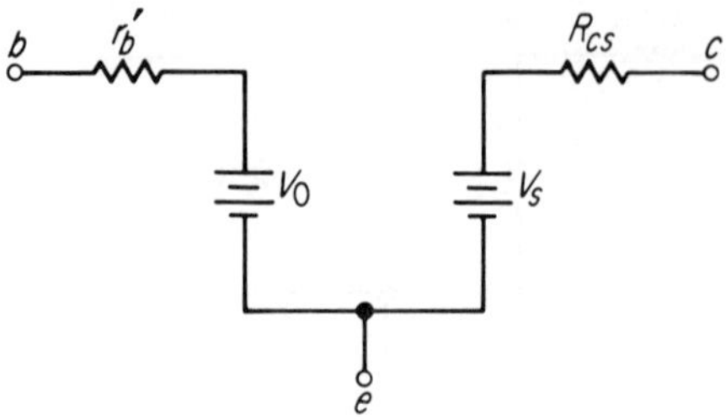

FIG. 7-10. Equivalent circuit for saturated transistor (*n-p-n* transistor).

The saturated transistor can be represented by the circuit shown in Fig. 7-10. It is often possible to neglect V_s, which may typically have a value of 0.1 V, and consider the drop across R_{cs} as the total voltage from collector to emitter. The transition from the active to the saturation region is not distinguished by a change in current direction as is the transition to the cutoff region. Consequently, a current-controlled switch cannot establish the correct parameters as the transistor enters the saturation region if placed in series with either terminal. Thus some method of reversing branch current at the proper point (as the transistor enters the saturation region) must be devised. A switch that controls the saturated parameters can then be placed in this branch and will change element values at the correct point.

At the boundary of the saturation region, both active-region and saturation-region equivalent circuits are valid. From the saturation model of Fig. 7-10 we can express the collector to emitter saturation voltage as

$$V_{cs} = V_s + I_{cs} R_{cs} \tag{7-1}$$

From the active-region model we can write the collector current just as the transistor reaches the saturation boundary as

$$I_{cs} = \beta I_b + V_{cs}/r_c \tag{7-2}$$

Both Eqs. 7-1 and 7-2 hold at the boundary, therefore we can combine these equations and solve for V_{cs} giving

$$V_{cs} = \frac{V_s + \beta I_b R_{cs}}{1 - R_{cs}/r_c} \approx V_s + \beta I_b R_{cs} \tag{7-3}$$

The approximate expression is quite accurate since R_{cs} is much smaller than r_c. Equation 7-3 gives us an expression that is valid only at the edge of the saturation region. This expression forms the basis of an automatic procedure for selecting the correct operating region of the transistor and the appropriate model to be used. When the transistor operates in the active region the collector voltage, calculated from the model of Fig. 7-9, will be greater than the voltage given by Eq. 7-3. As the base current is increased, lowering collector voltage of the active-region model, a point is reached where the collector voltage equals V_{cs}. Thus the saturation region has been reached and the model of Fig. 7-10 must now replace the active-region model. It is easy to show that as the transistor is driven further into the saturated region, the voltage V_{cs} given by Eq. 7-3 will continue to be larger than the actual voltage calculated from the saturation equivalent circuit. The problem reduces to one of monitoring the collector voltage of the circuit and comparing it to the voltage of Eq. 7-3. When the collector voltage is greater than V_{cs} the active-region model is used, but when V_{cs} exceeds the collector voltage the saturation model is used. The circuit of Fig. 7-11 demonstrates how this comparison of voltages and ex-

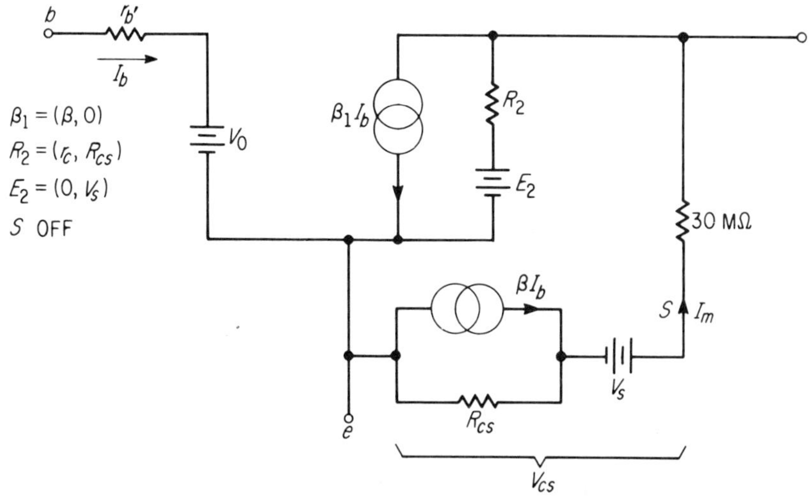

FIG. 7-11. Equivalent circuit valid for active and saturation regions (*n-p-n* transistor).

change of circuit models is automatically accomplished. The branch containing the switch has a large resistance and does not load the collector circuit. The current I_m is negative when the collector to emitter voltage is greater than V_{cs} and becomes positive if $\beta I_b R_{cs} + V_s$ is greater than the collector to emitter voltage. The current I_m then changes sign at the boundary of the saturation region and S controls β_1, R_2, and E_2 to switch between the active values and saturated values. The additional branches containing R_{cs}, the current generator βI_b, V_s, the 30-MΩ resistance, and the switch are present merely to generate a test voltage equal to V_{cs} and to sense the boundary of the saturation region.

EXAMPLE 7-2. A transistor which can be represented by the equivalent circuit of Fig. 7-11 has the following element values $\beta = 100$, $r_c = 20$ kΩ, $R_{cs} = 20$ Ω, $V_s = 0.2$ V, $r_b' = 200$ Ω and $V_0 = 0.6$ V. Sketch the output signal of the circuit in Fig. 7-12 resulting from the input shown.

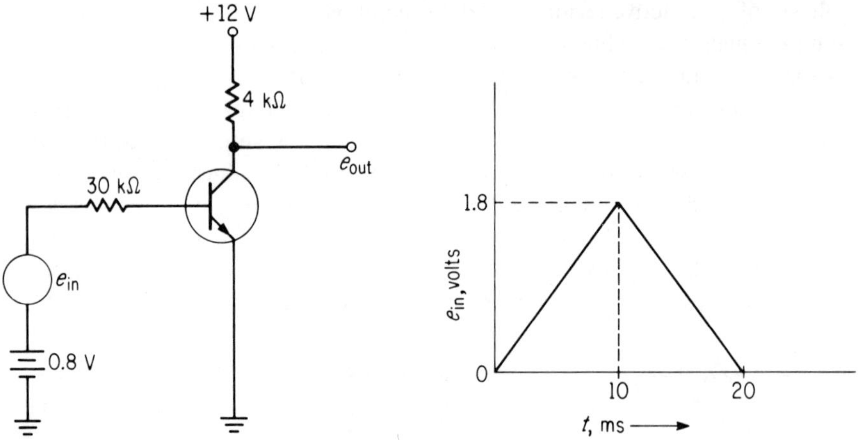

FIG. 7-12. Common-emitter switch.

Solution. When e_{in} is zero the transistor is in the active region. The collector voltage is found to be 7.8 V. The branch of the equivalent circuit which contains the switch senses a negative value of I_m, thus the switch is off, resulting in the active-region circuit displayed in Fig. 7-13.

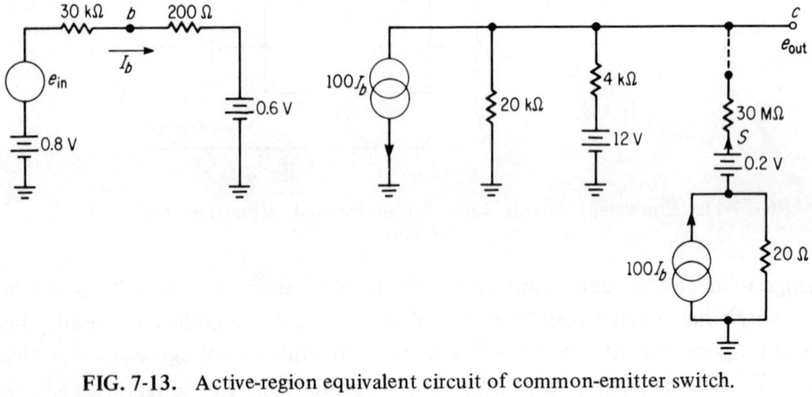

FIG. 7-13. Active-region equivalent circuit of common-emitter switch.

The base current can be written as

$$I_b = \frac{e_{in} + 0.2}{30.2 \times 10^3} \text{ amp}$$

The voltage V_{cs} generated by the sensing circuit is

$$V_{cs} = 0.2 + 100I_b \times 20 = (0.2 + 2 \times 10^3 I_b) \text{ V}$$

Using the active-region model the output voltage is found to be

$$e_{out} = 10 - 333 \times 10^3 I_b$$

The output voltage will be more positive than V_{cs} until I_b reaches a value of 29.3 μa. At this point e_{out} equals V_{cs} and a further increase in I_b causes the current through S to become positive. The switch then controls the appropriate element values to transform the circuit to the saturation equivalent circuit of Fig. 7-14.

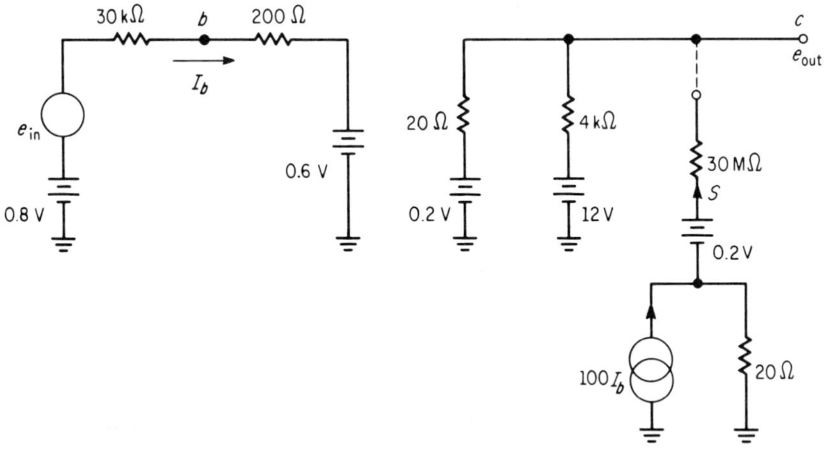

FIG. 7-14. Saturation equivalent circuit for common-emitter switch.

While the transistor is in the active region the output voltage is

$$e_{out} = 10 - 11.03\,(e_{in} + 0.2)$$

Saturation occurs when e_{in} reaches a value of 0.68 V and e_{out} is approximately 0.24 V. This condition is reached at 3.8 ms. The transistor remains saturated until e_{in} decreases to 0.68 V at which point the active region is again entered. This occurs at 16.2 ms. The output voltage is shown in Fig. 7-15.

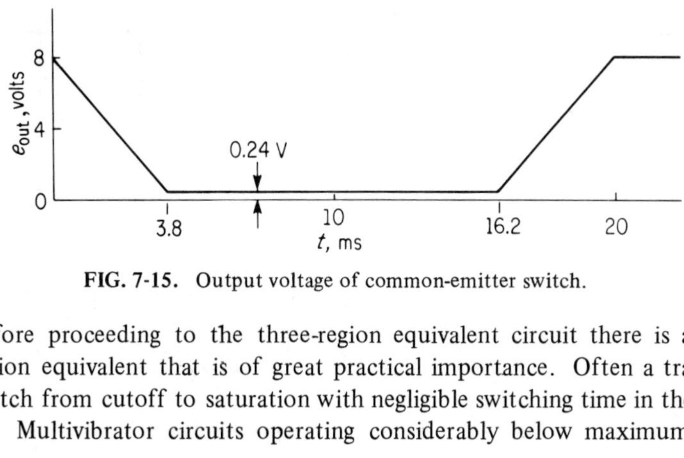

FIG. 7-15. Output voltage of common-emitter switch.

Before proceeding to the three-region equivalent circuit there is another two-region equivalent that is of great practical importance. Often a transistor will switch from cutoff to saturation with negligible switching time in the active region. Multivibrator circuits operating considerably below maximum pulse-repetition frequencies are examples of this type of circuit. It may be assumed that transformation between the cutoff-equivalent circuit and the saturation-equivalent circuit occurs instantaneously upon application of an appropriate input pulse [4]. The condition of negligible switching time is usually met if regeneration is present throughout the active region of a device. The equivalent circuit of Fig. 7-16 will be valid for this case. If I_b is negative the switch is OFF,

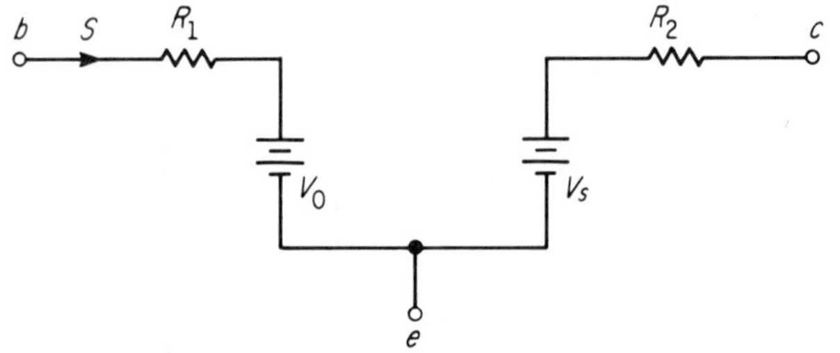

FIG. 7-16. Two-region equivalent circuit for use in cutoff and saturation regions.

resulting in very high resistances from base to emitter and collector to emitter. As I_b becomes positive the transistor parameters switch to the saturated values. This two-region equivalent should be used instead of the three-region model where possible to reduce the number of required elements. The voltage sources V_0 and V_s are not controlled, since the large resistances make the effects of these voltages negligible in the cutoff region.

The three-region equivalent circuit must represent the transistor in any of the three possible operating regions and must automatically switch between equivalent circuits at the appropriate points. Figure 7-17 depicts one possible three-region equivalent circuit.

There are three possible choices in specifying the controlled elements of the circuit. These choices relate to the three possible switch combinations. In cut-

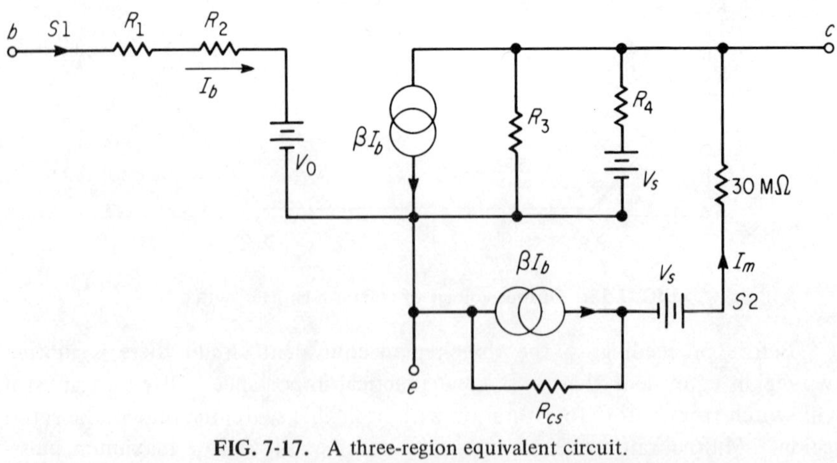

FIG. 7-17. A three-region equivalent circuit.

off both $S1$ and $S2$ are OFF; in the active region $S1$ is ON and $S2$ is OFF; in saturation $S1$ and $S2$ are both ON. In analyzing a transient problem with ECAP, it is general practice to specify the states of the switches corresponding to the actual state of the circuit at $t = 0$. Thus if the stage is initially cut off, the parameters would be specified relative to an initial switch combination of both $S1$ and $S2$ in the OFF condition. Table 7-1 summarizes the values of R_1, R_2, R_3, R_4, and β_1 that must prevail in the various operating regions.

TABLE 7-1. Three-Region Equivalent Circuit Parameters.

Parameter	Active Region	Cutoff Region	Saturation Region
R_1	$0.1\,\Omega$	$10\,\text{M}\Omega$	$0.1\,\Omega$
R_2	r_{b1}	r_{b2}	r_{b2}
R_3	r_c	$30\,\text{M}\Omega$	$30\,\text{M}\Omega$
R_4	$30\,\text{M}\Omega$	$30\,\text{M}\Omega$	R_{cs}
β_1	β	0	0
S_1	ON	OFF	ON
S_2	OFF	OFF	ON

The three combinations of element specifications are as follow:

$R_1 = (0.1, 10^7)$ controlled by S_1
$R_2 = (r_{b1}, r_{b2})$ controlled by $S1$ and $S2$
$R_3 = (r_c, 30\text{M}\Omega)$ controlled by $S1$ and $S2$
$R_4 = (30\text{M}\Omega, R_{cs})$ controlled by $S2$
$\beta_1 = (\beta, 0)$ controlled by $S1$ and $S2$
$S1$ ON
$S2$ OFF

Active region at $t = 0$

$R_1 = (10^7, 0.1)$ controlled by $S1$
$R_2 = (r_{b2}, r_{b1})$ controlled by $S1$ and $S2$
$R_3 = (30\text{M}\Omega, r_c)$ controlled by $S1$ and $S2$
$R_4 = (30\text{M}\Omega, R_{cs})$ controlled by $S2$
$\beta_1 = (0, \beta)$ controlled by $S1$ and $S2$
$S1$ OFF
$S2$ OFF

Cutoff region at $t = 0$

$R_1 = (0.1, 10^7)$ controlled by $S1$
$R_2 = (r_{b2}, r_{b1})$ controlled by $S1$ and $S2$
$R_3 = (30\text{M}\Omega, r_c)$ controlled by $S1$ and $S2$

$R_4 = (R_{cs}, 30\text{M}\Omega)$ controlled by $S2$
$\beta_1 = (0, \beta)$ controlled by $S1$ and $S2$
$S1$ ON
$S2$ ON

Saturation region at $t = 0$

When an element value is controlled by two switches, the first value listed in parentheses is in effect when both switches are in the specified state. If either switch changes state, the second value in parentheses is assumed by the element. When both switches change state, the element value reverts back to that given by the first figure in parentheses. For example, if the transistor stage is cut off at $t = 0$ then the second choice of element values would be used. The resistance R_2 has a value equal to r_{b2} at cutoff. In the active region $S1$ turns ON while $S2$ remains OFF, therefore R_2 takes on a value equal to r_{b1}. When saturation is reached, with $S1$ and $S2$ both ON, R_2 becomes equal to r_{b2}.

Using the equivalent circuit of Fig. 7-17 allows one to insert a value of input resistance for the active region (r_{b1}) that differs from that of the saturation region (r_{b2}). It may not always be necessary to do this in which case one of the input resistors can be omitted. Furthermore, the voltage V_s may often be negligible and the circuit can be further simplified.

Regardless of the initial state of the switches the equivalent circuit will automatically insert the appropriate element values as dictated by the exciting signal. It should be mentioned that when a current source is controlled by a switch, ECAP requires that the "from" branch resistance must also be controlled by the same switch. In this circuit R_2 is controlled by both S_1 and S_2 as also is β_1.

High-Frequency Analysis

The equivalent circuit of Fig. 7-17 is valid only when transistor switching times are unimportant. At high switching rates the various junction capacitances must be considered. While a switch can control inductors and capacitors as well as resistors it is general practice to change only resistive values in a transient solution. Changing capacitor or inductor values presents the additional problem of determining voltages or currents just after the switch transition. Thus it is difficult if not impossible to propose a two- or three-region equivalent circuit including reactive effects that simply switches capacitance values at the boundaries of the regions. Alternative methods can be used as dictated by the problem at hand.

If the circuit is to operate only in the active region, it is often possible to apply the equivalent circuit given in Fig. 6-21. This circuit is based on the assumption that linearized values of all element values are known. This is a poor assumption, since even if small signal element values are known the linearized values are functions of current or voltage swings in the circuit. The output signal is not known until the circuit is analyzed, requiring knowledge of the element values which depend on the output signal. Thus educated guesses are often made to provide information to analyze the circuit. Revised parameter values can then be based on the results of the first analysis to provide a more accurate equivalent circuit.

It is possible to develop a piecewise linear model for the transistor based on the charge-control concept [6]. The resulting model can be utilized for approximate analysis of high-speed switching circuits, but requires an enormous number of elements for each transistor stage.

Nonlinear Capacitance [7]

Rather than using an average value of capacitance to represent a nonlinear capacitor, a piecewise linear model reflecting the approximate nonlinear behavior can be constructed. The total charge of a capacitor is given by

$$q = Cv \tag{7-4}$$

where v is the applied voltage and C is the capacitance. If the capacitor is nonlinear, C is a function of applied voltage. The diode depletion region capacitance is an example of a voltage-variable element. Since the capacitor current is given by $i = dq/dt$, we can write

$$i = d[Cv]/dt = C(dv/dt) + v(dC/dt) \tag{7-5}$$

A plot of total charge as a function of applied voltage is shown in Fig. 7-18 for a capacitor obeying the equation $q = Kv^{1/2}$.

It is possible in ECAP to model nonlinear voltage-current characteristics, but no provisions are made for modeling charge-voltage characteristics directly. A different modeling procedure must be used to produce a circuit to represent this nonlinearity. The basic idea is to use a model which will yield an output voltage that depends on the applied voltage in the same way that the charge on the capacitor depends on applied voltage. For example, the capacitor represented by the characteristic of Fig. 7-18 contains a charge that is expressible as

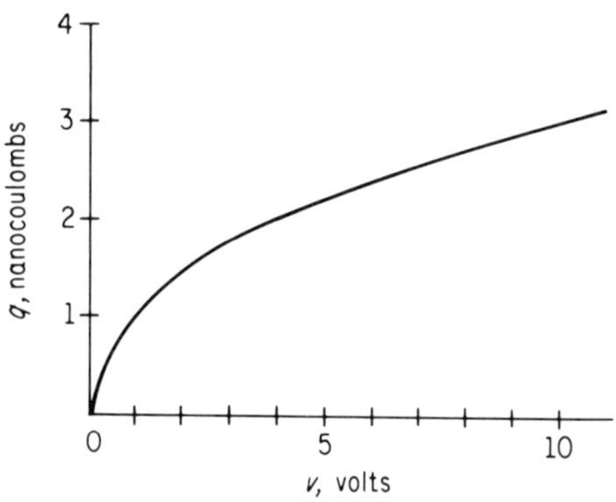

FIG. 7-18. Charge-voltage characteristic of a nonlinear capacitance.

$q = Kv^{1/2}$. A model is constructed so that the output voltage is given by

$$v_1 = K_1 v^{1/2}$$

The voltage v_1 is then applied to a capacitance of constant value C_1 resulting in a charge of

$$q_1 = C_1 v_1 = C_1 K_1 v^{1/2}$$

If C_1 and K_1 are selected such that $C_1 K_1 = K$, then

$$q_1 = q = Kv^{1/2}$$

The charge accumulation on C_1 is exactly the same as that of the nonlinear capacitance, thus the charging currents will be equal. If we can now cause this current to enter the node associated with the applied voltage, the voltage-current characteristic of that node will equal that of the nonlinear capacitor. To demonstrate this procedure a possible model of the capacitor represented by the characteristic of Fig. 7-18 will be developed.

Let us assume that a switch is closed at $t = 0$ on a series resistance and the uncharged nonlinear capacitor as indicated in Fig. 7.19. A voltage must be

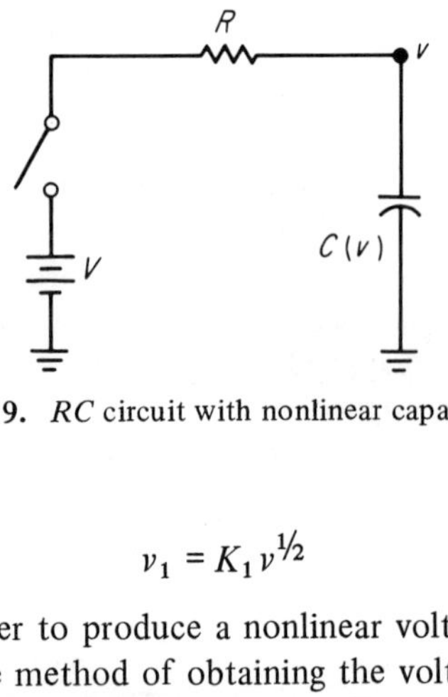

FIG. 7-19. *RC* circuit with nonlinear capacitance.

generated that varies as

$$v_1 = K_1 v^{1/2}$$

Since it is a simple matter to produce a nonlinear voltage-current characteristic with ECAP, a reasonable method of obtaining the voltage v_1 is to first obtain a current that varies as the square root of v then allow this current to develop the voltage v_1 across a resistance. Figure 7-20 shows schematically how this is done.

For the circuit of Fig. 7-20 the voltage v_1 is given by

$$v_1 = \beta Ri = \beta R K_2 v^{1/2}$$

where K_2 is a constant. The voltage v_1 should now be applied to a capacitor C_1, but we must avoid loading the resistor R since the entire current βi must flow through R. Figure 7-21 demonstrates one method to guarantee that the presence of C_1 does not decrease the current through R.

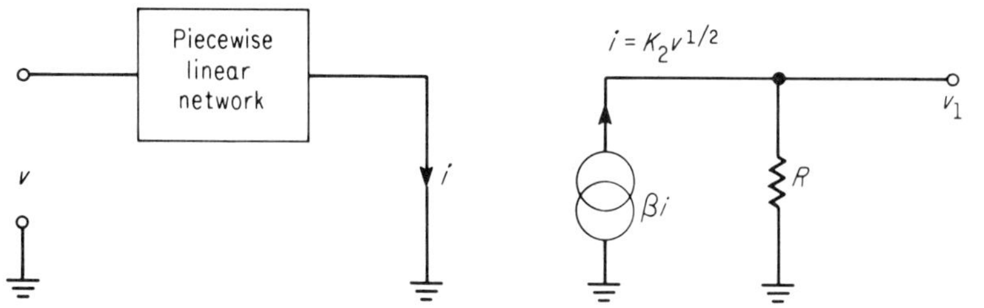

FIG. 7-20. Generation of nonlinear voltage v_1.

Any current drawn from node B by C_1 is replaced by the dependent source i_1. The charge q_1 is given by $C_1 v_1$ which is

$$q_1 = C_1 \beta R K_2 v^{1/2}$$

This charge is made to equal q by the appropriate selection of constants. For a given applied voltage v, the current i_1 equals the current that would flow if v were applied directly to the nonlinear capacitor. However, when v is applied to the nonlinear capacitor the current i_1 flows into the terminal at which voltage is applied. In the circuit of Fig. 7-21 the current i_1 does not flow into the

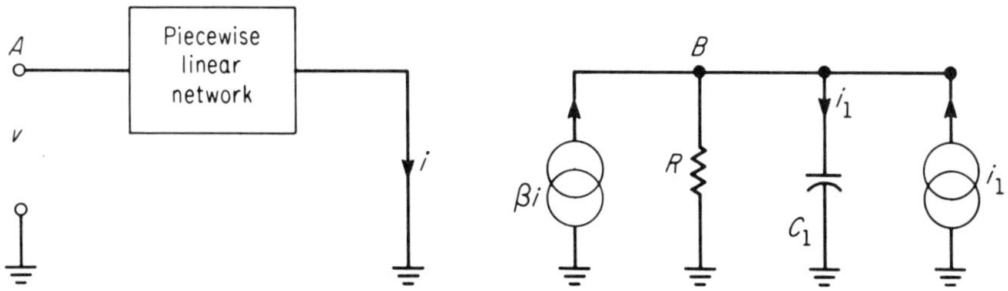

FIG. 7-21. Generation of nonlinear charge q_1.

terminal at which v is applied. We can add two dependent current generators to the input terminal to force i_1 to flow into node A. These additions are shown in Fig. 7-22.

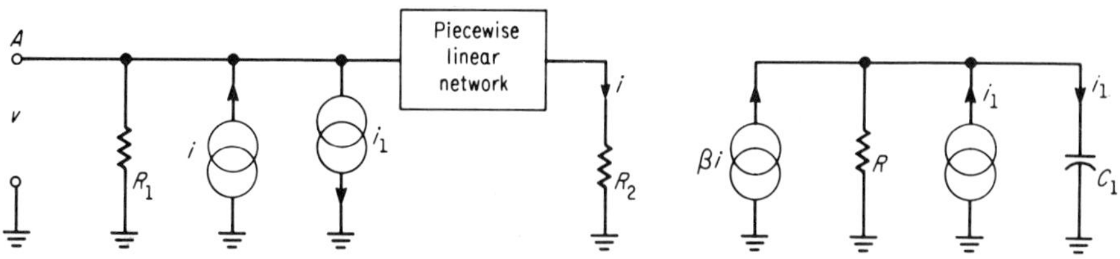

FIG. 7-22. Model of a nonlinear capacitance.

The resistance R_1 is a very large resistance and is added so that the dependent sources connected to node A can be associated with a finite branch resistance. Likewise, the resistor R_2 is added to give a finite resistance (very low resistance in this case) to this branch that controls two dependent sources. This resistance should be much smaller than those values used in constructing the piecewise linear network.

In summary, to create a terminal pair having the same voltage-current characteristic as a nonlinear capacitance we must

1. Evaluate the voltage-charge characteristic (q vs. v) of the capacitor.
2. Generate a voltage-current characteristic (i vs. v) that equals the voltage-charge characteristic (except for a multiplicative constant).
3. Generate an intermediate voltage v_1 that is always proportional to the current i.
4. Use the intermediate voltage to charge a linear capacitance which now has a current i_1 that varies with v in the same manner as the nonlinear capacitor current.
5. Force the input terminal current to equal i_1.

An example [6] will demonstrate the procedure necessary to develop the nonlinear-capacitance model.*

EXAMPLE 7-3. A nonlinear capacitance has a voltage-charge relationship given by

$$q = ve^{-0.1\,v} \times 10^{-9} \text{ f}$$

If this capacitance has an initial voltage of 10V and is shunted by a 1 kΩ resistance at $t = 0$, plot the discharge curve.

Solution. The piecewise linear network is found by plotting the voltage-charge characteristic and constructing a piecewise linear network giving a similar voltage-current characteristic. A possible network for this purpose is shown in Fig. 7-23.

In dealing with negative conductances or resistances we must remember that the resulting direction of current flow for a given polarity of applied voltage is opposite to that of a positive element. Figure 7-24 shows the complete model for this problem. The piecewise linear network approximates the characteristic

$$i = ve^{-0.1\,v} \text{ amp}$$

The charge on C_1 is then

$$q_1 = ve^{-0.1\,v} \times 10^{-9}$$

The relationship of charge on C_1 to applied voltage v is the same as the desired nonlinear relationship. The current i_1 will equal the current that would be drawn by an actual nonlinear capacitance. An initial charge of 10V results in a charge $q = 3.679 \times 10^{-9}$ coulomb. Thus, specifying q_1 to initially equal this quantity sets the initial conditions. To set $q_1 = 3.679 \times 10^{-9}$ coulomb requires an initial

*The ECAP program for this example is given in Table 8-13.

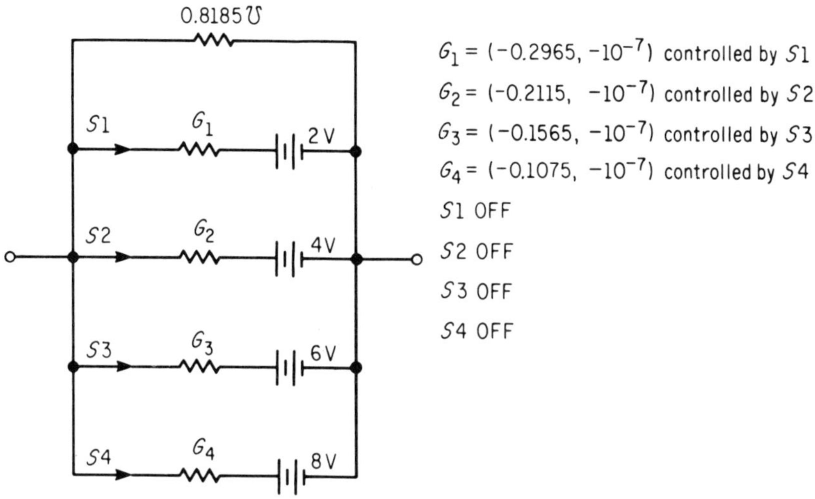

FIG. 7-23. Piecewise linear network.

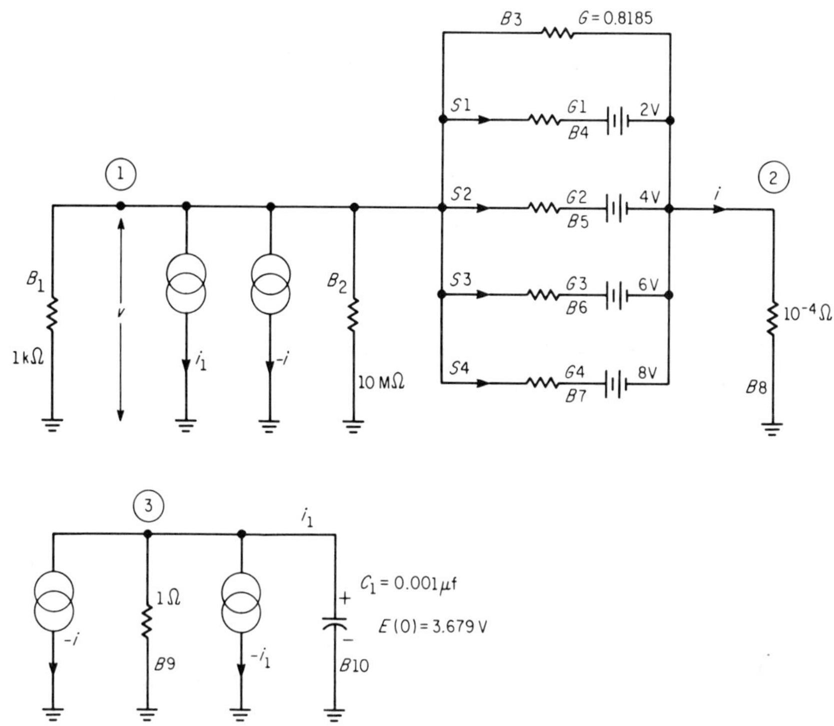

FIG. 7-24. Model for Example 7-3.

voltage on C_1 of 3.679 V. The results of execution of ECAP using this model are plotted in Fig. 7-25.

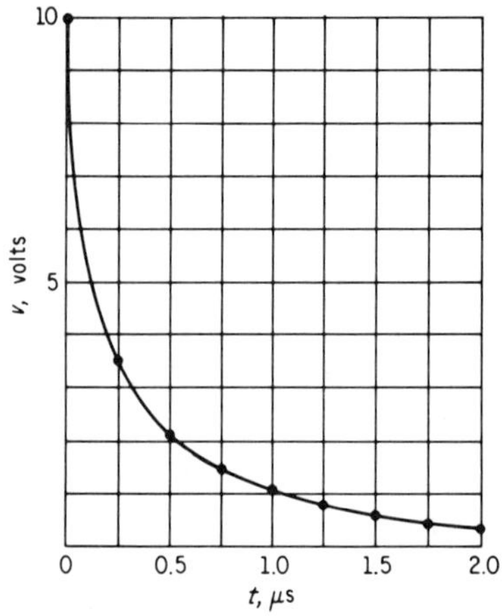

FIG. 7-25. Discharge curve of nonlinear capacitance.

Nonlinear resistances can be approximated as well as nonlinear capacitances by piecewise linear techniques. It follows that a general transistor model can theoretically be proposed based on these approximations to the nonlinear elements. A model such as this is not very practical however, due to the limitations on the total number of elements that can be handled by the computer. A transistor exhibits at least three nonlinear capacitances and two diodes. Just to model a single stage might require 100 elements, with multistage circuits requiring an inordinate number of elements. For switching circuits sufficient accuracy is generally obtained by using average values of these nonlinear transistor parameters as done in the previous section.

B. ITERATIVE TECHNIQUES FOR NONLINEAR PROBLEMS

Most electronic devices such as vacuum tubes or semiconductors possess nonlinear characteristics. In manually analyzing the large-signal behavior of a circuit involving such a device graphical techniques are used almost exclusively. The general technique used is that of load-line construction. Let us briefly review the theory of the load line.

Figure 7-26 shows a circuit consisting of a nonlinear device in series with a resistance and voltage source. If we are to calculate the voltage across the device and the current flow we must consider the factors that determine these values. The device does not entirely determine the current flow, nor does the external circuit consisting of R and E_{cc}. Rather, both the device and external circuit influence the current flow. The load-line approach separates the overall circuit characteristics into two sets of characteristics. One V-I characteristic, that of

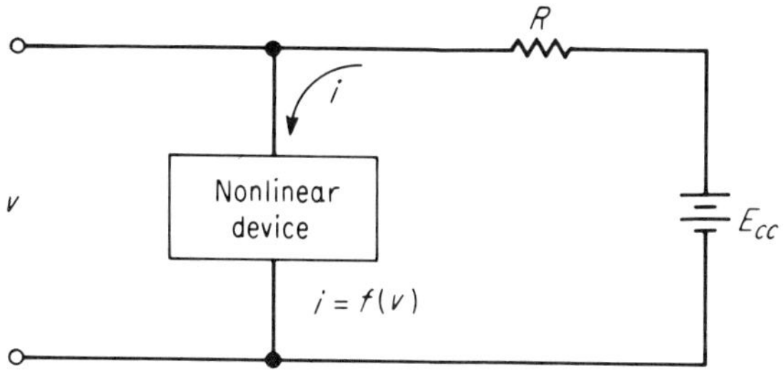

FIG. 7-26. Nonlinear circuit.

the external circuit, is completely linear and is called the *load line*. The other set of characteristics relates to the device behavior. We note that the current through the device equals that of the external circuit, as does also the device voltage.

The load line is a graphical expression of external circuit voltage in terms of current, while the device characteristics express device voltage in terms of current. The intersection of the load line with the appropriate device characteristic determines the simultaneous solution to the problem. It is only at this point that the device variables equal the external circuit variables. Figure 7-27 demon-

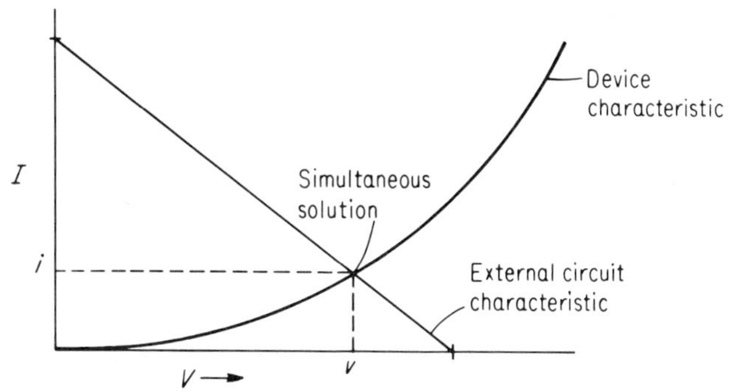

FIG. 7-27. Device characteristics with load line.

strates this graphical approach. The basic idea, which is extendible to iterative computer techniques, is to separate a nonlinear circuit into a portion described by linear equations and a portion described by nonlinear equations. For each terminal of separation we can equate the voltage or current expressible as a linear relationship to the voltage or current expressible as a nonlinear relationship. Unfortunately, the computer does not have the facility to examine graphical data and immediately locate points of intersection, thus iterative methods must be applied. This requires that an analytic expression for the nonlinear device be made available to the computer. An initial guess at the solution is made by the designer, and succeeding estimates by the computer are guided by some mathematical process which will hopefully lead the computer to the correct solution in minimal time. There are several numeric procedures available to guide iterative processes, but only the Newton-Raphson method will be discussed here.

The Newton-Raphson method considers the slopes of the characteristics at each point of the iterative process to guide the selection of the succeeding values of variables. To demonstrate this procedure, let us assume that the nonlinear device of Fig. 7-26 has a current given by $i_{nl} = v_{nl}^3$ ma. If the source is 10 V and $R = 1\,\text{k}\Omega$, then the external circuit current is $i_l = (10 - v)$ma. These characteristics are shown in Fig. 7-28. If an initial value of $v^0 = 1$ V is used (the super-

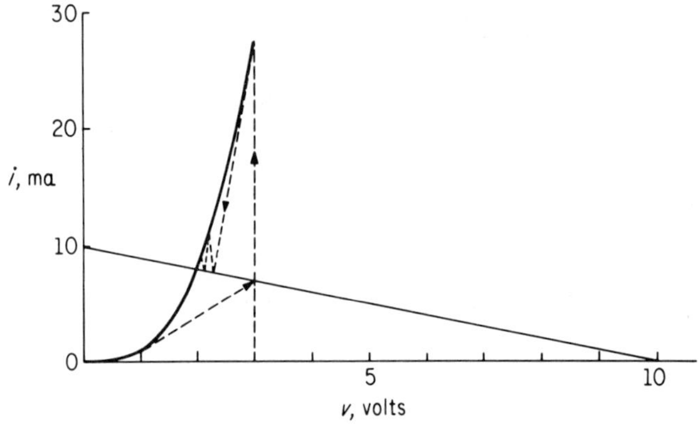

FIG. 7-28. Newton-Raphson iteration.

script corresponds to the number of iteration) the total difference in current or current error is

$$\Delta i^0 = i_l^0 - i_{nl}^0 = 8\,\text{ma}$$

Since the error is nonzero, a new estimate of the voltage must be made. A piecewise linear approximation to the nonlinear characteristic will allow the selection of a second voltage estimate. The slopes of both curves are evaluated at v^0. If the slope of the nonlinear curve remained constant with v, then the intersection of the corresponding piecewise linear approximation to the curve and the load line would be the final solution. Since this slope does not remain constant, the intersection serves only as a new estimate of the solution and is designated as v^1. The value of v^1 can be found from a consideration of the current error Δi^0. The change in Δi with respect to v evaluated at v^0 is

$$\frac{d\Delta i^0}{dv} = \left.\frac{di_l}{dv}\right|_{v^0} - \left.\frac{di_{nl}}{dv}\right|_{v^0}$$

$$= -1 - 3(v^0)^2 = -4\,\text{ma/V}$$

Assuming the slope di_{nl}/dv remains constant as v changes, the current error would be reduced from 8 ma to zero when

$$-\Delta i^0 = \left.\frac{d\Delta i}{dv}\right|_{v^0} \Delta v^0$$

Thus Δv^0 can be found as

$$\Delta v^0 = -\Delta i^0 \Big/ \left.\frac{d\Delta i}{dv}\right|_{v^0}$$

$$= -8/-4 = 2\,\text{V}$$

The initial estimate is $v^0 = 1\,\text{V}$, hence a second estimate is $v^1 = v^0 + \Delta v^0 = 3\,\text{V}$. This point corresponds to the intersection of the load line with the piecewise linear approximation to the curve at v^0. The current error at v^1 is now found to be

$$\Delta i^1 = i_l^1 - i_{nl}^1 = -20\,\text{ma}$$

The change in Δi^1 with respect to v at v^1 is

$$\frac{d\Delta i^1}{dv} = \left.\frac{di_l}{dv}\right|_{v_1} - \left.\frac{di_{nl}}{dv}\right|_{v_1} = -28\,\text{ma/V}$$

A new change in voltage can be determined to reduce the current error to zero, assuming a constant change of $-28\,\text{ma/V}$ as v varies. This value is

$$\Delta v^1 = -\Delta i^1 \Big/ \left.\frac{d\Delta i}{dv}\right|_{v_1}$$

$$= -5/7$$

The next voltage estimate is then $v^2 = v^1 + \Delta v^1 = 2.29\,\text{V}$. This process can be continued until the solution is arbitrarily close to the actual values of $v = 2\,\text{V}$ and $i = 8\,\text{ma}$.

The Newton-Raphson method can be easily formulated for use with the digital computer. A program for the solution of the preceding example is given in Table 7-2. This program is written in FORTRAN IV and was executed on the IBM 1132 computer.

TABLE 7-2. Newton-Raphson Iterative Program.

```
C     NEWTON-RAPHSON ITERATION PROGRAM
C     V IS THE INITIAL VOLTAGE ESTIMATE
      V = 0.1
C     VS IS THE SUPPLY VOLTAGE
      VS = 10.0
C     R IS THE LOAD RESISTANCE
      R = 1.0E3
```

TABLE 7-2 *(continued)*

```
C     K IS THE MAXIMUM NUMBER OF ITERATIONS ALLOWED
      K = 500
      REAL IR,IN
      WRITE (6,5)
    5 FORMAT (1H1, '     V        IR         IN',//)
      DO 11 I= 1,K
      IR = (VS-V)/R
      IN = 0.001*V**3.0
      ER = IR - IN
      WRITE (6,1) V,IR,IN
    1 FORMAT (1P3E11.3)
      IF (ABS(ER)-1.0E-6)10,2,2
    2 V1 =-ER/(-1.0/R-0.003*V*V)
      V = V + V1
   11 CONTINUE
      GO TO 13
   10 WRITE (6,12)
   12 FORMAT ('     SOLUTION HAS BEEN OBTAINED TO DESIRED
     1 TOLERANCE')
   13 CALL EXIT
      END
```

Table 7-3 shows the results of execution of the program.

TABLE 7-3. Solution to Nonlinear Problem.

V	IR	IN
1.000E-01	9.900E-03	1.000E-06
9.711E-00	2.893E-04	9.157E-01
6.486E-00	3.514E-03	2.729E-01
4.369E-00	5.631E-03	8.338E-02
3.034E-00	6.966E-03	2.793E-02
2.302E-00	7.698E-03	1.219E-02
2.036E-00	7.964E-03	8.434E-03
2.001E-00	7.999E-03	8.007E-03
2.000E-00	8.000E-03	8.000E-03
SOLUTION HAS BEEN OBTAINED TO DESIRED TOLERANCE		

There are certain disadvantages to the iterative method outlined in the previous paragraphs. There is no guarantee that convergence will take place. For example, the diode equation is $I = I_s(e^{40\,V} - 1)$ near room temperature. If this diode is the nonlinear device of Fig. 7-26 and the initial guess v^0 is zero volts, the next estimate of voltage will be approximately $v^1 = E_{cc}$. For E_{cc} = 10V, the diode current becomes $I = I_s(e^{400} - 1)$ a number too large to be handled by any digital computer. Furthermore, convergence is often slow when a voltage estimate larger than the correct solution results during the process. The exponential current dependence on voltage exhibited by the diode causes large current errors in this case. Both problems can be alleviated somewhat by modifying the Newton-Raphson procedure [2].

The modified Newton-Raphson procedure is especially effective for exponential nonlinearities as often encountered in semiconductor circuits. In this case a value of Δv is calculated as before. If Δv is negative the procedure continues with no modification. When Δv is positive a new current rather than a voltage estimate is determined by the intersection of the piecewise linear approximation and the load line. The corresponding voltage calculation is based on this value of current and the nonlinear equation. Table 7-4 shows a program applying the modified Newton-Raphson method to the solution of the circuit of Fig. 7-26 with the nonlinear element represented by the exponential equation $I = 2 \times 10^{-6} (e^{40V} - 1)$. Table 7-5 lists the results obtained by both the Newton-Raphson procedure and the modified procedure. These programs were executed on the IBM 1132 computer.

TABLE 7-4. Modified Newton-Raphson Procedure.

```
C      MODIFIED NEWTON-RAPHSON ITERATION PROGRAM
C      V IS THE INITIAL VOLTAGE ESTIMATE
       V = 0.1
C      VS IS THE SUPPLY VOLTAGE
       VS = 10.0
C      R IS THE VALUE OF RESISTANCE
       R = 1.0E3
C      K IS THE MAXIMUM NUMBER OF ITERATIONS ALLOWED
       K = 500
       REAL IR,IN
       WRITE (6,5)
     5 FORMAT (1H1,'     V      IR      IN',//)
       DO 11 I= 1,K
       IR = (VS-V)/R
       IN = (2.0E-6)*(EXP(40.0*V)-1.0)
       ER = IR - IN
       WRITE (6,1) V,IR,IN
     1 FORMAT (3E11.3)
       IF  (ABS(ER)-1.0E-6)  10,2,2
     2 V1 = -ER/ (-1.0/R- (80.0E-6)*(EXP(40.0*V)))
       V = V + V1
       IF  (V1)  11,11,3
     3 IR = (VS-V)/R
       V = (1.0/40.0)*ALOG (1.0+IR/2.0E-6)
    11 CONTINUE
       GO TO 13
    10 WRITE  (6,12)
    12 FORMAT ('    SOLUTION HAS BEEN OBTAINED TO DESIRED
      1 TOLERANCE')
    13 CALL EXIT
       END
```

The modified procedure requires only 3 iterations, whereas the simple

TABLE 7-5. Solution to Nonlinear Program.

Modified Newton-Raphson Results

V	IR	IN
0.100E 00	0.990E-02	0.107E-03
0.207E 00	0.979E-02	0.807E-02
0.212E 00	0.978E-02	0.978E-02

SOLUTION HAS BEEN OBTAINED TO DESIRED TOLERANCE

Newton-Raphson Results

V	IR	IN
0.100E 00	0.990E-02	0.107E-03
0.192E 01	0.807E-02	0.537E 28
0.189E 01	0.810E-02	0.197E 28
0.187E 01	0.812E-02	0.727E 27
0.184E 01	0.815E-02	0.267E 27
0.182E 01	0.817E-02	0.984E 26
0.179E 01	0.820E-02	0.362E 26
0.177E 01	0.822E-02	0.133E 26
0.174E 01	0.825E-02	0.490E 25
0.172E 01	0.827E-02	0.180E 25
0.169E 01	0.830E-02	0.663E 24
0.167E 01	0.832E-02	0.243E 24
0.164E 01	0.835E-02	0.897E 23
0.162E 01	0.837E-02	0.330E 23
0.159E 01	0.840E-02	0.121E 23
0.157E 01	0.842E-02	0.446E 22
0.154E 01	0.845E-02	0.164E 22
0.152E 01	0.847E-02	0.604E 21
0.149E 01	0.850E-02	0.222E 21
0.147E 01	0.852E-02	0.818E 20
0.144E 01	0.855E-02	0.301E 20
0.142E 01	0.857E-02	0.110E 20
0.139E 01	0.860E-02	0.407E 19
0.137E 01	0.862E-02	0.149E 19
0.134E 01	0.865E-02	0.551E 18
0.132E 01	0.867E-02	0.202E 18
0.129E 01	0.870E-02	0.746E 17
0.127E 01	0.872E-02	0.274E 17
0.124E 01	0.875E-02	0.100E 17
0.122E 01	0.877E-02	0.371E 16
0.119E 01	0.880E-02	0.136E 16
0.117E 01	0.882E-02	0.502E 15
0.114E 01	0.885E-02	0.184E 15
0.112E 01	0.887E-02	0.680E 14
0.109E 01	0.890E-02	0.250E 14
0.107E 01	0.892E-01	0.920E 13
0.104E 01	0.895E-02	0.338E 13
0.102E 01	0.897E-02	0.124E 13
0.999E 00	0.900E-02	0.458E 12
0.974E 00	0.902E-02	0.168E 12

TABLE 7-5 (*continued*)

Newton-Raphson Results		
V	IR	IN
0.949E 00	0.905E-02	0.620E 11
0.924E 00	0.907E-02	0.228E 11
0.899E 00	0.910E-02	0.839E 10
0.874E 00	0.912E-02	0.308E 10
0.849E 00	0.915E-02	0.113E 10
0.824E 00	0.917E-02	0.418E 09
0.799E 00	0.920E-02	0.153E 09
0.774E 00	0.922E-02	0.565E 08
0.749E 00	0.925E-02	0.208E 08
0.724E 00	0.927E-02	0.765E 07
0.699E 00	0.930E-02	0.281E 07
0.674E 00	0.932E-02	0.103E 07
0.649E 00	0.935E-02	0.381E 06
0.624E 00	0.937E-02	0.140E 06
0.599E 00	0.940E-02	0.515E 05
0.574E 00	0.942E-02	0.189E 05
0.549E 00	0.945E-02	0.698E 04
0.524E 00	0.947E-02	0.256E 04
0.499E 00	0.950E-02	0.944E 03
0.474E 00	0.952E-02	0.347E 03
0.449E 00	0.955E-02	0.127E 03
0.424E 00	0.957E-02	0.470E 02
0.399E 00	0.960E-02	0.173E 02
0.374E 00	0.962E-02	0.637E 01
0.349E 00	0.965E-02	0.234E 01
0.324E 00	0.967E-02	0.867E 00
0.299E 00	0.970E-02	0.322E 00
0.275E 00	0.972E-02	0.122E 00
0.252E 00	0.974E-02	0.487E-01
0.232E 00	0.976E-02	0.219E-01
0.218E 00	0.978E-02	0.125E-01
0.213E 00	0.978E-02	0.100E-01
0.212E 00	0.978E-02	0.979E-02
0.212E 00	0.978E-02	0.978E-02
SOLUTION HAS BEEN OBTAINED TO DESIRED TOLERANCE		

procedure requires 74. Obviously the modified procedure should be used where possible to reduce computer costs.

Programs that solve nonlinear equations such as SCEPTRE generally apply some modified form of Newton-Raphson procedure. Of course, a given network may contain several nonlinear elements, resulting in a more complex procedure. Although the theory is more involved, the number of iterations required does not drastically increase.

A general procedure for analysis of a network containing a number of nonlinearities has been developed [2]. To analyze the circuit of Fig. 7-26 the nonlinear element terminal voltage or current is equated to the voltage or current of the constant parameter portion of the network. Thus, the network is separated into a nonlinear section and a section describable by linear equations. A circuit with many nonlinearities can be treated in a similar manner. All nonlinear elements are withdrawn from the circuit. The *V-I* characteristics at all terminals from which nonlinear elements have been removed are described in terms of linear equations and in terms of the nonlinear device equations. The corresponding voltages or currents are equated at all terminals resulting in a matrix equation. This equation contains a nonlinear matrix (variable elements) and a constant element matrix. Iterative techniques are then applied to obtain a solution.

In equating terminal variables as expressed by linear and nonlinear equations we have a choice of variables to be used—that is, we can equate voltages or currents. One major consideration results from the fact that more accuracy in solving nonlinear equations often can be obtained by expressing the equation in the appropriate form. Usually that variable whose value varies most rapidly will be chosen as the terminal quantity. For example, the semiconductor diode equation

$$I = I_s[e^{qV/kT} - 1] \tag{7-6}$$

can be expressed as

$$V = \frac{kT}{q} \ln\left(\frac{I}{I_s} + 1\right) \tag{7-7}$$

The current changes very rapidly with voltage, whereas the opposite situation is not true. Thus the terminal variable to be equated to the linear expression for the corresponding variable should be current for the semiconductor diode. These are other cases, possibly in the same network, that will require use of the terminal voltage as the variable of interest. In formulating a general procedure provision must be made to allow expression of the equations in terms of current for some terminals and voltage for others.

Analysis for Nonlinear Resistive Networks*

Let us assume that we have a network with $m + n$ devices exhibiting nonlinear *V-I* characteristics. We will assume that, due to computational considerations, the first m terminal variables are chosen as currents and the next

*This section is based on the approach outlined by D. A. Calahan in *Computer-Aided Network Design*, New York: McGraw-Hill Book Company, 1968.

n variables are chosen as voltages. A vector made up of these terminal variables can be expressed as

$$\mathbf{n} = \begin{bmatrix} i_1 \\ i_2 \\ \cdot \\ \cdot \\ \cdot \\ i_m \\ v_{m+1} \\ v_{m+2} \\ \cdot \\ \cdot \\ \cdot \\ v_{m+n} \end{bmatrix}$$

This vector depends both on the nonlinear device characteristics and also on the characteristics of that section of the network containing only constant parameter elements. In terms of the nonlinear characteristics, **n** is given by

$$\mathbf{n} = - \begin{bmatrix} f_1(v_1) \\ \cdot \\ \cdot \\ \cdot \\ f_m(v_m) \\ f_{m+1}(i_{m+1}) \\ \cdot \\ \cdot \\ \cdot \\ f_{m+n}(i_{m+n}) \end{bmatrix} \tag{7-8}$$

The negative sign accounts for the fact that all currents are to be positive when they flow into the terminals of the linear network (out of the device terminals). This vector can also be expressed in terms of the linear part of the network as

$$\mathbf{n} = \begin{bmatrix} \mathbf{y} & \boldsymbol{\alpha} \\ \mathbf{t} & \mathbf{z} \end{bmatrix} \begin{bmatrix} v_1 \\ \cdot \\ \cdot \\ \cdot \\ v_m \\ i_{m+1} \\ \cdot \\ \cdot \\ \cdot \\ i_{m+n} \end{bmatrix} + \mathbf{Bu} \tag{7-9}$$

where $\mathbf{u}$ is the vector of independent sources, $\mathbf{y}$ is an $m \times m$ admittance matrix, α is an $m \times n$ current transfer matrix, $\mathbf{t}$ is an $n \times m$ voltage-transfer matrix, $\mathbf{z}$ is an $n \times n$ impedance matrix, and $\mathbf{B}$ is an $(m + n) \times s$ matrix where s is the number of independent sources. We will later consider methods of obtaining these matrices. The correct solution for $\mathbf{n}$ will occur when the difference between Eqs. 7-8 and 7-9 is zero. We can form this difference, which will be

$$\Delta\mathbf{n} = \begin{bmatrix} \mathbf{y} & \boldsymbol{\alpha} \\ \mathbf{t} & \mathbf{z} \end{bmatrix} \begin{bmatrix} v_1 \\ \cdot \\ \cdot \\ \cdot \\ v_m \\ i_{m+1} \\ \cdot \\ \cdot \\ \cdot \\ i_{m+n} \end{bmatrix} + \mathbf{Bu} + \begin{bmatrix} f_1(v_1) \\ \cdot \\ \cdot \\ \cdot \\ f_m(v_m) \\ f_{m+1}(i_{m+1}) \\ \cdot \\ \cdot \\ \cdot \\ f_{m+n}(i_{m+n}) \end{bmatrix}$$

$$= \begin{bmatrix} \mathbf{y} & \boldsymbol{\alpha} \\ \mathbf{t} & \mathbf{z} \end{bmatrix} \begin{bmatrix} \mathbf{v}_i \\ \mathbf{i}_j \end{bmatrix} + \mathbf{Bu} + \begin{bmatrix} \mathbf{f}(v_i) \\ \mathbf{f}(i_j) \end{bmatrix} \tag{7-10}$$

where $i = 1, 2, \ldots, m$ and $j = m + 1, m + 2, \ldots m + n$. If an initial estimate of $\mathbf{v}_i$ and $\mathbf{i}_j$ result in a zero value of error vector $\Delta\mathbf{n}$, these values of terminal variables are the correct values. In general, the initial estimate will result in a nonzero value of $\Delta\mathbf{n}$ which we can express as

$$\Delta\mathbf{n}^0 = \begin{bmatrix} \Delta\mathbf{i}_i^0 \\ \Delta\mathbf{v}_j^0 \end{bmatrix}$$

The error vector $\Delta\mathbf{n}$ is a function of several variables, thus the total change in this vector must be calculated using partial derivatives. Assuming that the derivative of $\Delta\mathbf{n}$ remains constant at the value found at $v_1^0, v_2^0, \ldots, v_m^0, i_{m+1}^0, \ldots, i_{m+n}^0$, then the change required in $\begin{bmatrix} \mathbf{v}_i \\ \mathbf{i}_j \end{bmatrix}$ to reduce the error to zero can be found by equating the required change $(-\Delta\mathbf{n}^0)$ to the expression for change in $\Delta\mathbf{n}$. This gives

$$-\Delta\mathbf{n}^0 = \begin{bmatrix} \left.\dfrac{\partial\Delta i_1}{\partial v_1}\right|_{v_1^0} \Delta v_1^0 + \left.\dfrac{\partial\Delta i_1}{\partial v_1}\right|_{v_2^0} \Delta v_2^0 + \cdots + \left.\dfrac{\partial\Delta i_1}{\partial i_{m+n}}\right|_{i_{m+n}} \Delta i_{m+n}^0 \\ \left.\dfrac{\partial\Delta i_2}{\partial v_1}\right|_{v_1^0} \Delta v_1^0 + \left.\dfrac{\partial\Delta i_2}{\partial v_2}\right|_{v_2^0} \Delta v_2^0 + \cdots + \left.\dfrac{\partial\Delta i_2}{\partial i_{m+n}}\right|_{i_{m+n}^0} \Delta i_{m+n}^0 \\ \vdots \\ \left.\dfrac{\partial\Delta v_{m+n}}{\partial v_1}\right|_{v_1^0} \Delta v_1^0 + \left.\dfrac{\partial\Delta v_{m+n}}{\partial v_2}\right|_{v_2^0} \Delta v_2^0 + \cdots + \left.\dfrac{\partial\Delta v_{m+n}}{\partial i_{m+n}}\right|_{i_{m+n}^0} \Delta i_{m+n} \end{bmatrix}$$

This change can also be written as

$$-\Delta\mathbf{n}^0 = \mathbf{J}(\Delta\mathbf{n}^0) \begin{bmatrix} \Delta\mathbf{v}_i^0 \\ \Delta\mathbf{i}_j^0 \end{bmatrix} \tag{7-11}$$

where $\mathbf{J}(\Delta\mathbf{n}^0)$ is the Jacobian of $\Delta\mathbf{n}$ evaluated at $v_1^0, v_2^0, \ldots, i_{m+n}^0$. The Jacobian is defined by

$$\mathbf{J}(\Delta n) = \begin{bmatrix} \dfrac{\partial\Delta i_1}{\partial v_1} & \dfrac{\partial\Delta i_1}{\partial v_2} & \cdots & \dfrac{\partial\Delta i_1}{\partial i_{m+n}} \\ \vdots & \vdots & & \vdots \\ \dfrac{\partial\Delta v_{m+n}}{\partial v_1} & \dfrac{\partial\Delta v_{m+n}}{\partial v_2} & \cdots & \dfrac{\partial\Delta v_{m+n}}{\partial i_{m+n}} \end{bmatrix} \tag{7-12}$$

We can now solve for $\begin{bmatrix} \Delta\mathbf{v}_i^0 \\ \Delta\mathbf{i}_j^0 \end{bmatrix}$ from Eq. 7-11, giving

$$\begin{bmatrix} \Delta \mathbf{v}_i^0 \\ \Delta \mathbf{i}_j^0 \end{bmatrix} = -[\mathbf{J}(\Delta \mathbf{n}^0)]^{-1} \Delta \mathbf{n}^0 \tag{7-13}$$

The next estimate for $\begin{bmatrix} \mathbf{v}_i \\ \mathbf{i}_j \end{bmatrix}$ is then

$$\begin{bmatrix} \mathbf{v}_i^1 \\ \mathbf{i}_j^1 \end{bmatrix} = \begin{bmatrix} \mathbf{v}_i^0 \\ \mathbf{i}_j^0 \end{bmatrix} + \begin{bmatrix} \Delta \mathbf{v}_i^0 \\ \Delta \mathbf{i}_j^0 \end{bmatrix} \tag{7-14}$$

Equation 7-10 can now be used to calculate a new error vector $\Delta \mathbf{n}^1$ and the process is repeated until the length of the error vector is reduced below some prescribed value.

We must now consider methods of evaluating the matrices $\mathbf{z}$, $\mathbf{t}$, α, and $\mathbf{y}$ in order to complete the procedure. The terminal equations of the linear network, after removal of the $m + n$ nonlinearities, are given by the matrix expression of Eq. 7-9 which we can write as

$$\begin{bmatrix} i_1 \\ \cdot \\ \cdot \\ \cdot \\ i_m \\ v_{m+1} \\ \cdot \\ \cdot \\ \cdot \\ v_{m+n} \end{bmatrix}$$

$$= \begin{bmatrix} y_{11}v_1 + \cdots + y_{1m}v_m + \alpha_{1(m+1)}i_{m+1} + \cdots + \alpha_{1(m+n)}i_{m+n} \\ \cdot \quad \cdot \quad \cdot \quad \cdot \\ \cdot \quad \cdot \quad \cdot \quad \cdot \\ \cdot \quad \cdot \quad \cdot \quad \cdot \\ y_{m1}v_1 + \cdots + y_{mm}v_m + \alpha_{m(m+1)}i_{m+1} + \cdots + \alpha_{m(m+n)}i_{m+n} \\ t_{(m+1)1}v_1 + \cdots + t_{(m+1)m}v_m + z_{(m+1)(m+1)}i_{m+1} + \cdots + z_{(m+1)(m+n)}i_{m+n} \\ \cdot \quad \cdot \quad \cdot \quad \cdot \\ \cdot \quad \cdot \quad \cdot \quad \cdot \\ \cdot \quad \cdot \quad \cdot \quad \cdot \\ t_{(m+n)1}v_1 + \cdots + t_{(m+n)m}v_m + z_{(m+n)(m+1)}i_{m+1} + \cdots + z_{(m+n)(m+n)}i_{m+n} \end{bmatrix} + \mathbf{Bu}$$

Any element of the hybrid matrix

$$\begin{bmatrix} \mathbf{y} & \boldsymbol{\alpha} \\ \mathbf{t} & \mathbf{z} \end{bmatrix}$$

can be found by assuming that all currents and voltages associated with the row in which the element occurs are equal to zero, except the variable associated with the element in question. In general,

$$z_{ab} = \frac{v_a}{i_b}\Bigg|_{\substack{v_i = i_j = u_k = 0, \quad i_b \neq 0 \\ i = 1 \text{ to } m \\ j = m+1 \text{ to } m+n \\ k = 1 \text{ to } s}} \tag{7-15}$$

$$t_{cd} = \frac{v_c}{v_d}\Bigg|_{\substack{v_i = i_j = u_k = 0, \quad v_d \neq 0 \\ i = 1 \text{ to } m \\ j = m+1 \text{ to } m+n \\ k = 1 \text{ to } s}} \tag{7-16}$$

$$\alpha_{ef} = \frac{i_e}{i_f}\Bigg|_{\substack{v_i = i_j = u_k = 0, \quad i_f \neq 0 \\ i = 1 \text{ to } m \\ j = m+1 \text{ to } m+n \\ k = 1 \text{ to } s}} \tag{7-17}$$

$$y_{gh} = \frac{i_g}{v_h}\Bigg|_{\substack{v_i = i_j = u_k = 0, \quad v_h \neq 0 \\ i = 1 \text{ to } m \\ j = m+1 \text{ to } m+n \\ k = 1 \text{ to } s}} \tag{7-18}$$

We can apply this same method to calculate the elements of **B**.

EXAMPLE 7-4. Calculate the output voltage for the circuit of Fig. 7.29. For this circuit assume that the current through diode D_1 obeys the equation

$$i_{D_1} = 10^{-8}(e^{40v_{D1}} - 1)$$

while i_{D_2} is

$$i_{D_2} = 2 \times 10^{-8}\ (e^{40v_{D2}} - 1)$$

and $\alpha = 0.99$.

Solution. The circuit of Fig. 7-30 results from replacing the transistor by its equivalent circuit. The variables to be equated at the terminals to which the nonlinear elements are connected will be current in this example. Thus we need

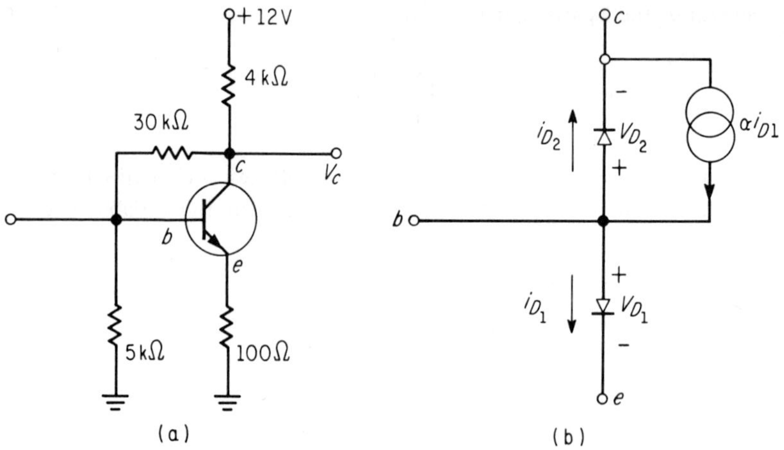

FIG. 7-29. Collector-base bias stage. (a) Actual circuit. (b) Transistor model for active region.

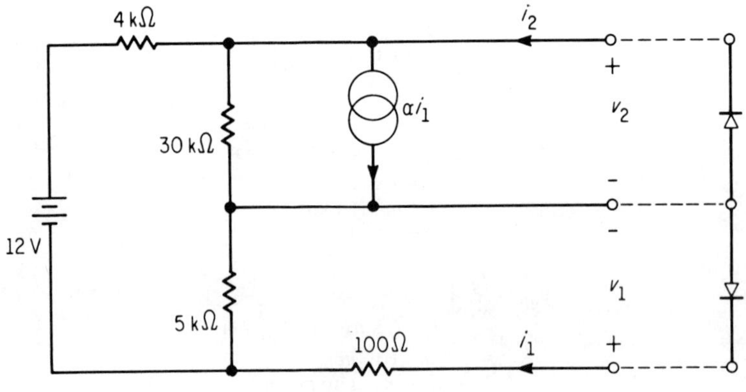

FIG. 7-30. Equivalent circuit.

only evaluate **y** and **B** to apply Eq. 7-10. Using Eq. 7-18, **y** is found to be

$$\mathbf{y} = \begin{bmatrix} 4.306 \times 10^{-4} & -2.392 \times 10^{-4} \\ 1.871 \times 10^{-4} & 4.051 \times 10^{-5} \end{bmatrix}$$

and

$$\mathbf{B} = \begin{bmatrix} 2.392 \times 10^{-4} \\ -7.176 \times 10^{-6} \end{bmatrix}$$

Equation 7-10 becomes

$$\Delta \mathbf{n} = \begin{bmatrix} 4.306 \times 10^{-4} & -2.392 \times 10^{-4} \\ 1.871 \times 10^{-4} & 4.051 \times 10^{-5} \end{bmatrix} \begin{bmatrix} v_1 \\ v_2 \end{bmatrix} + \begin{bmatrix} 2.392 \times 10^{-4} \\ -7.176 \times 10^{-6} \end{bmatrix} [V] - \begin{bmatrix} 10^{-8}\,(e^{-40v_1} - 1) \\ 2 \times 10^{-8}\,(e^{-40v_2} - 1) \end{bmatrix}$$

The negative signs preceding the nonlinear terms and appearing in the exponentials arise from the definitions of terminal variables in Fig. 7-30. A modified Newton-Raphson scheme is used to solve this problem. The program is shown in Table 7-6. Table 7-7 shows the results of this program along with the results of a simple Newton-Raphson analysis which required 55 iterations to complete. The modified program requires only 6 steps. The FORTRAN IV programs were run on the IBM 1132 computer.

TABLE 7-6. Modified Newton-Raphson Program.

```
C       MODIFIED NEWTON-RAPHSON ITERATION FOR A TWO PORT
C       NETWORK
        FUNCTION FI(J,I)
        REAL I
        IF(J-1),1,1,2
      1 FI=(-1.0/40.0)*ALOG(I/(-1.0E-8)+1.0)
        GO TO 3
      2 FI=(-1.0/40.0)*ALOG(I/(-2.0E-8)+1.0)
      3 RETURN
        END
C          H IS THE HYBRID MATRIX STORED IN COLUMN FORM
C          V IS THE VECTOR OF TERMINAL VOLTAGES
C          U IS THE VECTOR OF SOURCE VOLTAGES
C          B IS THE (M+N)*S MATRIX
C          F IS THE VECTOR OF NON-LINEAR ELEMENTS CREATED BY
C              THE STATEMENT FUNCTIONS F1,F2
C          THE FUNCTION FI EXPRESSES THE NON-LINEAR VECTOR F
C              AS A FUNCTION OF THE CURRENT I
C          JA IS THE JACOBIAN MATRIX CREATED BY THE
C              STATEMENT FUNCTIONS JA1, JA2
C          SUBROUTINES MPRD AND MINV ARE THE
C              SUBROUTINES FOR MATRIX MULTIPLICATION AND
C              MATRIX INVERSION
C       ********************************
        REAL JA(4),JA11,JA12,JA21,JA22,II(2)
        DIMENSION H(4),V(2),B(4),F(2),DELTN(2),X(2),Y(2),L(2),
      1     M(2)
        DIMENSION DELTV(2),U(4)
        DATA IN,I0/8,5/
        F1(V1)=-1.0E-8 *(EXP(40.0*(-V1))-1.0)
        F2(V2)=-2.0E-8*(EXP(40.0*(-V2))-1.0)
        JA11(V1)=+4.0E-7*EXP(+40.0*(-V1))+4.306E-4
        JA22(V2)=+4.0E-7*EXP(+40.0*(-V2))+4.051E-5
        JA12=-2.392E-4
        JA21=1.871E-4
        READ(IN,1) (H(I),I=1,4)
        READ(IN,1) (V(I),I=1,2)
        READ(IN,1) (B(I) ,I=1,2)
        READ(IN,1) (U(I) ,I=1,1)
      1 FORMAT (E10.3)
        DO 777 I=1,300
        IK=I
```

TABLE 7-6 (*continued*)

```
      F(1)=F1 (V(1))
      F(2)=F2 (V(2))
      CALL MPRD(H,V,X,2,2,0,0,1)
      CALL MPRD(B,U,Y,2,2,0,0,1)
      DO 2 J=1,2
    2 DELTN(J)=X(J)+Y(J)+F(J)
      JA(1)=JA11(V(1))
      JA(2)=JA21
      JA(3)=JA12
      JA(4)=JA22(V(2))
      CALL MINV(JA,2,D,L,M)
      DO 21 J=1,4
   21 JA(J)=-JA(J)
      CALL MPRD(JA,DELTN,DELTV,2,2,0,0,1)
      WRITE(I0,12) I,V
   12 FORMAT('    ,I3,E15.6,5X,E15.6)
      IF(ABS(DELTN(1))-ABS(DELTN(2))) 8,9,9
    8 XX=DELTN(2)
      GO TO 10
    9 XX=DELTN(1)
   10 IF(ABS(XX)-1.0E-6) 4,3,3
    3 CONTINUE
   55 DO 5 J=1,2
    5 V(J)=V(J)+DELTV(J)
      CALL MPRD (H,V,X,2,2,0,0,1)
      DO 60 J=1,2
   60 II(J)=X(J)+Y(J)
      DO 61 J=1,2
      IF(DELTV(J)) 62,62,61
   62 V(J)=FI(J ,II(J))
   61 CONTINUE
  777 CONTINUE
    4 VC=V(2)-V(1)+100.0*F1(V(1))
   11 WRITE(I0,6) IK,V(1) ,V(2)
    6 FORMAT(' ',T20,'I',T40,'V1',T60, 'V2',/('0',19X,I3,13X,
     1 E15.6,4X,E15.6))
      WRITE(I0,7) VC
    7 FORMAT('0',T7,'VC'/(' ',E15.6))
      CALL EXIT
      END
```

TABLE 7-7. Results of Programs: (a) Modified Newton-Raphson; (b) Simple Newton-Raphson.

(a)

1	0.100000E-03	0.100000E-03
2	-0.103551E 00	0.920347E 01
3	-0.204825E 00	0.134278E 00
4	-0.288117E 00	0.618675E 01
5	-0.303691E 00	0.158867E 00
6	-0.303811E 00	0.352833E 01

TABLE 7-7 (*continued*)

(a)

I	V1	V2
6	-0.303811E 00	0.352833E 01

VC
0.364258E 01

(b)

1	0.100000E-03	0.100000E-03
2	-0.156606E 01	0.917821E 01
3	-0.154106E 01	0.924278E 01
4	-0.151606E 01	0.912731E 01
5	-0.149106E 01	0.901185E 01
6	-0.146606E 01	0.889638E 01
7	-0.144106E 01	0.878092E 01
8	-0.141606E 01	0.866545E 01
9	-0.139106E 01	0.854998E 01
10	-0.136606E 01	0.843452E 01
11	-0.134106E 01	0.831905E 01
12	-0.131606E 01	0.820359E 01
13	-0.129106E 01	0.808812E 01
14	-0.126606E 01	0.797265E 01
15	-0.124106E 01	0.785719E 01
16	-0.121606E 01	0.774172E 01
17	-0.119106E 01	0.762626E 01
18	-0.116606E 01	0.751079E 01
19	-0.114106E 01	0.739532E 01
20	-0.111606E 01	0.727986E 01
21	-0.109106E 01	0.716439E 01
22	-0.106606E 01	0.704893E 01
23	-0.104106E 01	0.693346E 01
24	-0.101606E 01	0.681800E 01
25	-0.991062E 00	0.670253E 01
26	-0.966061E 00	0.658706E 01
27	-0.941061E 00	0.647160E 01
28	-0.916061E 00	0.635613E 01
29	-0.891061E 00	0.624067E 01
30	-0.866061E 00	0.612520E 01
31	-0.841061E 00	0.600974E 01
32	-0.816061E 00	0.589427E 01
33	-0.791061E 00	0.577880E 01
34	-0.766061E 00	0.566334E 01
35	-0.741061E 00	0.554787E 01
36	-0.716060E 00	0.543241E 01
37	-0.691060E 00	0.531694E 01
38	-0.666060E 00	0.520148E 01
39	-0.641060E 00	0.508601E 01
40	-0.616060E 00	0.497054E 01
41	-0.591060E 00	0.485508E 01
42	-0.566060E 00	0.473962E 01
43	-0.541061E 00	0.462415E 01

TABLE 7-7 *(continued)*

(b)

44	-0.516062E 00	0.450869E 01
45	-0.491067E 00	0.439325E 01
46	-0.466079E 00	0.427784E 01
47	-0.441113E 00	0.416253E 01
48	-0.416206E 00	0.404750E 01
49	-0.391465E 00	0.393323E 01
50	-0.367177E 00	0.382105E 01
51	-0.344095E 00	0.371444E 01
52	-0.324003E 00	0.362165E 01
53	-0.310094E 00	0.355740E 01
54	-0.304527E 00	0.353169E 01
55	-0.303821E 00	0.352843E 01

I	V1	V2
55	-0.303821E 00	0.352843E 01

VC
0.364262E 01

We note that the method proposed on the preceding pages suffers two limitations; namely, (1) no guarantee of convergence, and (2) the necessity of a matrix inversion at each iteration. Either of these limitations can result in excessive computational costs. Modifications of the process can result in a practical procedure [2], however, there are other methods that may reduce the consequence of the limitations mentioned. Branin and Wang [1] have reported a method of d-c analysis of nonlinear networks that does not require the inversion of the Jacobian at each step. Rather this matrix is updated at each step of the iterative process. Furthermore, the probability of divergence is reduced to zero by this scheme.

State Equations for Nonlinear Networks*

Nonlinear transient or a-c problems can be treated in much the same way that d-c nonlinear problems are handled. A great deal more effort is required since an iterative solution must be carried out at each time step of the problem.

The derivation of the necessary equations starts by separating the nonlinear elements from the network and numbering terminals at each point of separation. All capacitors and inductors are withdrawn from the network, and the withdrawal of these elements define additional terminals. Furthermore, all independent sources define terminal variables; thus we have three distinct classes of terminals–independent source terminals, reactive element terminals, and nonlinear element terminals. Also, connected to these terminals will be the remainder of the network which will consist of linear resistive elements as shown in Fig. 7-31.

Let us assume that there are s independent voltage sources, t independent

*This section is also based on the approach outlined by D. A. Calahan in *Computer-Aided Network Design,* New York: McGraw-Hill Book Company, 1968.

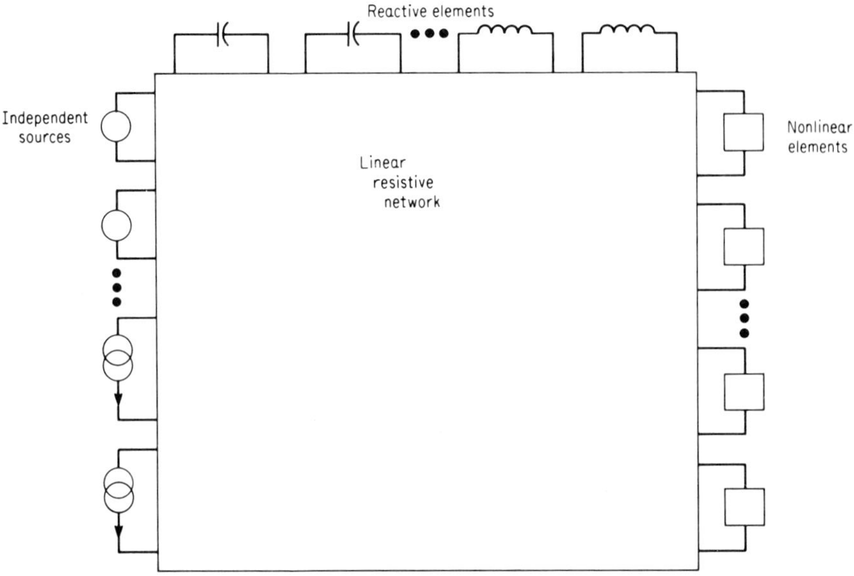

FIG. 7-31. Separation of classes of elements for nonlinear network.

current sources, p capacitors, r inductors, m nonlinear devices expressible as functions of voltage, and n nonlinear devices expressible in terms of current. We can write any set of terminal variables in terms of all terminal variables. Thus we can express capacitor currents and inductor voltages in terms of the state vector, the source vector, and the vector of variables appearing at the nonlinear device terminals. If we define

$$\begin{bmatrix} C_1 \dfrac{dv_{c1}}{dt} \\ \cdot \\ \cdot \\ \cdot \\ C_p \dfrac{dv_{cp}}{dt} \\ L_1 \dfrac{di_{L1}}{dt} \\ \cdot \\ \cdot \\ \cdot \\ L_r \dfrac{di_{Lr}}{dt} \end{bmatrix} = \mathbf{M}\dot{\mathbf{x}} \tag{7-19}$$

where

$$\mathbf{M} = \begin{bmatrix} C_1 & 0 & \cdot & \cdot & \cdot & \cdot & 0 \\ \cdot & & & & & & \cdot \\ \cdot & & & & & & \cdot \\ \cdot & & & & & & \cdot \\ 0 & \cdot & \cdot & C_p & \cdot & \cdot & 0 \\ 0 & \cdot & \cdot & \cdot & L_1 & \cdot & 0 \\ \cdot & & & & & & \cdot \\ \cdot & & & & & & \cdot \\ \cdot & & & & & & \cdot \\ 0 & \cdot & \cdot & \cdot & \cdot & 0 & L_r \end{bmatrix} \tag{7-20}$$

we can write

$$\mathbf{M}\dot{\mathbf{x}} = \mathbf{A}'\mathbf{x} + \mathbf{B}'\mathbf{u} + \mathbf{G}'\mathbf{g} \tag{7-21}$$

In this case the vector $\mathbf{g}$ consists of the variables at the terminals to which the nonlinear elements are connected. Explicitly,

$$\mathbf{g} = \begin{bmatrix} v_{g1} \\ \cdot \\ \cdot \\ \cdot \\ v_{gm} \\ i_{g(m+1)} \\ \cdot \\ \cdot \\ i_{g(m+n)} \end{bmatrix} \tag{7-22}$$

It is also possible to write the nonlinear terminal variables

$$\mathbf{h} = \begin{bmatrix} i_{g1} \\ \cdot \\ \cdot \\ \cdot \\ i_{gm} \\ v_{g(m+1)} \\ \cdot \\ \cdot \\ \cdot \\ v_{g(m+n)} \end{bmatrix}$$

in terms of **x**, **u**, and **g** as

$$\mathbf{h} = \mathbf{E}\mathbf{x} + \mathbf{F}\mathbf{u} + \mathbf{K}\mathbf{g} \tag{7-23}$$

Equation 7-21 can be rewritten as

$$\dot{\mathbf{x}} = \mathbf{A}\mathbf{x} + \mathbf{B}\mathbf{u} + \mathbf{G}\mathbf{g} \tag{7-24}$$

These two equations contain three unknowns, namely **h**, **x**, and **g**. The vector **h** can be eliminated by applying the nonlinear relations

$$\mathbf{h} = -\begin{bmatrix} f_1(v_{g1}) \\ \cdot \\ \cdot \\ \cdot \\ f_m(v_{gm}) \\ f_{m+1}(i_{g(m+1)}) \\ \cdot \\ \cdot \\ \cdot \\ f_{m+n}(i_{g(m+n)}) \end{bmatrix} = -f(g) \tag{7-25}$$

Making this substitution in Eq. 7-23 eliminates the unknown **h**, but results in a nonlinear equation. There are now only two unknowns, **x** and **g** and theoretically these vectors can be found from Eqs. 7-23 and 7-24.

We must realize, of course, that the two equations must be solved at each point in time rather than only one time as in the d-c analysis case. If the initial state vector is known, then this value can be used in Eq. 7-23 to find **g**. Another possibility is to evaluate an initial value of **g** (which may be relatively easy in many cases) and solve Eq. 7-23 for the corresponding value of $\mathbf{x}_0$. As pointed out by Calahan, some care must be taken when evaluating the quiescent conditions. For an oscillator quiescent conditions result in a time-varying state vector in which $\dot{\mathbf{x}} \neq 0$. Thus if quiescent conditions are to be calculated from Eq. 7-24 by setting $\dot{\mathbf{x}}$ to zero, one must guarantee that this condition is actually met.

Calahan suggests a method of solution of the nonlinear state equations. If $\mathbf{x}(nT)$, $\mathbf{g}(nT)$, and $\mathbf{u}(nT)$ are known and $\mathbf{u}(t)$ can be approximated as a constant for $nT \leqslant t \leqslant (n+1)T$, then Eq. 7-24 can be numerically integrated to find $\mathbf{x}[(n+1)T]$. We must assume here that $\mathbf{g}(nT)$ is also constant over this time interval. When $\mathbf{x}[(n+1)T]$ is found a value of $\mathbf{g}[(n+1)T]$ can be calculated. This procedure can be repeated to obtain succeeding values of **x** and **g**.

In applying these analytical methods to the solution of nonlinear networks we must bear in mind the tremendous number of calculations made by the computer. Matrix equations are solved by iteration at each point in time of interest. Newer methods for more efficient solution are constantly sought. A second problem that arises is that analytical expressions for all nonlinearities are required. These expressions are not always obtainable from theoretical considerations and approximations are often in order. Because of these problems the piecewise linear approach may be more amenable to a great number of nonlinear problems.

BIBLIOGRAPHY

1. Branin, F. H., Jr., and H. H. Wang, "A Fast Reliable Iteration Method for D-C Analysis of Nonlinear Networks," *Proc. IEEE*, Vol. 55, No. 11 (November 1967), pp. 1819–1826.
2. Calahan, D. A., *Computer-Aided Network Design*, preliminary edition, McGraw-Hill Book Company, New York, 1968.
3. Comer, D. J., *Introduction to Semiconductor Circuit Design*, Addison-Wesley Publishing Company, Reading, Mass., 1968.
4. Comer, D. T., *Large-Signal Transistor Circuits*, Prentice-Hall, Inc., Englewood Cliffs, N.J., 1967.
5. Gray, P. E., D. DeWitt, A. R. Boothroyd, and J. F. Gibbons, *Physical Electronics and Circuit Models of Transistors,* SEEC, Vol. 2, John Wiley & Sons, Inc., New York, 1964.
6. Jensen, R. W., and M. D. Lieberman, *IBM Electronic Circuit Analysis Program,* Prentice-Hall, Inc., Englewood Cliffs, N.J., 1968.
7. 1620 Electronic Circuit Analysis Program (ECAP) (1620-EE-02X), User's Manual, IBM Publication H20-0170-1, IBM Corporation, White Plains, N.Y.

PROBLEMS

7-1 Plot the *V-I* characteristics of a practical diode that can be represented as a diode obeying the equation

$$I_d = 10^{-6}\,(e^{40V_d} - 1)$$

plus a series ohmic resistance of 10 Ω. What slope is approached at high currents?

7-2 Specify all values for the ECAP two-break-point model of the diode of Prob. 7-1.

7-3 Plot the output voltage waveform of the circuit shown if $r_b' = 1{,}000\ \Omega$, $r_c = 30\ \text{k}\Omega$, $V_0 = 0.5\,V$, and $\beta = 80$. Use the equivalent circuit of Fig. 7-9.

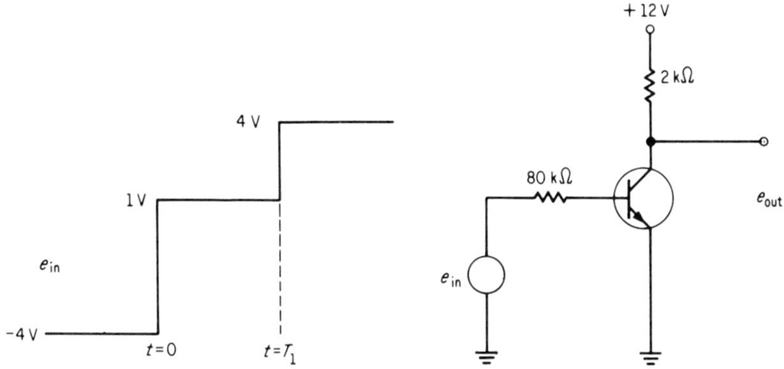

PROB. 7-3

7-4 Propose an ECAP model for a diode that obeys the equation

$$I = 10^{-5}\,(e^{40V} - 1)$$

Account for the reverse leakage current in your model.

7-5 Propose an ECAP model for a tunnel diode with piecewise linear characteristics as shown.

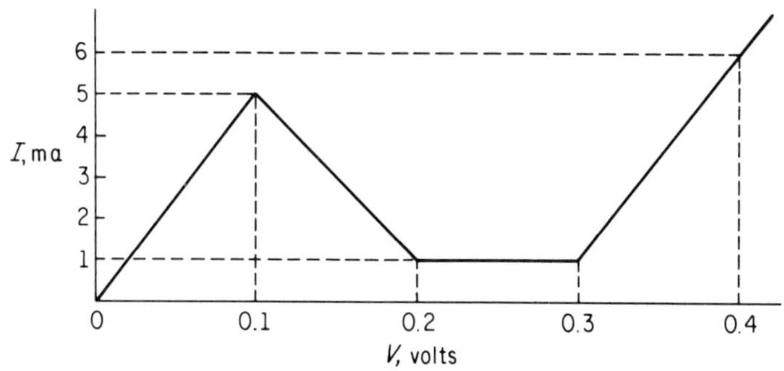

PROB. 7-5

7-6 Propose an ECAP model for a unijunction transistor. Sketch the resulting V-I characteristics.

7-7 Propose an ECAP model for a triode that always operates in the active region. Sketch the resulting V-I characteristics.

7-8 Propose an ECAP model for all operating regions of the pentode. Sketch the resulting V-I characteristics.

7-9 Propose an ECAP model for the junction field-effect transistor. Sketch the resulting V-I characteristics.

7-10 A vacuum-tube diode obeys the equation $I = 10^{-4}\,V^{3/2}$, where I is expressed in amps and V in volts. Propose an ECAP model for this diode that never departs from the true current by more than 10 percent in the range of 1 ma to 100 ma.

7-11 Find a nonlinear capacitance model which has the characteristic given by Fig. 7-18.

7-12 Plot the V-I characteristic of the circuit of Fig. 7-23. Assume that the applied voltage is positive when it opposes the branch source voltages.

7-13 Write a Newton-Raphson program to solve for the operating point of the

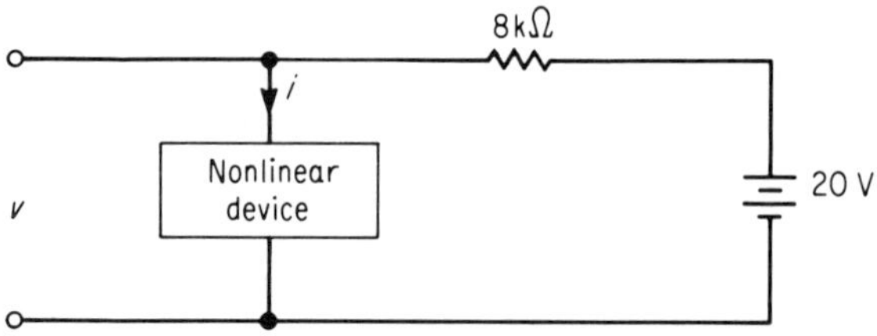

PROB. 7-13

circuit shown. The nonlinear device exhibits the relationship

$$i = 10^{-5}\, v^{6} + 10^{-4}\, v^{4}$$

7-14 Write a modified Newton-Raphson program to find the operating point of the circuit in Prob. 7-13.

7-15 Write a modified Newton-Raphson program to find the operating point of the circuit in Prob. 7-13 if the nonlinear relationship is

$$i = 0.5 \ln v$$

The modification should minimize the number of iterations required.

7-16 Write a modified Newton-Raphson program to solve Example 7-4 if the 30 kΩ resistor is changed to 60 kΩ.

CHAPTER **8**

ECAP APPLICATIONS* and SCEPTRE

The Electronic Circuit Analysis Program has been mentioned many times throughout this text. ECAP is one of the more popular commercially available circuit analysis programs. This chapter will discuss some capabilities and programming details of the program.

ECAP performs d-c, a-c, and transient analysis. Under control of this program, network equations are formulated and solved by machine requiring only topological information along with element values of the network. In d-c and a-c analysis, parameter modification can be effected, but automatic comparison of calculated results to desired results is not possible.

The three types of analysis must be done separately although the transient analysis portion of the program will accept a harmonic input. To analyze a circuit for both d-c and a-c response requires separate analyses along with a manual superposition of results.

A. THE STANDARD ECAP BRANCH

All analysis done by ECAP is based on a standard branch. This branch was discussed in some detail in Chapter 3. There are some additional constraints imposed on the configuration by ECAP and these points will be reviewed. The branch used for this program is shown in Fig. 8-1.

The current source I_i is an independent source while I_{di} is a dependent source. The voltage source E_i is an independent source. The branch variables are e_i and i_i, while V and J_i are element variables. In comparing this branch to the branch of Fig. 3-11 we note that the dependent current source does not appear in Fig. 3-11. As shown later in Chapter 3, the effects of the dependent generator can be expressed as a transadmittance or a transimpedance. Thus the impedance Z or the admittance of the branch can account for this generator. Note that this approach is consistent with ECAP in that the element current is considered to be made up of the current through the self-impedance or self-admittance plus the dependent current.

*This chapter considers only some of the many features of ECAP. For a full discussion of the program and applications see an *ECAP User's Manual* or R. W. Jensen and M. D. Lieberman, *IBM Electronic Circuit Analysis Program,* Englewood Cliffs, N. J. : Prentice-Hall, Inc., 1968.

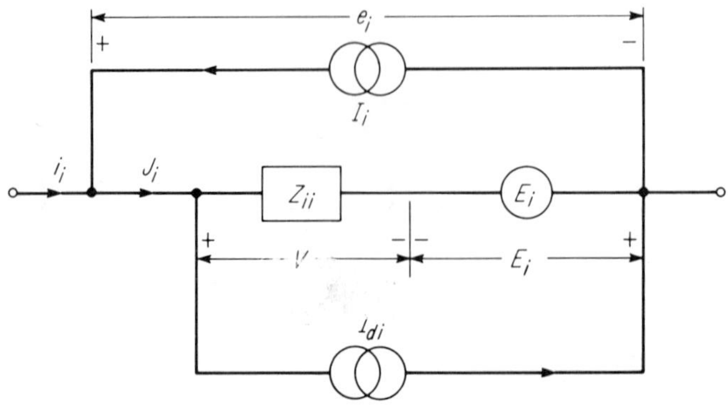

FIG. 8-1. Standard ECAP branch.

The values of all elements in the standard branch except for Z_{ii} can be zero. As explained in Chapter 3 a finite value of passive element must appear in each branch to guarantee that the analysis can be carried out. We should also note that more than one dependent source can be applied to the standard branch although all dependent sources must be represented by current sources. The fact that ECAP does not recognize dependent voltage sources does not limit the program generality. Chapter 3 demonstrates the method that can be used for converting a dependent voltage source to a dependent current source.

There are certain networks having a configuration that does not allow direct representation by standard branches. Generally some slight modification can be performed to allow standard branches to accurately represent the network. Two important nonstandard configurations have been considered in Chapter 3 under the section entitled Special Cases.

The units assumed by ECAP are shown in Table 8-1.

TABLE 8-1. Units Assumed by ECAP.

Variable	Units	Variable	Units
Current	amperes	Inductance	henries
Voltage	volts	Time	seconds
Resistance	ohms	Frequency	Hertz
Conductance	mhos	Phase angle	degrees
Capacitance	farads		

The formats of numerical values that are acceptable are

342.1, 3.421E2, 3.421E+2, 3.421E+02, 3421.E−1, 3421.E−01, 3421E−1

The E symbol indicates that the preceding number is to be multiplied by ten raised to the power specified after the symbol. When this is a positive number, the plus sign can be deleted. Note that if no decimal is used, the program assumes a decimal to the right of the least significant digit. The maximum sizes of the mantissas and exponents depend on the type of computer used.

B. D-C ANALYSIS

The d-c analysis program recognizes the standard branch containing only independent d-c sources, dependent d-c current sources, and resistance or conductance. The dependent current source can be specified in terms of a current transfer or a transconductance. Output quantities that can be calculated by ECAP using the d-c program are node voltages, element voltages, branch voltages, element currents (note that element current is the sum of the self-impedance current plus dependent source current), branch currents, and element power losses. Further capabilities include calculation of sensitivities and partial derivatives of output variables with respect to element value changes, standard deviation analysis, and calculation of various other quantities of interest such as the nodal-admittance matrix.

A typical problem is prepared for analysis by dividing the network into standard branches, numbering branches consecutively (starting with number 1), and numbering all nodes consecutively (starting with zero for the reference node). Figure 8-2 shows a network to be analyzed, while Fig. 8-3 shows the equivalent circuit which is appropriately subdivided into standard branches. The quiescent collector voltage is to be found.

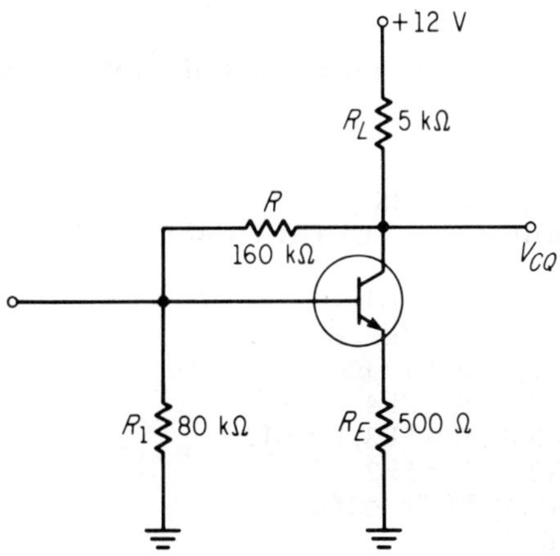

FIG. 8-2. A transistor amplifier.

A piecewise linear model for the transistor is used with values of $\beta = 100$, $r_c = 30$ kΩ, $r_b' = 2$ kΩ, and $V_0 = 0.5$ V. The program to calculate node voltages is shown in Table 8-2 along with the output.

The program first specifies the type of analysis to be performed; in this case d-c analysis. This information starts in column 7. Any number of comment cards can be included before or after the program specification card. Branch information is then listed starting with B1 and proceeding consecutively to the

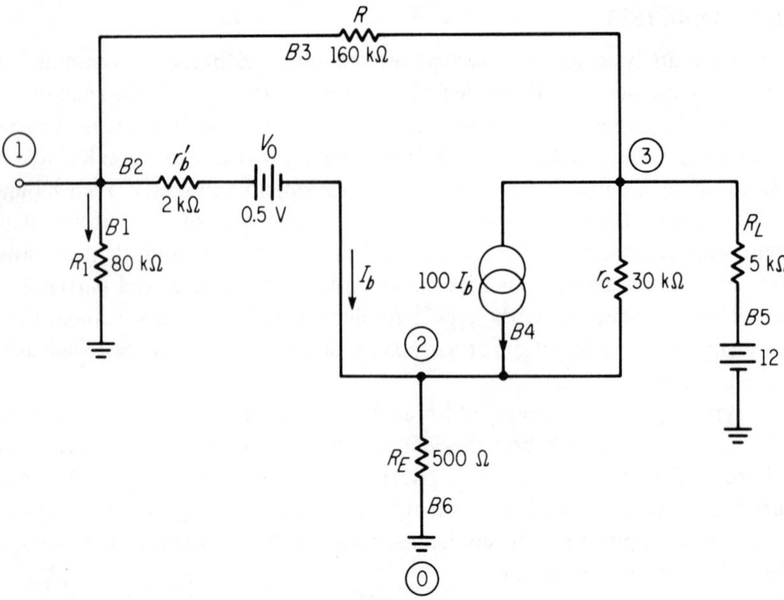

FIG. 8-3. The equivalent circuit, showing standard branches.

TABLE 8-2. ECAP Program for Quiescent Voltage Analysis.

```
Column No.

1          7
           DC ANALYSIS
C          QUIESCENT VOLTAGE CALCULATION
C
B1         N(1,0), R = 8E4
B2         N(1,2), R = 2E3, E = -0.5
B3         N(1,3), R = 1.6E5
B4         N(3,2), R = 3E4
B5         N(3,0), R = 5E3, E = -12.
B6         N(2,0), R = 5E2
T1         B(2,4), BETA = 100.
           PRINT, NV
           EXECUTE
NODE VOLTAGES
NODES                        VOLTAGES
  1 - 3      0.12039154D 01   0.67992230D 00   0.51255323D 01
R = 0.16000000E 06
```

highest branch number in the circuit. The branch number must be placed within columns 1 to 5. The nodal connectivity information for each branch starts in column 7 with the letter *N*. The two numbers in parentheses specify the two nodes to which the branch connects and the direction defining positive current.

Positive current flows from the first node specified to the second node. Nodal information is followed by a comma and a finite value of resistance. Independent voltage-source information follows the resistance value and independent source current would follow the voltage source value if a current source appears in the branch. Commas always separate the data subgroups.

Dependent current-source cards follow the branch cards and are identified by the letter *T* followed by the number of the source. Dependent sources are also numbered consecutively starting with number 1. This information is located in columns 1 to 5. Note that the number following the letter *T* does not necessarily correspond to the branch number. Branch information for the current source starts in column 7 with the letter *B*. The first number in parenthesis specifies the branch current to which the dependent source is proportional, while the second number corresponds to the branch that contains the source. The strength of the source is then specified in terms of the word BETA which is the current gain of the source. The value of BETA can be negative depending on the chosen directions of positive current in both the "from" branch and the "to" branch. Current source information can also be specified in terms of a transconductance. In this case the letters *GM* are used instead of BETA. The current is then equal to the value of *GM* times the voltage appearing across the resistance of the "from" branch.

There are only four circuit elements recognized by the ECAP d-c analysis program; resistors (or conductances), independent voltage sources, independent current sources, and dependent current sources. All of these elements except independent current sources are used in the program of Table 8-2.

After the topology of the network is described by *B*-cards (branches) and *T*-cards (dependent sources), an output specification card determines the output block of data to be printed. In some cases a solution control card will be inserted before the output specification card. These cases will be discussed more fully in later paragraphs. When no solution control card is included, the output quantities available are node voltages, branch voltages, element voltages, branch currents, element currents, element power loss, and the nodal-admittance matrix. The output block indicators used in the program are listed in Table 8-3. Any or all of the symbols for output quantities can be used in a given problem. Commas must be used after the word PRINT and between all output quantities called for.

There are other possible outputs in the d-c analysis program requiring the use of solution control cards. Sensitivities and partial derivatives can be calculated; worst-case analysis and standard deviation analysis can be done.

The sensitivity of some dependent variable *A* with respect to change in some independent variable *X* is defined as

$$S_X^A = \frac{\dfrac{\partial A}{A}}{\dfrac{\partial X}{X}} = \frac{X}{A}\frac{\partial A}{\partial X} \tag{8-1}$$

TABLE 8-3. Outputs For D-C Analysis Program.

Output Quantity	Program Symbol
Node voltages	NV or VOLTAGES
Branch voltages	BV
Element voltages	CV
Branch currents	BA
Element currents	CA or CURRENTS
Element power loss	BP
Nodal admittance and impedance matrix	MI or MISCELLANEOUS

The variable A could be gain, quiescent voltage, quiescent current, or any other dependent quantity of interest while X might be an element value such as resistance or bias voltage. The value of sensitivity indicates the fractional change that will take place in the dependent variable when the independent variable exhibits a given fractional change.

In ECAP d-c analysis the sensitivities of all node voltages with respect to every circuit parameter are calculated when specified by a solution control card. The calculation of the partial derivatives is based on an exact formula that has been derived in the literature [3]. Sensitivities are then found from the partial derivative information by applying Eq. 8-1.

In worst-case analysis, some understanding of the solution method is required to determine if the results are a true worst-case analysis. In production circuits resistors will have certain tolerances, power supplies are subject to variations, and active elements have properties that vary widely from one device to another. Furthermore, temperature changes can result in parameter variations that will modify the overall circuit performance. For circuits that are to be produced in quantity and allow interchange of components, the performance of the circuit must not be too strongly dependent on any component. Stability of circuit performance with respect to parameter changes must be such that the performance of one circuit is nearly equal to that of a similar circuit. One method used to guarantee performance is to calculate the extreme values of some output quantity based on a worst-case combination of element values. The output quantity, of course, will depend differently on each element value. An increase in one resistor may cause the output voltage to increase, while decreasing a second resistor may also result in an increased value of output voltage. Thus, before an extreme value of output voltage can be calculated, the dependence of this quantity on each parameter must be determined.

The direction of variation of any node voltage with respect to change in any element value is given by the appropriate partial derivative. If the partial derivative $\partial v_i/\partial R_j$ has a positive value, then an increase in R_j also increases v_i. If the derivative is negative then R_j must be decreased to increase v_i. A worst-case analysis of the node voltages starts by calculating the voltages for nominal element values. The partial derivatives of the voltages with respect to all element values are then found. To calculate the minimum value of a given node voltage, the signs of the partial derivatives of this voltage with respect to all element values are considered. All elements having positive derivatives are set to minimum values and all elements with negative derivatives are set to maximum values. The node voltage of interest is now recalculated using these extreme element values. The resulting voltage is the minimum value. The maximum value of voltage is found by setting all elements to the opposite extremes and recalculating this node voltage. This procedure can be repeated for all node voltages required.

The method outlined is subject to the fulfillment of a basic assumption. The dependent variables must be monotonic functions of all circuit parameters over the ranges of interest. That is, the sign of the partial derivatives must not change as the parameters change from nominal values to extreme values. For many circuits this assumption will be valid, but there are other circuits wherein the derivatives will change sign. If a derivative does change sign it may still be possible to find a worst case solution by using certain modifications [3], but in general a worst-case analysis will not be possible with ECAP. In order to verify that a worst-case condition has resulted, all derivatives are calculated at the nominal parameter values, at the parameter values resulting in a minimum node voltage, and at the parameter values resulting in a maximum node voltage. The signs of each derivative at these three sets of parameter values are compared. If a sign change has occurred, this fact is printed out to inform the designer that true worst-case conditions have not been calculated. The possibility exists of two sign changes occurring, but is very remote. If this happened the program would yield a solution that is not worst-case, but would go undetected.

Also related to production circuit analysis is the standard deviation analysis. With nominal value plus maximum and minimum for each variable parameter specified, the standard deviation of node voltages can be calculated. The assumptions upon which this calculation is based are not too realistic [3], thus one might consider the results as approximations rather than accurate deviations.

An example of a worst-case analysis of the circuit of Fig. 8-3 with R = 300 kΩ (nominally) is presented in Table 8-4. In this analysis all resistances are assumed to have a 10 percent tolerance, the current gain of the transistor is assumed to vary from 50 to 200, and the base emitter voltage drop is assumed to change from 0.4 to 0.6 V.

The letter D preceding the exponents in the WCMAX voltages of Table 8-4 has the same meaning as the letter E, but indicates that double precision arithmetic was used in the calculation.

For the circuit of Fig. 8-3 the quiescent output voltage varies between 5.3 and 8.6 V as the components vary from one extreme to another. While the combination of element values to give the worst case may not occur in a given circuit, production circuits generally are based on worst-case design to allow for this possibility.

We should note that any particular nodes of interest can be specified by

TABLE 8-4. Worst-Case Analysis.

```
Column No.
1           7
C           WORST CASE ANALYSIS
C           RESISTOR TOLERANCES ARE 10 PER CENT
            DC ANALYSIS
C
B1          N(1,0), R = 8E4(.1)
B2          N(1,2), R = 2E3, E = -0.5(.2)
B3          N(1,3), R = 3.00E5(.1)
B4          N(3,2), R = 3E4
B5          N(3,0), R = 5E3(.1), E = -12.
B6          N(2,0), R = 5E2(.1)
T1          B(2,4), BETA = 100(50,200)
            WORST CASE
            PRINT, WO
            EXECUTE
```

WORST CASE SOLUTIONS FOR NODE VOLTAGES

NODE	WCMIN	NOMINAL	WCMAX
1	0.76796103E 00	0.10051003E 01	0.12713031D 01
2	0.31551850E 00	0.49001592E 00	0.71536248D 00
3	0.52931423E 01	0.70370216E 01	0.86466578D 01

worst-case analysis. If node 3 were the node of interest, printout for this node only would result if the statement WORST CASE, 3 replaces the statement WORST CASE. When no node is specified, analysis is done and results printed for all node voltages. It should also be mentioned that the print statement is not actually required when worst-case, sensitivity, or standard deviation analysis is done. However, if a nominal printout of some variable other than node voltages is desired, it must be specified by a print statement.

There are two formats for specification of tolerance information. The nominal value must first be specified in both cases. For fractional variation, the fraction of variation is specified. A 10 percent variation of an 80 kΩ resistance can be specified as $R = 8E4(.1)$. The variation is then plus and minus 10 percent of the nominal value giving $R_{min} = 72$ kΩ and $R_{max} = 88$ kΩ. The second format specifies the nominal value with the algebraic minimum and maximum values in parentheses. The specification of current gain in Table 8-4 is given as BETA = 100(50,200). It should be noted that if a current direction associated with this generator were redefined causing the nominal value of β to be -100, then the statement would be changed to BETA = -100(-200,-50) to cause the algebraic minimum value to appear in the appropriate position.

Another very important capability of ECAP is that of parameter modification. This feature allows an element value to take on several discrete values while output quantities are calculated for each element value. Parameter modification is controlled by a command card containing the statement MODIFY. The element value to be modified along with the specified values follow the MODIFY card. An example of the use of the modify feature is given in Table 8-5. The circuit of Fig. 8-3 is analyzed with the resistor R varying from 100 to 400 kΩ in steps of 30 kΩ. This calculation was used to find the correct value of R to set the quiescent output voltage at approximately 7 V.

After the modify statement follows the card (or cards) specifying the branch (or branches) to be changed. In this program the resistance of the third branch is varied. Note that nodal information is not included on this branch modification card. The resistance is specified as R = 1E5(10)4E5 which means that the resistance is varied from 100 kΩ to 400 kΩ in 10 steps. The resistance increments will be equal to

$$\frac{4\text{E}5 - 1\text{E}5}{10} = 3\text{E}4 = 30 \text{ k}\Omega$$

A nominal solution is first obtained using whatever value of resistance is given in the program for branch 3. The resistance is then changed to the first value given following the modify statement and successive solutions are obtained for each new resistance value.

Rather than changing a resistance through several values, a single value change can also be effected. In fact, after a modify statement the values of parameters in several branches can be changed. Voltage and current sources can also be changed or added to a branch with this statement.

Several modify statements can be used in a given program, however only one statement can appear before the first execute statement. Succeeding modify statements must be followed by execute statements. The information that will be printed out when a modify statement is used will be the originally requested information of the program. Additional information can be requested after the modify statement. In Table 8-5 the print command is included after the modify statement, but the same output would have been obtained if this command appeared before the modify statement. Parameters not affected by a modify statement remain equal to the values specified by the program preceding the modify statement. For example, if one modify statement changes the resistance in branch 3 to 10 Ω, a succeeding modify statement that does not change branch 3 will use 10 Ω for this resistance.

One precaution that should be observed relates to the maximum spread of values in resistances connected to a given node. Inaccuracies can arise from the inversion of a matrix with widely spread values. The term $[\mathbf{AYA}^t]^{-1}$ required in Eq. 3-28 can lead to inaccuracies if the elements of $\mathbf{Y}$ vary widely. Secondly, the term $\mathbf{I}-\mathbf{YE}$ in the same equation can lead to a certain amount of inaccuracy that is dependent on resistor spread. With the IBM 1620 or CDC 3200, the suggested spread in values should not exceed 10^6 or 10^7 while the IBM System 360 recommends that the spread does not exceed 10^7 or 10^8. Thus the ratio of the largest resistance to the smallest resistance of a node should be kept within this limit.

TABLE 8-5. Parameter Modification Program and Output.

```
Column No.
1          7
C          DETERMINATION OF BIAS RESISTOR
C          V3 IS TO BE APPROXIMATELY 7 VOLTS
           DC ANALYSIS
B1         N(1,0), R = 8E4
B2         N(1,2), R = 2E3, E = -0.5
B3         N(1,3), R = 1E5
B4         N(3,2), R = 3E4
B5         N(3,0), R = 5E3, E = -12.
B6         N(2,0), R = 5E2
T1         B(2,4), BETA = 100.
           MODIFY
B3         R = 1E5(10)4E5
           PRINT, NV
           EXECUTE
NODE VOLTAGES
NODES                          VOLTAGES
  1 - 3    0.12902023D 01   0.76234295D 00   0.42959329D 01
R = 0.10000000E 06
NODE VOLTAGES
NODES                          VOLTAGES
  1 - 3    0.12902023D 01   0.76234295D 00   0.42959329D 01
R = 0.13000000E 06
NODE VOLTAGES
NODES                          VOLTAGES
  1 - 3    0.12306901D 01   0.70549724D 00   0.48681095D 01
R = 0.16000000E 06
NODE VOLTAGES
NODES                          VOLTAGES
  1 - 3    0.12039154D 01   0.67992230D 00   0.51255323D 01
R = 0.19000000E 06
NODE VOLTAGES
NODES                          VOLTAGES
  1 - 3    0.11333454D 01   0.61251417D 00   0.58040242D 01
R = 0.22000000E 06
NODE VOLTAGES
NODES                          VOLTAGES
  1 - 3    0.10930310D 01   0.57400612D 00   0.61916243D 01
R = 0.25000000E 06
NODE VOLTAGES
NODES                          VOLTAGES
  1 - 3    0.10570819D 01   0.53966775D 00   0.65372549D 01
R = 0.28000000E 06
NODE VOLTAGES
NODES                          VOLTAGES
  1 - 3    0.10248255D 01   0.50885665D 00   0.68473819D 01
```

TABLE 8-5 *(continued)*

```
R = 0.31000000E 06
NODE VOLTAGES
NODES                       VOLTAGES
 1 - 3     0.99572054D 00   0.48105580D 00   0.71272094D 01
R = 0.34000000E 06
NODE VOLTAGES
NODES                       VOLTAGES
 1 - 3     0.96932682D 00   0.45584467D 00   0.73809702D 01
R = 0.37000000E 06
NODE VOLTAGES
NODES                       VOLTAGES
 1 - 3     0.94528236D 00   0.43287755D 00   0.76121442D 01
R = 0.40000000E 06
NODE VOLTAGES
NODES                       VOLTAGES
 1 - 3     0.92328676D 00   0.41186749D 00   0.78236196D 01
```

The program automatically sums the current at each node to see if this value is approximately zero as it should be. If it exceeds 1 ma, a warning message is printed out showing the current error at that node. The error current which must be exceeded before a warning occurs can be changed by using the statement 1ERROR = I, where I is the numerical value of the desired error current. This statement should appear after the topological information.

C. A-C ANALYSIS

The a-c analysis program recognizes the standard branch containing independent a-c voltage or current sources, dependent a-c current sources, resistances or conductances, capacitances, self-inductances, or mutual inductances. The dependent current source can be specified in terms of a current gain or a transconductance, but if an L or C is a "from" branch, the dependent source must be expressed as a transconductance. Output quantities that can be calculated by ECAP using the a-c program are node voltages, element voltages, branch voltages, element currents (note that element current is the sum of the self-impedance current plus the dependent source current), branch currents, and element power losses. These quantities are all expressed in terms of rms values and phase angles. The nodal-admittance matrix can also be calculated. While parameter modification can be used in the a-c program, neither worst-case, sensitivity, or standard deviation analyses can be performed.

A transistor amplifier having the equivalent circuit shown in Fig. 8-4 is analyzed at a frequency near 1 mHz to demonstrate an a-c program. The program and results are shown in Table 8-6.

The term MI included in the print statement causes the nodal-admittance matrix and an equivalent (node) current vector to appear as output information.

TABLE 8-6. Program and Results of Analysis of a Transistor Amplifier.

```
Column No.
1           7
C           HYBRID PI EQUIVALENT CIRCUIT
            AC ANALYSIS
C
B1          N(1, 0), R = 2.0E2, I = 1.0E-6
B2          N(1, 0), R = 7.85E2
B3          N(1, 0), C = 1.3E-9
B4          N(2, 0), R = 1.0E4
B5          N(1, 2), C = 1.26E-11
T1          B(2, 4), BETA = 150
            1ERROR = 1.0E-6
            FREQUENCY = .99045494 E06
            PRINT, NV, CV, CA, BA, MI
            EXECUTE
        ROW   COL
        NODE NODE                 NODAL ADMITTANCE MATRIX
REAL     1    1- 2                0.62738841D-02    0.0
IMAG                              0.81685844D-02  -0.78412442D-04
REAL     2    1- 2                0.19108272D 00    0.99999990D-04
IMAG                             -0.78412442D-04    0.78412442D-04
        NODES                    EQUIVALENT CURRENT VECTOR
REAL    1- 2  0.99999943D-06               0.0
IMAG          0.0                          0.0
FREQ =  0.99045494E 06
        NODES       NODE VOLTAGES
MAG     1- 2  0.77999102E-05   0.11728562E-01
PHA          -0.51923676E 02   0.89951889E 02
 BRANCHES           ELEMENT VOLTAGES
MAG     1- 4  0.77999102E-05   0.77999102E-05   0.77999102E-05   0.11728562E-01
PHA          -0.51923676E 02  -0.51923676E 02  -0.51923676E 02   0.89951889E 02
MAG     5- 5  0.11734698E-01
PHA          -0.90024506E 02
 BRANCHES           ELEMENT CURRENTS
MAG     1- 4  0.38999541E-07   0.99361870E-08   0.63102561E-07   0.92014653E-06
PHA          -0.51923676E 02  -0.51923676E 02   0.38076294E 02  -0.24553489E-01
MAG     5- 5  0.92014653E-06
PHA          -0.24553489E-01
 BRANCHES           BRANCH CURRENTS
MAG     1- 4  0.97642987E-06   0.99361870E-08   0.63102561E-07   0.92014653E-06
PHA          -0.17819821E 03  -0.51923676E 02   0.38076294E 02  -0.24553489E-01
MAG     5- 5  0.92014653E-06
PHA          -0.24553489E-01
```

Data for the frequency response of a circuit can be obtained by using the modify statement, as will be demonstrated in connection with the problem to be discussed.

The manual analysis of high-frequency transistor amplifiers is somewhat difficult unless some simplifications to the equivalent circuit are made [2]. The

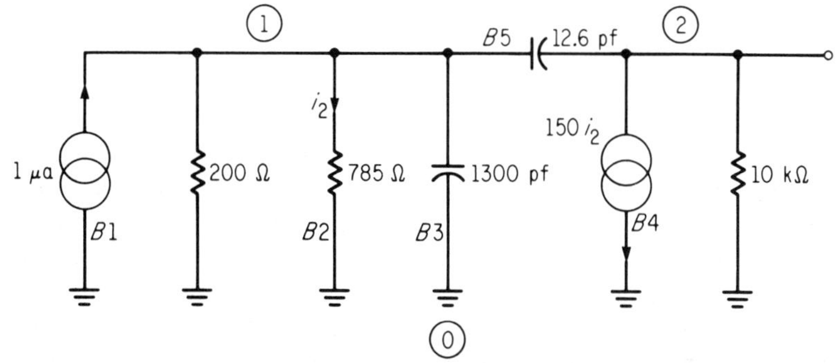

FIG. 8-4. High-frequency transistor equivalent circuit.

hybrid π circuit often used as the starting point for high-frequency analysis is shown in Fig. 8-5. ECAP can be used to analyze this circuit quite readily, but for manual analysis the circuit is generally simplified by reflecting the effects of $C_{b'c}$ to the input. This results in the circuit of Fig. 8-6 [1]. In the circuit of

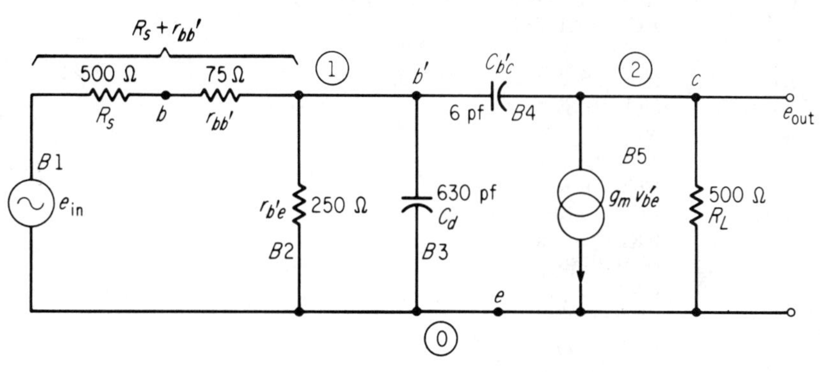

FIG. 8-5. Hybrid π equivalent circuit.

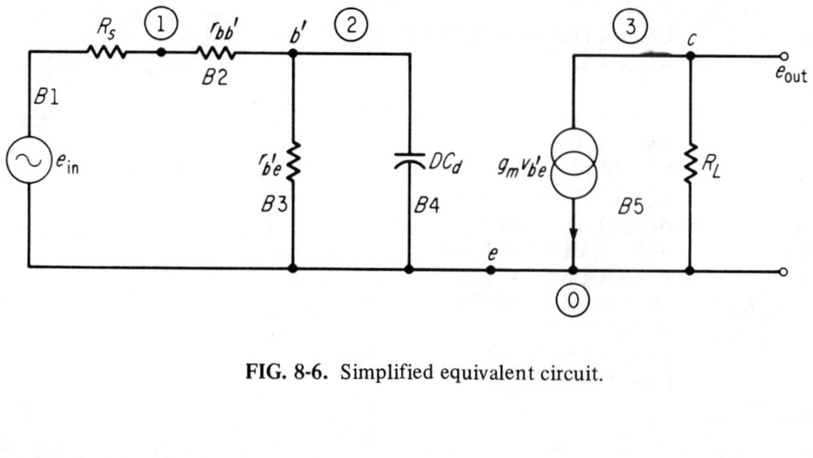

FIG. 8-6. Simplified equivalent circuit.

Fig. 8-6, the output loop is isolated from the input loop and the stage is easier to analyze. The D factor is given by

$$D = 1 + R_L \omega_t C_{b'c} \tag{8-2}$$

where ω_t is the current gain-bandwidth of the transistor. It must be noted that the loading effect of $C_{b'c}$ on the output circuit is assumed negligible to use the unilateral circuit of Fig. 8-6. For the conditions prevailing in most high-frequency stages it has been shown [2] that the magnitude response resulting from the circuit of Fig. 8-6 closely approximates that of the more accurate circuit of Fig. 8-5, at least up to the corner frequency (3-db point). This fact can be demonstrated by ECAP. The program of Table 8-7a corresponds to the accurate circuit, while

TABLE 8-7. Programs for High-Frequency Analysis: (a) Accurate Hybrid π Circuit. (b) Unilateral Model.

(a)

```
Column No.

1           7
C           ACCURATE HYBRID PI CIRCUIT
            AC ANALYSIS
C
B1          N(1, 0), R = 575., E = -0.1
B2          N(1, 0), R = 250.
B3          N(1, 0), C =630.E-12
B4          N(1, 2), C = 6.E-12
B5          N(2, 0), R = 500.
T1          B(2, 5), BETA = 100.
            FREQ = 1000.
            MODIFY
            FREQ = .5E3(1.585)200.E6
            PRINT, NV
            EXECUTE
```

(b)

```
Column No.

1           7
C           UNILATERAL CIRCUIT
            AC ANALYSIS
C
B1          N(0, 1), R = 0.5E3, E = 1.0E-1/0
B2          N(1, 2), R = 75.0
B3          N(2, 0), R = 250.0
B4          N(2, 0), C = 0.1817E-8
B5          N(3, 0), R = 5.0E2
T1          B(3, 5), BETA = 100.0
            FREQUENCY = 1.0E8
            PRINT, VOLTAGES
            MODIFY
            FREQUENCY = 1.0E3(2)2.0E8
            EXECUTE
```

Table 8-7b shows the program for the unilateral model. A source resistance of 500 Ω is assumed along with parameter values of $r_{bb'} = 75\ \Omega$, $r_{b'e} = 250\ \Omega$, $R_L = 500\ \Omega$, $C_d = 630$ pf, $C_{b'c} = 6$ pf, and $g_m = 0.4$ mho. The D factor for this case is 2.88.

The output voltages as functions of frequency are shown in Fig. 8-7.

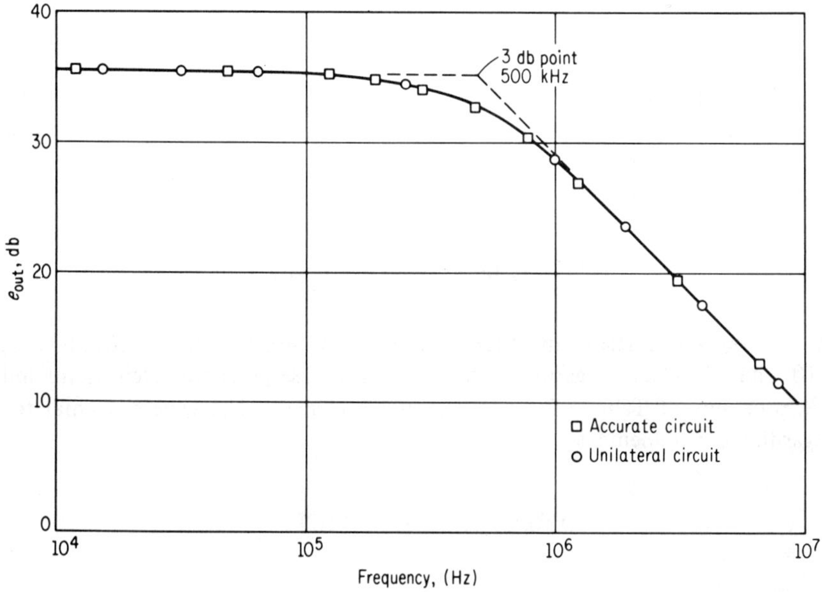

FIG. 8-7. Frequency responses.

In entering data pertaining to independent sources, a phase angle can be specified. The data card for branch one of the unilateral circuit analysis shows the format for phase specification. The format $E = 10/-80$ means an a-c signal of 10 V rms value at an angle of −80 degrees. If the slash and phase-angle information are deleted, the program assumes a zero phase angle.

The two programs indicate that the print command can be placed before or after the modify statement. If a print command is given before the modify statement, the results of each parameter modification are printed out according to the original print command. A second print command can be placed after the modify statement to cause additional data to be printed out after each parameter modification.

When parameter modification is used in d-c or a-c analysis a nominal solution must be completed before any modification takes place. If frequency is the parameter to be modified, a nominal frequency (not necessarily included in the frequency range of interest) must be specified before the modify statement is used. After this solution has been effected any number of modify statements can be executed.

There are two possible formats to specify frequency changes in addition to the possibility of specifying a single new frequency. These two formats are useful for frequency-response calculations. For linear increments in frequency the format is

FREQUENCY = 2E4(+10)12E4

The first number specified is the lowest frequency, the last number is the highest frequency, and the figure in parentheses indicates the number of equal increments between the low and high frequencies. The above specification would result in increments of

(12E4-2E4)/10 = 1E4 = 10kHz

A solution would result at frequencies of 20, 30, 40, 50, 60, 70, 80, 90, 100, 110, and 120 kHz. Logarithmic frequency response plots are often useful and the programs of Table 8-7 demonstrate this format. The general format for logarithmic increments is

FREQUENCY = 1E3(2)2E8

The first term indicates the starting frequency. The term in parentheses is a multiplicative constant. A new solution frequency is determined by multiplying this term by the frequency of the preceding solution. In this case the first solution is performed at a frequency of 1 kHz, the next at 2 kHz, the next at 4 kHz, and the next at 8 kHz. The frequency progression continues until the upper frequency, specified by the last number, is equaled or exceeded. In the above case the frequency increase is in octaves, but a noninteger value of multiplicative constant can also be used. If a logarithmic response containing k increments is desired, the constant can be found to be

$$\left(\frac{f_{\text{high}}}{f_{\text{low}}}\right)^{1/k-1}$$

Note the absence of the plus sign in parentheses of the logarithmic frequency format.

Element values can be modified in the a-c analysis program also. Passive element modification is accomplished the same way as in the d-c case. To vary an independent voltage or current source, both magnitude and phase of the new value must be included. In cases where only magnitude is changed the phase-angle specification still must be given. When a magnitude is changed in steps, a phase angle is required at each step. For example, to change a source of 1 V rms at 90° to 5 V at 90° in four steps, the appropriate format is E = 1.0(4)5.0/90(4)90.

Inductors and capacitors are specified in the same way as resistors noting that the units are henrys and farads respectively. For problems involving transformers one must decide whether the ideal transformer model of Chapter 3 or a mutual-inductance concept is to be applied. For midfrequency work where transformer reactances can be neglected and the coupling coefficient approximates unity, the ideal model can be applied. For more critical work, an M card can be used to denote mutual inductance between two inductances. After expressing branch and dependent source information in the usual way, the mutual inductance is specified. An example of the format for an M card is

```
M1      B(6, 7), L = -0.99999
```

The branches between which coupling is effective is specified in parentheses–in this case branch 6 and 7. These branches must both contain self-inductances. The power-driver stage of Fig. 8-8 is analyzed to find the power output at 2,000 Hz. For this circuit the primary resistance of the transformer R_p is 10 Ω, the secondary resistance R_s is 1 Ω, the primary inductance is 4 h, the secondary inductance is 0.25 h, and the mutual inductance is 0.999994 H giving a coefficient of coupling of 0.999994. The load resistance is 4 Ω, R_E is 10 Ω, C_E is 100 μf, $\beta = 25$, and the transistor input impedance is 100 Ω. We assume an input signal of 0.3 V rms which drives the transistor to maximum output. The output resistance of the transistor is neglected, and the 10-mΩ resistance is included simply to provide the branch with a finite resistance. The program and results are given in Table 8-8.

There are certain types of circuits that cannot be completely analyzed by ECAP, but useful information regarding these circuits can be obtained from the program. Complementary stages are examples of circuits that are used in a-c amplifiers while the transistors operate as nonlinear elements. As far as a-c analysis is concerned, however, the two complementary stages can often be replaced by a single stage that is assumed to amplify both positive-going and negative-going signals equally.

Figure 8-9 contains the schematic of the power-amplifier section of a phonograph amplifier. The two output transistors are complementary emitter followers. The network consisting of the 8,200-Ω and 1,500-Ω resistances and the 10-μf capacitor is a bias network that can be neglected for a-c analysis. The

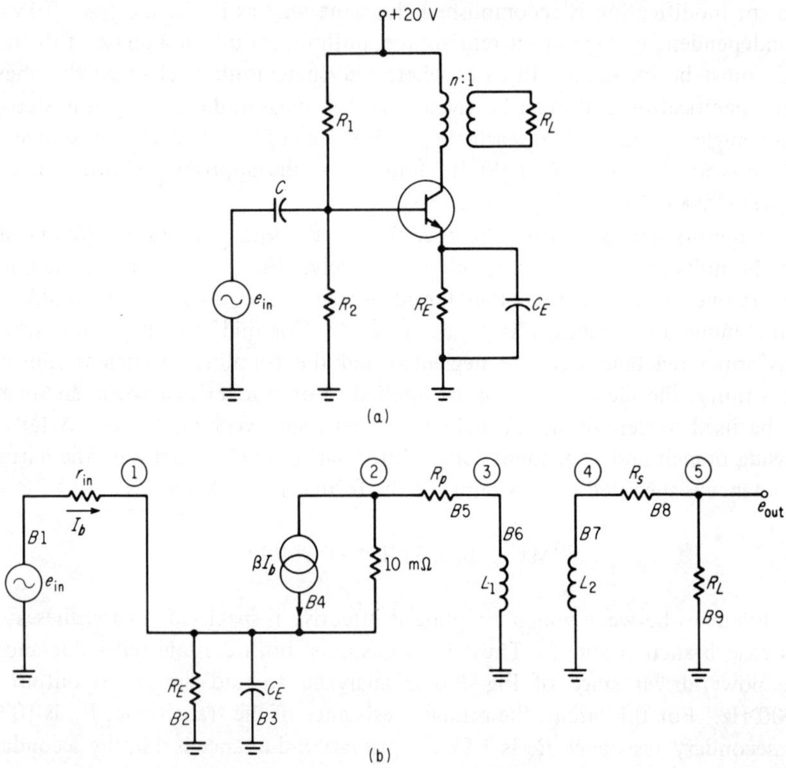

FIG. 8-8. (a) Power-driver stage. (b) Equivalent circuit.

33-Ω resistance is also negligible; thus the equivalent circuit can be represented as in Fig. 8-10. The frequency variation of output power is plotted in Fig. 8-11. The information for this plot was obtained from the program of Table 8-9.

The complementary stages are represented by a single transistor in this case. The midband power delivered to the speaker is approximately 2 watts, while the low-frequency half-power point occurs near 40 Hz. The high frequency half-power point cannot be found for this circuit unless the high-frequency models for the transistors are used.

Oscillator circuits cannot be completely analyzed by ECAP, since amplitude limiting of the output signal due to nonlinear distortion cannot be considered. It is possible to calculate loop gain and phase shift around the loop. When the phase shift results in positive feedback, a gain of unity or greater indicates that oscillations are highly probable at that frequency. One must exercise some caution when breaking the loop to simulate the correct impedances at the point where the loop is broken. The RC phase-shift oscillator of Fig. 8-12 is analyzed by breaking the loop at point *a*. The emitter follower is loaded by the

TABLE 8-8. Power-Driver Stage Analysis.

```
Column No.
1        7
C        ANALYSIS OF POWER DRIVER WITH TRANSFORMER
C
         AC ANALYSIS
C
B1       N(0,1),R=100,E=0.30/0
B2       N(1,0),R=10
B3       N(1,0),C=100E-6
B4       N(1,2),R=10E6
B5       N(2,3),R=10
B6       N(3,0),L=4.0
B7       N(4,0),L=0.25
B8       N(4,5),R=1
B9       N(5,0),R=4
T1       B(1,4), BETA=-25
M1       B(6,7),L=-0.999994
C
         FREQ=2000
         PRINT,NV,CV,BP
         EXECUTE
FREQ =  0.19999998E 04
        NODES         NODE VOLTAGES
MAG     1- 4   0.59669651E-01   0.65095339E 01   0.57862864E 01   0.14465294E 01
PHA           -0.74014160E 02  -0.16813538E 03  -0.16808177E 03   0.11523365E 02
MAG     5- 5   0.11572237E 01
PHA            0.11523365E 02
 BRANCHES          ELEMENT VOLTAGES
MAG     1- 4   0.28931063E 00   0.59669651E-01   0.59669651E-01   0.65140934E 01
PHA            0.11435932E 02  -0.74014160E 02  -0.74014160E 02   0.11341112E 02
MAG     5- 8   0.72326976E 00   0.57862864E 01   0.14465294E 01   0.28930593E 00
PHA           -0.16856404E 03  -0.16808177E 03   0.11523365E 02   0.11523365E 02
MAG     9- 9   0.11572237E 01
PHA            0.11523365E 02
 BRANCHES             BRANCH POWER
MAG     1- 4   0.83700614E-03   0.35604648E-03  -0.54210109E-19  -0.47114420E 00
MAG     5- 8   0.52311897E-01   0.41848963E 00  -0.41848963E 00   0.83697915E-01
MAG     9- 9   0.33479172E 00
```

three-stage *RC* network. The equivalent resistance between point *b* and ground is approximately 2 kΩ, thus the equivalent circuit of Fig. 8-13 can be used to calculate the loaded gain of the stage.

The gain from node 1 to node 7 will be equivalent to the loaded loop gain of the oscillator. A frequency response is first performed by ECAP over a broad range of frequencies. Results of this program indicate that positive feedback (zero phase shift) exists near a frequency of 100 kHz. The circuit is then re-analyzed over a narrow range of frequencies that includes 100 kHz. Equal frequency increments of 150 Hz are used in this case. The program used is reported in Table 8-10.

At a frequency of 101,650 Hz the voltage at node 7 is about 1.02 V with

TABLE 8-9. Program Used to Obtain Frequency Response of Power Amplifier

Column No.

```
1           7
C           FREQUENCY RESPONSE OF POWER AMPLIFIER
C
            AC ANALYSIS
C
B1          N(0,1),C=.1E-6,E=.1
B2          N(1,0),R=470E3
B3          N(1,2),R=600
B4          N(0,2),R=1E5
B5          N(2,3),R=150
B6          N(3,0),R=22
B7          N(4,3),R=1E5
B8          N(4,5),R=150
B9          N(0,5),R=1E5
B10         N(5,6),R=2.2
B11         N(6,7),C=250E-6
B12         N(7,4),R=1E3
B13         N(7,0),R=25
T1          B(3,4), BETA=90
T2          B(5,7), BETA=90
T3          B(8,9), BETA=90
            FREQUENCY=5
            MODIFY
            FREQUENCY=10(1.8739)10000
            PRINT,NV,CA,BP
            EXECUTE
```

a phase shift of 0.031 degree. At a frequency of 101,800 Hz the voltage at this node is still 1.02 V with a phase shift of −0.056 degree. Positive feedback exists at some frequency between 101,650 Hz and 101,800 Hz with a gain of slightly greater than unity. The circuit would then be expected to oscillate near these frequencies. The amplitude of oscillations cannot be predicted, since the variation of loop gain with distortion and the variation of distortion with output amplitude must be known for this calculation.

D. TRANSIENT ANALYSIS

The transient-analysis program recognizes the standard branch containing independent voltage and current sources, resistances, capacitors, or inductors. In addition, current-controlled switches are recognized by this program, allowing synthesis of piecewise linear circuit models or waveforms. Time-dependent solutions are provided for node voltages, element voltages, branch voltages, element currents, and instantaneous element power dissipation. The time interval between each response calculation along with the total time interval of interest must be specified. The printout of results at specific times is controlled by a statement card. The transient analysis program does not have the capa-

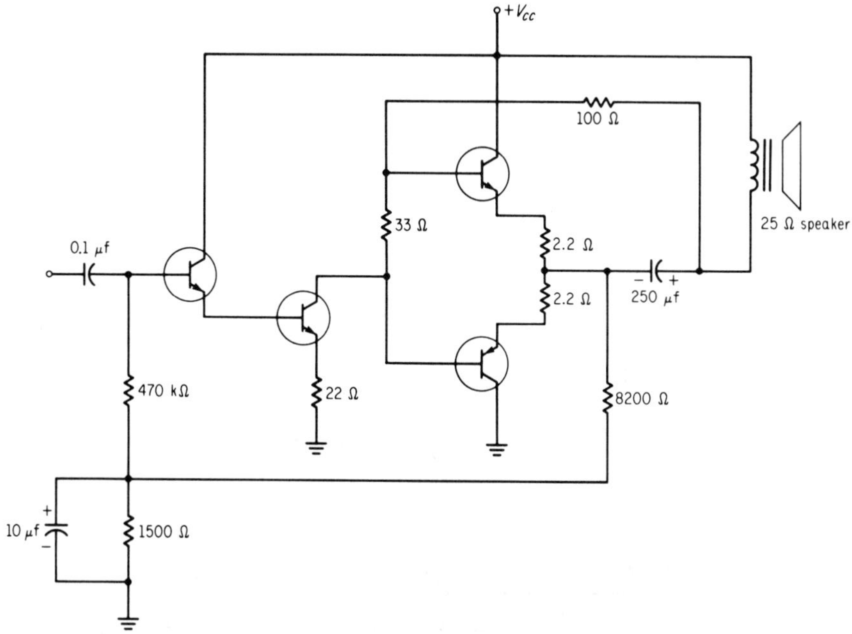

FIG. 8-9. Power amplifier.

bilities to modify parameters or perform worst-case, sensitivity, or standard deviation calculations. On some computers (namely the IBM System/360) mutual inductance cards can be used, but for other computers the M card is not allowed in the transient mode.

To introduce the time-control cards and demonstrate the importance of proper time-step selection, a simple *RC* charging circuit is analyzed. A 1-mΩ resistance and an initially uncharged 1 μf series capacitor are placed across a 10-V d-c source with 10 Ω internal resistance at $t = 0$ (Fig. 8-14). The program and results are shown in Table 8-11.

ECAP solves transient problems by a low-accuracy numerical integration scheme which avoids numerical instability problems, but requires a very small time step for usable results [3]. The ECAP User's Manual [8] suggests that a time step equal to the time constant of an *RC* circuit divided by 100 results in sufficient accuracy for most problems. For the program of Table 8-11 a time step equal to $\tau/800$ is used. At the end of 1 sec, which corresponds to a time constant, the capacitor voltage is given as 6.3188715. This same program with different time steps was used to generate the data in Table 8-12.

The accuracy decreases as the time step increases, but computing time also decreases as the time step increases. Thus the largest time step resulting in sufficiently accurate results is desirable. In many circuits it is difficult to select the appropriate time step. For oscillatory circuits a time step equal to the

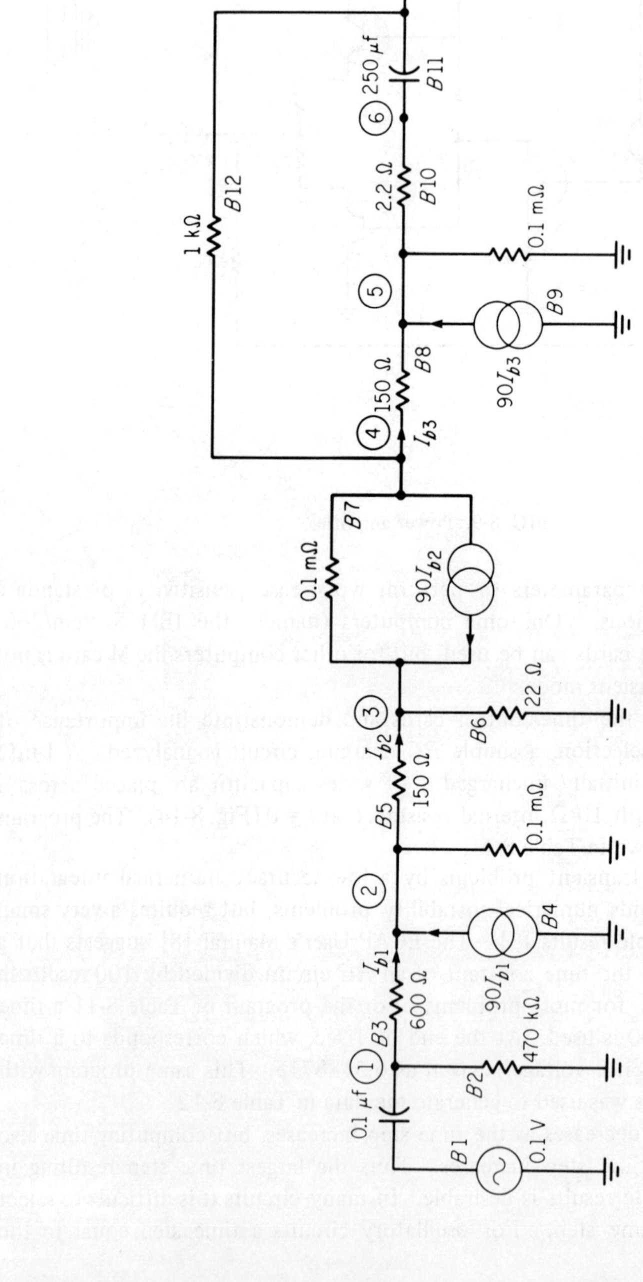

FIG. 8-10. Equivalent circuit of power amplifier.

TABLE 8-10. Program for Analysis of an Oscillator.

Column No.	
1	7
C	RC OSCILLATOR
C	
	AC ANALYSIS
C	
B1	N(0,1),R=0.01,E=1/0
B2	N(1,2),C=325E-12
B3	N(2,0),R=2E3
B4	N(2,3),C=325E-12
B5	N(3,0),R=2E3
B6	N(3,4),C=325E-12
B7	N(4,0),R=6.2E4
B8	N(4,5),R=2613
B9	N(4,0),R=1E4
B10	N(0,6),R=758
B11	N(6,5),R=1E7
B12	N(5,0),R=500
B13	N(5,0),C=1E-6
B14	N(6,7),R=1513
B15	N(0,7),R=1E7
B16	N(7,0),R=5E3
B17	N(7,8),C=325E-12
B18	N(8,0),R=2E3
B19	N(8,9),C=325E-12
B20	N(9,0),R=2E3
B21	N(9,10),C=325E-12
B22	N(10,0),R=2E3
T1	B(8,11), BETA=100
T2	B(14,15), BETA=100
	FREQ=1000
	MODIFY
	FREQ=97,600(+50)105,100
	PRINT,NV
	EXECUTE

reciprocal of the natural frequency divided by 1,000 may be appropriate. A method that can be used to check accuracy of results is to successively run the program with time steps equal to one-half the previous value. When the results converge, no further reduction of time is necessary. This process must be done manually and presents one serious disadvantage of the ECAP program. Automatic selection of the appropriate time step would improve the efficiency of the transient program considerably.

Since the time step must be very small compared to the smallest circuit time constant, many calculations are required to complete a solution. Output data at each time step are not required as much of this information is redundant.

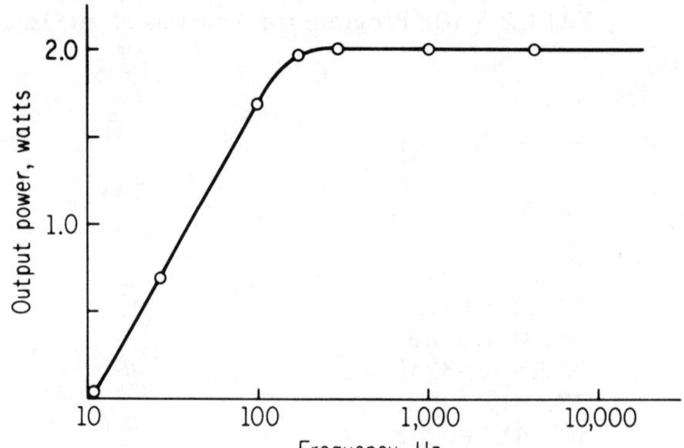

FIG. 8-11. Output power vs. frequency.

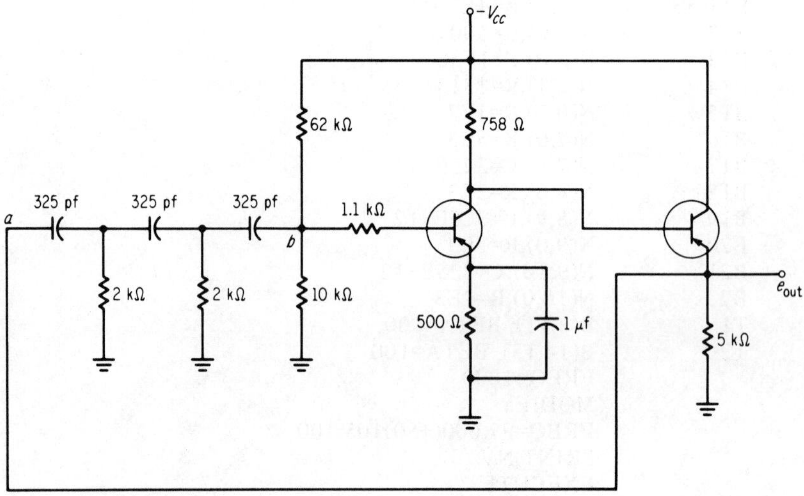

FIG. 8-12. *RC* oscillator.

The output interval statement, following the time-step statement, specifies the number of time steps between printouts. In the program of Table 8-11 solutions are obtained at $t = 0$ and at the end of each time interval of 0.00125 sec. The results are printed out at $t = 0$ and at 200 times the time step (0.25 sec), 400 times the time step (0.5 sec), and every 0.25 sec until the final time is reached. If the output interval statement is deleted, the program assumes a value of unity for this parameter resulting in printout at each time step. The final time statement determines the total time over which calculations are to be performed. The final time specified depends on the problem at hand, but if steady-state conditions are to be approached, the final time must equal from three to five

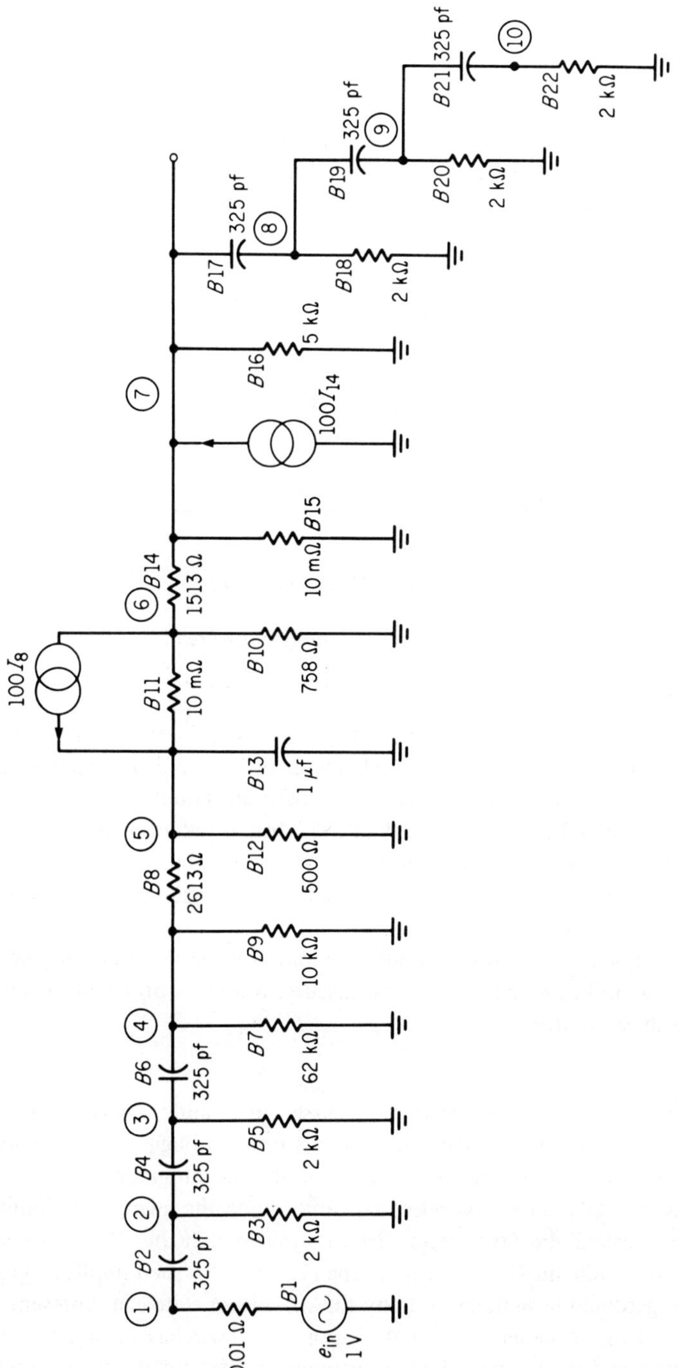

FIG. 8-13. Equivalent circuit of oscillator.

times the longest circuit time constant. Both time-step and final-time statements must be included in all transient-analysis programs.

Initial Conditions

Initial voltages on capacitors or currents through inductors can be specified on the branch cards containing the appropriate elements. For example, if the capacitor of Fig. 8-14 had a voltage at $t = 0$ of -6 V referenced to the ground

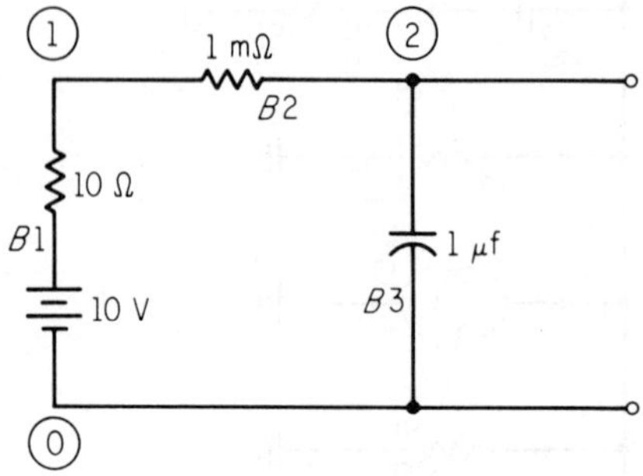

FIG. 8-14. *RC* circuit.

node, the B3 card of Table 8-11 would be modified to read

B3 N(2,0), C=1E−6, E0=6.0

The sign of the initial voltage is determined by comparing the polarity to that of a voltage source in the branch. The character following E in this specification can be zero or the letter O. Initial current for an inductor is specified in the same way except IO is used rather than EO. Initial inductor current is positive when flowing in the same direction as positive branch current. When no initial condition is specified for a reactive element, the program assumes this initial condition to be zero.

In a complex circuit initial conditions may not be obvious nor easy to calculate. An equilibrium solution, to be discussed later, is provided by ECAP to determine these conditions.

Switches

Current-controlled switches were discussed in Chapter 7 with little mention of the programming details involved. When current through a branch containing a switch is positive, the switch is assumed by the program to be in a state defined as the ON state. Negative current causes the switch to change to a second state called the OFF state. In a normal switch the ON state results in zero resistance while the OFF state is an open circuit. No such implication should be made regarding the switches used by ECAP. Either state can represent a large or small resistance, depending on the program. Switches must be numbered consecutively and the branch being monitored for direction of current flow must

TABLE 8-11. Program and Results for Analysis of an *RC* Network.

```
Column No.

1         7
C
          TRANSIENT ANALYSIS INTRODUCTION
C
B1        N(0,1),R=10,E=10
B2        N(1,2),R=1E6
B3        N(2,0),C=1E-6
C
          TIME STEP=.00125
          OUTPUT INTERVAL = 200
          FINAL TIME=1.00125
          PR,NV
          EXECUTE

   T = 0.0
NODES           NODE VOLTAGES
  1- 2    0.99998999E 01    0.99998886E-07
   T = 0.2499953E 00
NODES           NODE VOLTAGES
  1- 2    0.99999218E 01    0.22107573E 01
   T = 0.4999891E 00
NODES           NODE VOLTAGES
  1- 2    0.99999390E 01    0.39327698E 01
   T = 0.7499829E 00
NODES           NODE VOLTAGES
  1- 2    0.99999523E 01    0.52740870E 01
   T = 0.9999767E 00
NODES           NODE VOLTAGES
  1- 2    0.99999628E 01    0.63188715E 01
   T = 0.1002476E 01
NODES           NODE VOLTAGES
  1- 2    0.99999628E 01    0.63280563E 01
```

be listed. When a switch changes states, element values in any branch or branches can be changed. The controlled branches are listed on the *S* card along with the switch number and the switch branch. A typical *S* card might appear as

S1 B=4,(3,4,5),OFF

In this example, S1 appears in branch 4 and controls either passive elements, independent sources, or both in branches 3, 4, and 5. If the B3 card is given as

B3 N(2, 3),R=(10,9E4)

then the resistance of branch 3 will be 10Ω when S1 is OFF and 90 kΩ when S1

TABLE 8-12. Accuracy Comparison for Various Time Steps.

Time interval, sec		Capacitor Voltage at t = 1 Sec, Volts
0.00125	(τ/800)	6.3188715
0.0125	(τ/80)	6.2982950
0.025	(τ/40)	6.2756577
0.25	(τ/4)	5.9039660

is ON. The state listed on the *S* card means that when the switch takes on this state the elements in the controlled branches take on a value given by the first number in parentheses on the corresponding branch cards.

To demonstrate the use of *S* cards, the program for calculating the discharge of the nonlinear capacitor in Example 7-3 is given in Table 8-13

TABLE 8-13. Program for Discharge of a Nonlinear Capacitor.

```
Column No.

1     7
C     NONLINEAR CAPACITOR DISCHARGE
      TRANSIENT ANALYSIS
C
B1    N(1,0),R=1E3
B2    N(1,0),R=1E7
B3    N(1,2),G=0.8185
B4    N(1,2),G=(-0.2965,-1E-7),E=-2
B5    N(1,2),G=(-0.2115,-1E-7),E=-4
B6    N(1,2),G=(-0.1565,-1E-7),E=-6
B7    N(1,2),G=(-0.1075,-1E-7),E=-8
B8    N(2,0),R=1E-4
B9    N(3,0),R=1
B10   N(3,0),C=1E-9,EO=-3.679
T1    B(10,2), BETA=1
T2    B(8,2), BETA=-1
T3    B(8,9), BETA=-1
T4    B(10,9), BETA=-1
S1    B=4,(4),OFF
S2    B=5,(5),OFF
S3    B=6,(6),OFF
S4    B=7,(7),OFF
      TIME STEP=1E-8
      OUTPUT INTERVAL=2
      FINAL TIME=2E-9
      PRINT,NV,CA
      EXECUTE
```

The switch S1 in Table 8-13 appears in branch 4. Thus, when current through this branch is positive S1 is ON and when this current becomes negative S1 changes to the OFF state. The B4 card specifies two values of negative conductance associated with branch 4. The first value specified (−0.2965) will apply when the state of the switch corresponds to that specified on the S1 card (OFF).

The conductance of branch 4 will then be −0.2965 mhos when S1 is OFF and 10^{-7} mhos when S1 is ON. Equivalent information could be specified by reversing the order of conductances appearing on the B4 card and revising the S1 card to read

S1 B=4,(4),ON

The choice as to which initial switch state to specify depends on the circuit. The specified state should correspond to the actual initial state of the switch in the circuit. In the program of Table 8-13 each switch changes only the conductance of its own branch. Elements of other branches can be controlled by a switch as will be demonstrated by the next example. Furthermore it is not necessary for the switch to control any elements in its own branch.

Switches not only can control resistances, capacitances, inductances, and independent sources, but values of dependent-current generators can also be changed as a switch changes state. To control the value of BETA or GM, the "from" branch must be controlled by the appropriate switch. While the element values of the "from" branch may be unaffected this branch must be listed as a controlled branch of the *S* card. If only one element value is listed on a branch card, and if this branch is controlled by a switch, the element value is unaffected as the switch changes state. Table 8-14 includes a partial program with a controlled dependent-current source.

TABLE 8-14. Partial Program with Switched Element Values.

Column No.

1	7
B3	N(0,1),R=(1E7,1)
B4	N(2,1),R=10,E=(0.6,0)
B5	N(3,4),R=(6,800),I=(0.1,0.2)
B6	N(4,0),R=100
T1	B(6,8), BETA=(50,200)
S1	B=3,(3,4,5,6),ON

The switch S1 controls branches 3, 4, 5, and 6. Note that branch 5 contains two element values controlled by the switch. Branch 6 is the "from" branch for the dependent source. While the resistance of this branch is unaffected as S1 changes state, the value of BETA for the source changes from 50 to 200 as S1 changes to the OFF state.

A branch-element value can be controlled by more than one switch. The first element value specified on the *B* card will be in effect when all controlling switches are in the states specified on the *S* cards. When one switch or an odd number of switches change state, the element takes on the second value listed on the *B* card. When the state of two switches or an even number of switches differ from the states listed on the *S* cards, the original element value will be reassumed. Table 8-15 shows a branch value controlled by three switches and the resistance values corresponding to the various possible switch conditions.

TABLE 8-15. Element Value Controlled by Three Switches.

B2	N(3,2),R=(10,6E6)
S1	B=2,(2),ON
S2	B=4,(2),OFF
S3	B=5,(2),ON

Switch Conditions			Resistance Value
S1	S2	S3	
ON	OFF	ON	10Ω
OFF	OFF	ON	6MΩ
ON	ON	ON	6MΩ
ON	OFF	OFF	6MΩ
OFF	ON	ON	10Ω
OFF	OFF	OFF	10Ω
ON	ON	OFF	10Ω
OFF	ON	OFF	6MΩ

The next example contains elements controlled by two switches. In Chapter 7 a three-region equivalent circuit was proposed for both low and high-frequency switching. The circuit of Fig. 8-15 is to be analyzed using the low-frequency equivalent circuit. This assumes that the inherent transistor switching

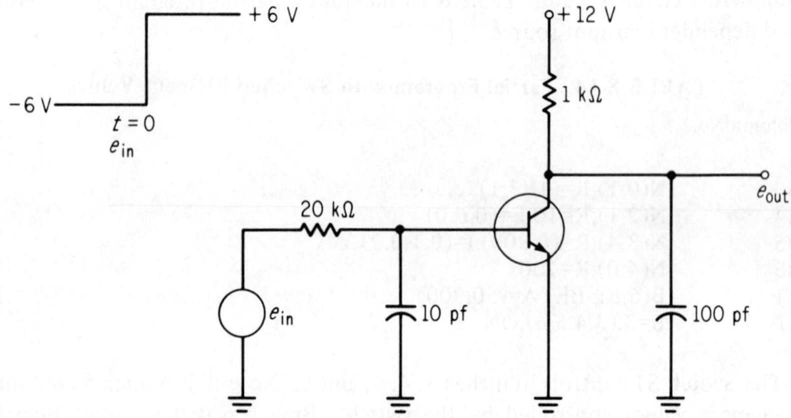

FIG. 8-15. Transistor switching circuit.

time constants are small compared to the circuit time constants. This assumption is probably not accurate in this case, but the example is useful for demonstration purposes. The equivalent circuit for this stage is shown in Fig. 8-16. For this problem the parameters are taken as follows:

Average input resistance in active region, 1,600 Ω
Average input resistance in saturation, 250 Ω
Average current gain, 100
Base-emitter voltage drop, 0.6 V

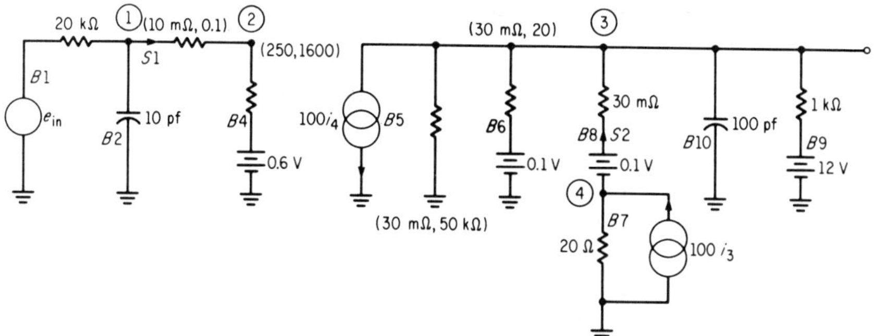

FIG. 8-16. Equivalent circuit.

Saturation resistance, 20 Ω
Saturation voltage, 0.1 V
Average collector resistance in active region, 50 kΩ

A step function input can be handled quite easily by ECAP. If the initial voltages on the capacitors are known prior to the step, we can assume that a constant input signal of +6 V is applied to the input. Thus, we must find the steady state voltages that appear across the capacitors when the input voltage equals −6 V. These voltages are then entered as initial conditions on the capacitors and the input is taken as +6 V. The response then proceeds with $t = 0$ occurring at the transition of the step.

An equilibrium solution is available to calculate steady-state conditions for transient problems. The transient program is written and the solution control card EQUILIBRIUM is inserted before the EXECUTE card. A d-c analysis is then performed which replaces all capacitors with large resistances (10 mΩ) and inductors with small resistances (0.01 Ω). If switches have been included in the equivalent circuit, the equilibrium control card assumes that the switches are in the states specified on the *S* cards. Care must be taken to insure that the switch condition assumed by the equilibrium solution is the correct condition. No switching takes place during the equilibrium solution. The results of this solution allow initial conditions to be specified for the actual transient solution. Of course, the correct input level must be used for the equilibrium calculations. In the circuit of Fig. 8-16 the input voltage should be taken as −6 V. The equilibrium program is given in Table 8-16 along with the results.

Note that the assumed switch condition, that is both S1 and S2 OFF, is the actual steady-state condition. The initial voltage on the 10-pf capacitor equals the negative of the voltage at node 2 according to the manner in which branch direction has been defined. The initial voltage on the 100-pf capacitor equals the negative of the voltage at node 3. The transient program including these initial conditions and the new input voltage is given in Table 8-17 with a summary of results.

TABLE 8-16. Equilibrium Program and Results.

```
Column No.
1        7
         TRANSIENT ANALYSIS OF SWITCHING TIME
C
B1       N(0,1),R=2E4,E=-6
B2       N(1,0),C=10E-12
B3       N(1,2),R=(1E7,0.1)
B4       N(2,0),R=(250,1600),E=-0.6
B5       N(3,0),R=(3E7,5E4)
B6       N(3,0),R=(3E7,20),E=-0.1
B7       N(0,4),R=20
B8       N(4,3),R=3E7,E=0.1
B9       N(0,3),R=1E3,E=12
B10      N(3,0),C=1E-10
T1       B(4,5), BETA=(0,100)
T2       B(3,7), BETA=100
S1       B=3,(3,4,5),OFF
S2       B=8,(4,5,6),OFF
         TIME STEP=1E-9
         OUTPUT INTERVAL=10
         FINISH TIME=5E-7
         PRINT,NV
         EQUILIBRIUM
         EXECUTE
STEADY STATE SOLUTION
NODES        NODE VOLTAGES
 1- 4 -0.59999816E 01  0.59983499E 00 -0.11999996E 00 -0.13120299E-02
```

TABLE 8-17. Transient-Analysis Program and Some Results.

```
Column No.
1        7
         TRANSIENT ANALYSIS OF SWITCHING TIME
C
B1       N(0,1),R=2E4,E=+6
B2       N(1,0),C=10E-12,EO=+6
B3       N(1,2),R=(1E7,0.1)
B4       N(2,0),R=(250,1600),E=-0.6
B5       N(3,0),R=(3E7,5E4)
B6       N(3,0),R=(3E7,20),E=-0.1
B7       N(0,4),R=20
B8       N(4,3),R=3E7,E=0.1
B9       N(0,3),R=1E3,E=12
B10      N(3,0),C=1E-10,EO=-12
T1       B(4,5), BETA=(0,100)
T2       B(3,7), BETA=100
S1       B=3,(3,4,5),OFF
S2       B=8,(4,5,6),OFF
         TIME STEP=1E-9
         OUTPUT INTERVAL=10
         FINISH TIME=5E-7
         PRINT,NV
         EXECUTE
```

TABLE 8-17 *(continued)*

```
T = 0.0
NODES            NODE VOLTAGES
1 - 4   -0.59999924E 01    0.59983492E 00    0.11999998E 02   -0.13120300E-02
T = 0.4999984E-07
NODES            NODE VOLTAGES
1 - 4   -0.38277016E 01    0.59988922E 00    0.11999606E 02   -0.87758363E-03
T = 0.9999934E-07
NODES            NODE VOLTAGES
1 - 4   -0.12844286E 01    0.599995282E 00   0.11999247E 02   -0.36894274E-03
SWITCH 1 IS ON
T = 0.1599986E-06
NODES            NODE VOLTAGES
1 - 4    0.59998655E 00    0.59998655E 00    0.11999046E 02   -0.87774524E-05
T = 0.1799985E-06
NODES            NODE VOLTAGES
1 - 4    0.89168721E 00    0.89166898E 00    0.98337679E 01    0.36459237E 00
T = 0.1999983E-06
NODES            NODE VOLTAGES
1 - 4    0.97068399E 00    0.97066087E 00    0.63306799E 01    0.46332973E 00
T = 0.2099982E-06
NODES            NODE VOLTAGES
1 - 4    0.98475504E00     0.98473096E 00    0.46123734E 01    0.48091626E 00
SWITCH 2 IS ON
T = 0.2366492E-06
NODES            NODE VOLTAGES
1 - 4    0.99733317E 00    0.99717426E 00    0.59664029E 00    0.31773891E 01
T = 0.2699975E-06
NODES            NODE VOLTAGES
1 - 4    0.66669685E 00    0.66667020E 00    0.33333349E 00    0.53336138E 00
T = 0.5009951E-06
NODES            NODE VOLATGES
1 - 4    0.66669297E 00    0.66666633E 00    0.33333325E 00    0.53333020E 00
```

Only selected output points are listed in Table 8-17. There are some problems in the transient-analysis program that may be removed in the future. The necessity of a separate equilibrium solution before the transient solution decreases the efficiency of the program. Automatic evaluation of the initial conditions with automatic substitution into the transient program would require the analysis program to be run only once, thus minimizing turnaround time of the computer. It would also be advantageous if the appropriate switch condition could be determined by the program rather than manually.

It was noted in Chapter 7 that regenerative circuits such as multivibrators can be analyzed using a two-region equivalent circuit if transistor switching time

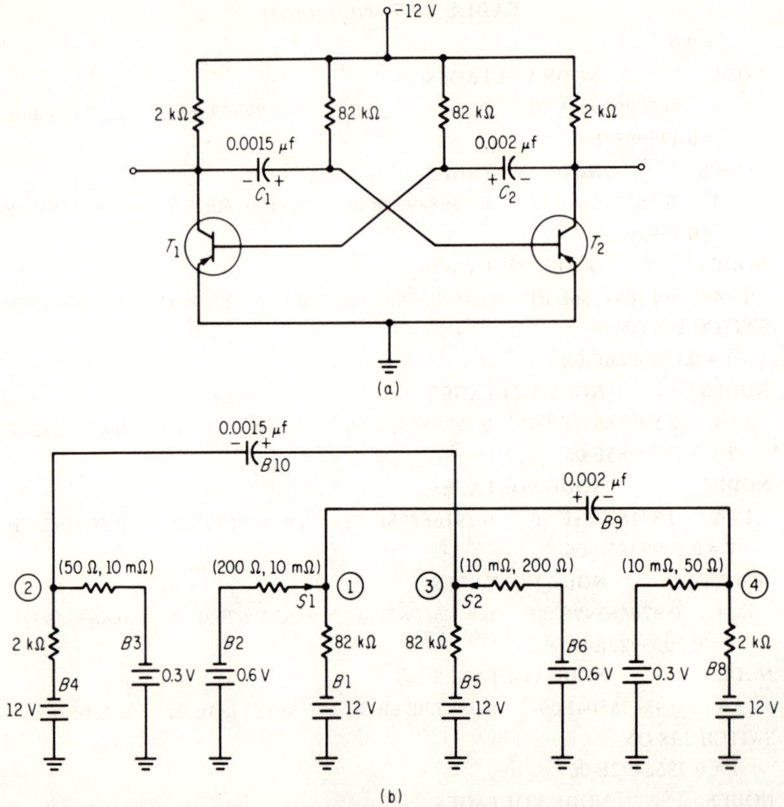

FIG. 8-17. Astable multivibrator. (a) Actual circuit. (b) Equivalent circuit.

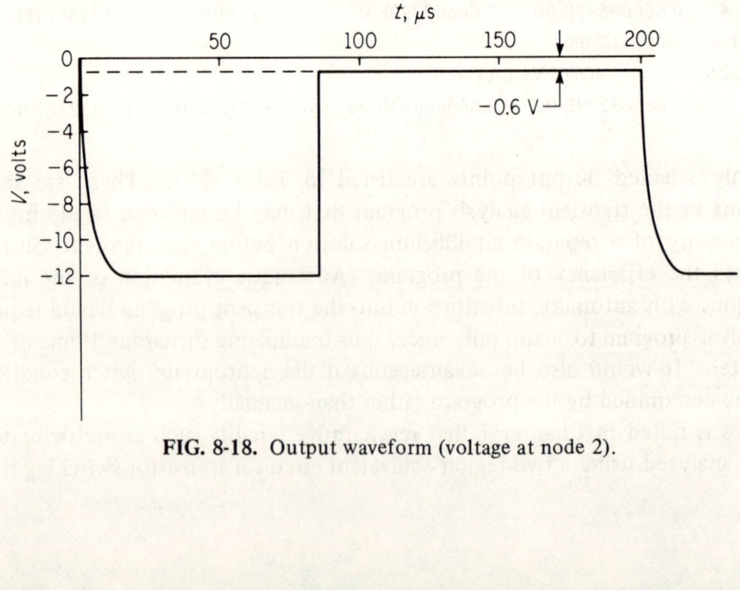

FIG. 8-18. Output waveform (voltage at node 2).

TABLE 8-18. Program for Astable Analysis.

```
Column No.

1          7
           TRANSIENT ANALYSIS
C
B1         N(1,0),R=82E3,E=12
B2         N(0,1),R=(200,10E6),E=-0.6
B3         N(2,0),R=(50,10E6),E=0.3
B4         N(2,0),R=2E3,E= 12
B5         N(3,0),R=82E3,E=12
B6         N(0,3),R=(10E6,200),E=-0.6
B7         N(4,0),R=(10E6,50),E=0.3
B8         N(4,0),R=2E3,E=12
B9         N(4,1),C=2E-9,EO=11
B10        N(2,3),C=15E-10,EO=12
S1         B=2,(2,3),ON
S2         B=6,(6,7),OFF
           TIME STEP=1E-7
           OUTPUT INTERVAL=20
           FINISH TIME=4E-4
           PRINT,NV
           EXECUTE
```

is negligible. For the astable circuit of Fig. 8-17 it is assumed that the time required to switch across the active region of the transistor is small compared to external circuit time constants. Initial voltages are assumed on C_1 and C_2 to be 12 V and 11 V respectively. Transistor T_1 is assumed to be ON while T_2 is assumed OFF.

The input impedance of transistors when in the active region is taken to be 200 Ω, the saturation resistance is 50 Ω, the saturation voltage is 0.3 V, and the forward-biased base to emitter voltage is 0.6 V. The program is shown in Table 8-18, while Fig. 8-18 plots one cycle of the voltage at node 4.

The -12-V power supply is connected to four resistors in Fig. 8-17a. In Fig. 8-17b, four -12-V supplies are shown each associated with one of these four resistors in order to conform to a standard branch. An alternative method is to insert a small resistance in series with the power supply to create a finite impedance in this branch. Each of the four resistors can then make up an acceptable branch.

Time-Dependent Waveforms

All transient circuits analyzed to this point have either required no input signal or have been excited by a step function occurring at $t = 0$. This type of step function requires only that the correct initial conditions are specified corresponding to the d-c input signal present for t less than zero. The equilibrium solution is available for evaluating the initial conditions of complex circuits.

Other types of input can be applied in the transient analysis. Time-dependent sources are directly available in the program and other types of function can be generated using waveform-synthesis techniques. There are three types of time-dependent waveform that can be specified in the program; sinusoidal, non-periodic, and periodic. Both current and voltage sources can be used. The data card for these sources must always appear after the data card for the branch in which the source appears. An example of the specification of input signals is shown in Table 8-19.

TABLE 8-19. Source Specifications.

Column No.

1	7
E12	SIN(0.16666666E-1),52,12,.083333
I33	(10),2E-3,4E-3,4E-3,2E-3,0,0
E9	P(8),0,2,2,0,-2,-2,0

The first card causes a sinusoidal voltage source to appear in branch 12. The period as indicated in parentheses is 16.666666 ms. The first number after the parentheses is the amplitude of the signal which, in this case, is 52 V. The next number is the d-c offset voltage, thus rather than a sine wave swinging about the zero volt reference, this sine wave would be centered at +12 V. The last number specifies the time delay in seconds. In this example the sine wave is delayed one-half period. The period must fall in the range between 3 and 126 times the specified time step of the program.

The second card corresponds to a nonperiodic current source in branch 33. The number in parentheses is the number of time steps between each successive source value, while the following numbers are successive source values. The first is the value occurring at $t = 0$. From Table 8-19 the initial source value is 2 ma. Ten time steps later the source takes on a value of 4 ma which it maintains until 20 time steps. At 30 time steps the value is 2 ma and at 40 and 50 time steps

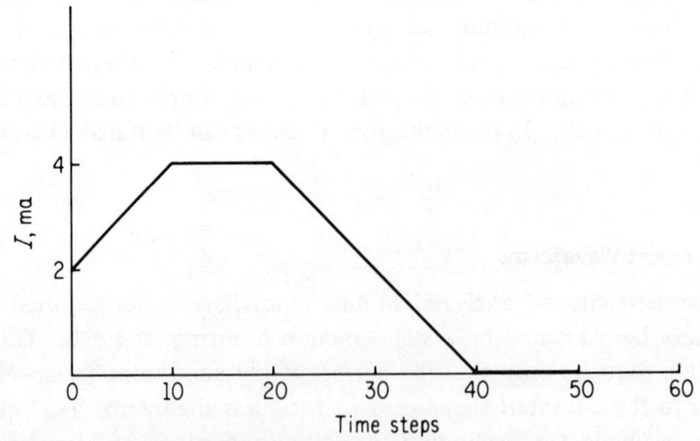

FIG. 8-19. Nonperiodic waveform specified in Table 8-19.

the value is zero. Straight-line segments are assumed to join the points resulting in the waveform of Fig. 8-19 for this card. The maximum number of source variables in the list is 125. Continuation cards can be used to list all source values if one card is not sufficient. An asterisk should appear in column six of the continuation cards.

The third card specifies a periodic voltage source in branch 9. In this case there are 8 time steps between successive source values. For a periodic waveform the last source value must equal the first. Figure 8-20 shows the source voltage for this card.

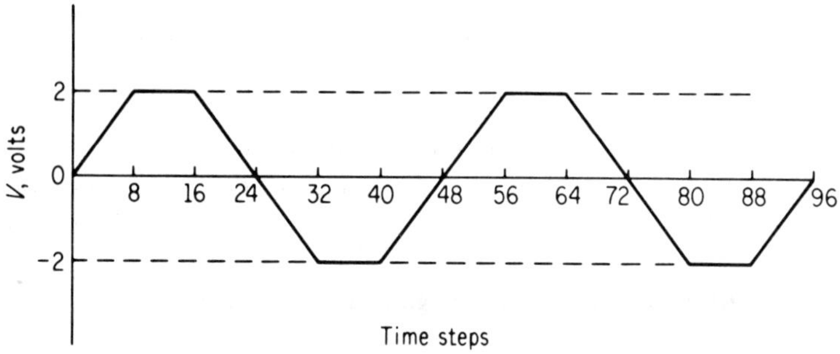

FIG. 8-20. Periodic waveform specified in Table 8-19.

Waveforms can be generated by combining constant or time-dependent sources with switches and other ECAP circuit elements. Waveforms with essentially zero rise time can be generated, although this type of waveform cannot

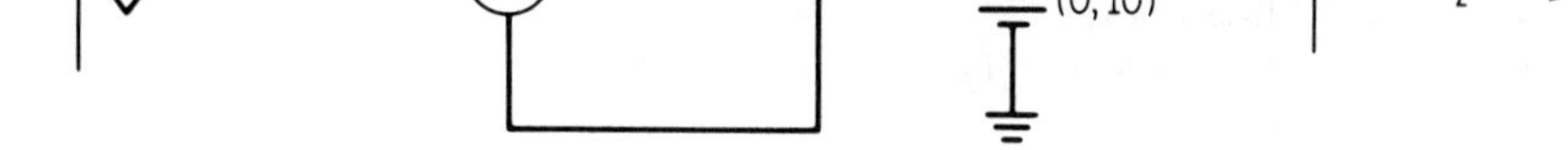

FIG. 8-21. Pulse-train generation.

be specified as a time-dependent source. Figure 8-21 indicates a circuit capable of generating a pulse train with minimum rise time. The switch controls the voltage source which will abruptly change value each time the applied current reverses polarity.

An example of the generation of a full-wave rectified waveform is given in the analysis of the power-supply filter of Fig. 8-22. The Zener diode can be represented by a 20-V source in series with an 8-Ω resistance. The resistance of the inductor plus forward diode and transformer resistance is 50 Ω.

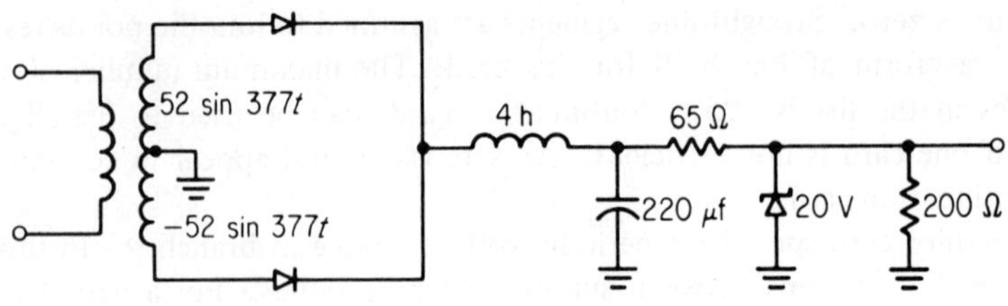

FIG. 8-22. *L*-section filter with Zener regulation.

The equivalent circuit used to analyze this circuit is shown in Fig. 8-23. The portion of the circuit enclosed by dashed lines is a buffered waveform generator which synthesizes a full-wave rectified signal of 52 volts amplitude. The current of branch 2 is sinusoidal as also is the current of branch 4. However, the current in branch 4 is displaced by one-half period. The positive portions of these waveforms are caused to flow through the 10-kΩ resistor of branch 5 developing a 52-V full-wave rectified signal. To avoid loading of the circuit by the filter, a

TABLE 8-20. Program for Regulated Filter Analysis.

```
Column No.
1      7
C      FULL WAVE RECTIFIER - CHOKE INPUT FILTER - ZENER DIODE
       TRANSIENT ANALYSIS
C
B1     N(0,1),R=1
E1     SIN(0.16666666E-1),52,0,0
B2     N(1,0),R=1E4
B3     N(0,2),R=1
E3     SIN(0.16666666E-1),52,0,0.83333333E-2
B4     N(2,0),R=1E4
B5     N(0,3),R=1E4
T1     B(2,5), BETA=(1,0)
T2     B(4,5), BETA=(1,0)
S1     B=2,(2),ON
S2     B=4,(4),ON
B6     N(3,4),R=(50,1E6)
S3     B=6,(6),ON
T3     B(6,5), BETA=1
B7     N(4,5),L=4,IO=.1
B8     N(5,0),C=220E-06,EO=-26.5
B9     N(5,6),R=65
B10    N(6,0),R=8,E=-20
B11    N(6,0),R=200
       TIME STEP=0.17361111E-3
       OUTPUT INTERVAL=12
       FINAL TIME=.16666666
       PRINT,NV
       EXECUTE
```

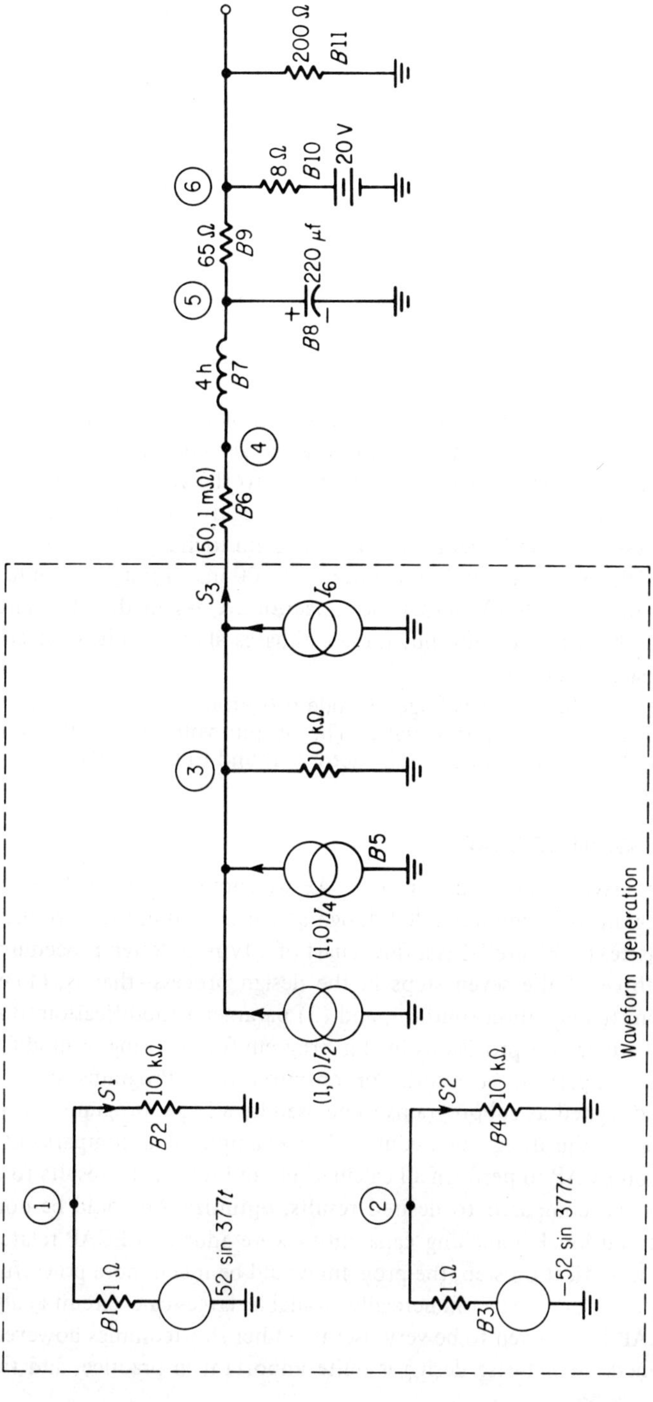

FIG. 8-23. Equivalent circuit of regulated power supply.

TABLE 8-21. Output Voltage of Power Supply.

Angle (referred to 60-Hz fundamental), Degrees	Voltage Volts
0	20.105
45	20.099
90	20.096
135	20.101
180	20.105
225	20.099
270	20.096
315	20.100
360	20.105

current generator proportional to the current drawn through branch 6 is placed across node 3. The waveform generator now has zero output impedance. Branch 6 contains a 50-Ω resistor when current flows to the filter to simulate resistive losses. If current through this branch becomes negative the resistance changes to 1 mΩ, representing diode behavior. The program to analyze this circuit is listed in Table 8-20. Note that an initial current of 100 ma through the inductor and an initial voltage of 26.5 V across the capacitor are assumed. This is not essential to complete the program, but these estimates allow steady state conditions to be approached sooner.

Table 8-21 shows the voltage at node 6 (output) over the last period (referenced to the 60-Hz input signal). The output voltage is fairly constant at approximately 20.1 V and the ripple factor is found to be 0.00016.

E. DISCUSSION OF ECAP

Before leaving the discussion of ECAP we should examine the relationship of this program to computer-aided design as considered in the introductory section of this text. Figure I-1 is a flow chart of a typical design procedure. ECAP performs three of the seven steps in the design process—that is, (1) equation formulation, (2) equation solution, and (3) parameter modification to a limited extent. There are no provisions in this program for choosing or modifying configuration, comparison of results, or optimization. Programs such as ECAP should be designed as subprograms to be used by a larger program that performs other steps in the design procedure. For example, if a comparison program could call on ECAP to perform all calculations and return the results to the main program to be compared to desired results, optimization could be done. Certainly if conditional branching capabilities were added to ECAP related to the parameter modification step, the program would be much more powerful.

As one of the first commercially available large-scale circuit-analysis programs, ECAP has proven to be very useful. Other shortcomings however, should be mentioned. Worst-case design is quite important in practice, and the extension to the a-c and transient-analysis programs of this feature would represent a significant improvement

Modeling of devices, which is sometimes tedious and quite redundant, is an area of potential improvement. Rather than modeling with circuit elements, a "device deck" could be used to characterize the device. Measured parameters or characteristics, stored on punched cards or tape, could be called by the main program when required. The "device deck" could even include extreme values of parameters to relate to worst-case design problems. SCEPTRE includes this stored-model feature.

There are a few disadvantages in the present transient-analysis program. The integration scheme is quite inefficient, requiring far more calculations than neccessary for many circuits. This problem is one of universal consequence, as indicated in Chapter 5. Automatic selection of an appropriate time step could be implemented, however, and this feature could decrease the design time significantly over the presently used trial-and-error approach to time-step selection. A variable time interval could also lead to more efficient analysis. Some networks, such as astable multivibrators, remain in a given state for long periods of time. Recalculation of voltages at each time step is redundant for this case. If the time step could be varied either by input specification or automatically, fewer calculations would be required of the computer.

ECAP requires that a-c and d-c solutions be found separately and does not have the capability to superimpose the results. If results could be automatically superimposed, some design-time savings could result. One must always consider results in light of operation in the appropriate region. Examination of a-c results alone does not always indicate to the designer when the active region of a device is exceeded. In addition to machine superposition of results, the computer could check to see if all devices operate within a prescribed region.

F. THE SCEPTRE PROGRAM

SCEPTRE, a transient program developed by the IBM Electronics Systems Center under sponsorship of the United States Air Force [4,5,6] represents several improvements over the transient section of ECAP.

SCEPTRE performs d-c and transient analysis, but has no provision for a-c analysis. The method of equation formulation is similar to that outlined in Chapter 4. A proper tree is selected and equations are constructed and manipulated, with the state equation being the final equation. This equation is now solved by one of three possible numerical integration methods. Each of these methods offers a different accuracy-speed combination. Unfortunately, all suffer from numerical-instability problems, resulting in excessive computer time for widely separated eigenvalues. For many problems however, the integration process is carried to completion in a reasonably short time. The integration step size is variable, with automatic selection of this size based on a suitable convergence criterion [7].

A very important feature of this program is that of a stored-model capability. ECAP requires that each device be modeled on paper before the data are entered into the computer. The program referred to as NET-1 allows a diode and transistor model to be stored on tape and called to the main program when necessary. Certain key parameters for the diode or transistor can be specified as input information. While this represents an improvement over ECAP, SCEPTRE goes one step further in allowing the user to propose and store models for any device made up of passive elements, mutual inductances, and voltage or current sources. Models for almost any device can be stored on tape and used at any point in the circuit analysis. This approach saves the designer a great deal of time and is also a step in the direction of automatic parameter measurement. It was mentioned earlier that one possible approach to the modeling problem is to develop a measuring scheme whereby the device parameters are automatically measured and inserted into the program. With the SCEPTRE approach this type of proposal becomes more feasible.

In the SCEPTRE program, initial conditions can be evaluated with the d-c analysis section and then automatically entered into the transient section. It will be recalled that ECAP requires two separate solutions to evaluate a transient response when initial conditions must be found. An equilibrium solution is executed and the results are manually entered into the transient solution program. With the automatic evaluation of initial conditions SCEPTRE eliminates the necessity of the equilibrium solution.

Parameter variation is allowed in SCEPTRE. ECAP allows this feature to be used with the d-c and a-c sections of the program, but not with transient response. With the somewhat more efficient integration routine of SCEPTRE parameter variation can be effected in the transient mode of the program. The user must exercise some judgment in applying this feature, as computer costs can mount rapidly when dealing with transient analysis.

Dependent voltage and current sources are allowed in SCEPTRE, whereas only dependent current sources are recognized by ECAP. Conversion of voltage sources to current sources is thus a requirement that is eliminated with SCEPTRE.

Another very significant feature of SCEPTRE is the ability to accept equations or tables that describe variable element values. A variable resistance, for example, can be described by an equation that reflects the resistance-current relationship of the resistor. Alternately a table can list piecewise linear values for the element. This feature not only leads to more accurate analysis but reduces the time required by the designer to model some of these nonlinear effects. The table approach is very effective in specifying the excitation of the network. Almost any transient waveform, regardless of complexity, can be entered in terms of a table of signal strength at appropriate time intervals. The program assumes that a linear segment appears between each pair of listed points.

As implied by the preceding paragraph, nonlinear analysis can be performed by SCEPTRE. The number of nonlinear elements affects the solution time since

Newton-Raphson iterative techniques are applied. Thus if a nonlinear resistance can be represented by an average value or a piecewise linear resistance with little loss of accuracy, the user is well advised to do so. On the other hand, when a very accurate solution is warranted the additional computer cost may be quite worthwhile.

Rather than describe the detailed format involved in programming SCEPTRE, a few simple programs will now be considered to demonstrate its application. The two-stage logic circuit of Fig. 8-24 is to be analyzed for various amplitudes of input signal [5].

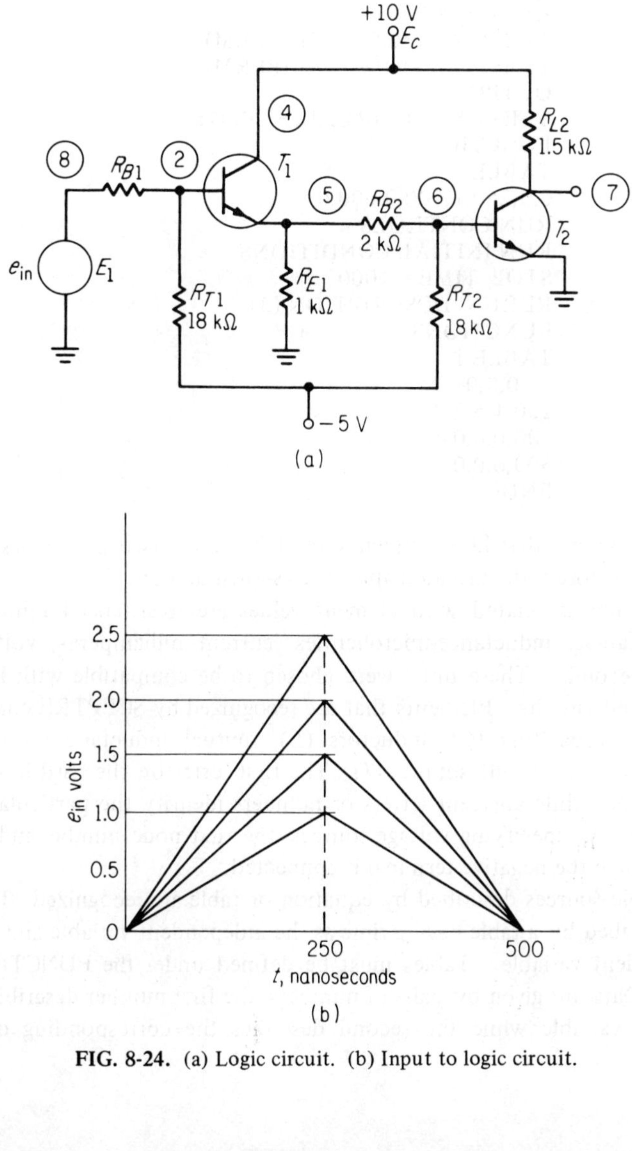

FIG. 8-24. (a) Logic circuit. (b) Input to logic circuit.

TABLE 8-22. SCEPTRE Program for Circuit of Fig. 8-24.

```
CIRCUIT DESCRIPTION
LOGIC CIRCUIT WITH SPIKE INPUT
ELEMENTS
EC,1-4 = 10
ET,3-1 = 5
E1,1-8=TABLE 1(TIME)
RB1,8-2 = 2
RT1,2-3 = 18
RE1,5-1 = 1
RB2,5-6 = 2
RT2,6-3 = 18
RL2,4-7 = 1.5
T1,2-5-4 = MODEL X1 (PERM)
T2,6-1-7 = MODEL X1 (PERM)
OUTPUTS
VCET1,VCET2,IRL2,IRE1,PLOT
FUNCTIONS
TABLE 1
0,0,250,1,500,0,600,0
RUN CONTROLS
RUN INITIAL CONDITIONS
STOP  TIME = 1000
RERUN  DESCRIPTION (3)
FUNCTIONS
TABLE 1
0  ,0,0,0
250,1.5,2,2.5
500,0,0,0
600,0,0,0
END
```

If we assume that high-frequency models for the two transistors have previously been stored, the program appears as shown in Table 8-22.

The units associated with element values are: resistance-kilohms, capacitance-picofarads, inductance-microhenries, current-milliamperes, voltage-volts, time-nanoseconds. These units were chosen to be compatible with high-speed transistorized circuits. Elements that are recognized by SCEPTRE consist of resistors (*R*), capacitors (*C*), inductors (*L*), mutual inductances (*M*), voltage sources (*E*), and current sources (*J*). The first letter on the card indicates the element type, while adjacent letters or numbers identify the particular element referred to. In specifying voltage sources, the first node number indicates the node to which the negative terminal is connected.

Variable sources described by equation or table are recognized. The source E_1 is described by a table having time as the independent variable and voltage as the dependent variable. Tables must be defined under the FUNCTIONS subheading. Data are given by pairs of numbers, the first number describing the independent variable while the second describes the corresponding dependent variable. The third number again refers to the independent variable and the fourth to the corresponding dependent variable. This format continues until the curve has been completely described. A piecewise linear curve is assumed by the program between each pair of points. Note that the description of E_1 in Table 8-22 lists the voltage at 600 nanoseconds as zero volts, even though the voltage at 500 nanoseconds was also zero volts. The slope at $t = 500$ was nonzero; thus the program would have incorrectly calculated the value of E_1 by extrapolation at times later than 500 nanoseconds if this additional point was not specified.

In specifying the transistor models, it was assumed that a stored model (MODEL X1) had been entered prior to execution of this program. Temporary models can be stored as part of the program to be used only by a single program. Permanent models are generally reserved for those devices that find widespread usage. The model stored for the transistors in this case might be the Ebers-Moll model mentioned previously. The circuit designation for the first transistor of the circuit is T_1. The three numbers following the circuit designation determine the configuration of the transistor. The first number is the node to which the base connects; the second is the node to which the emitter connects; and the third specifies the collector connections. For T_1, the base is connected to node 2, the emitter to node 5, and the collector to node 4. The reader is referred to the User's Manual [6] for more detail regarding the stored-model capability.

The output quantities will be printed in tabular form and can also be plotted if specified. All quantities appearing on the punched card that contains the instruction PLOT will be plotted. If only one variable of several outputs need be plotted, this variable and the word PLOT can be included on a separate card. Voltage and current associated with any passive element can be given as output quantities as can voltage and current of a source. Any output, however, must consist of some dependent variable that is a function of some independent variable. Element values or parameters dependent on some other quantity can be given as output information.

The specification of VECTI as an output quantity means that the voltage across the capacitor *CE* of transistor T_1 is required. This capacitor is included in the model for T_1. The correct format to specify current through R_{L_2} as an output quantity is IRL2.

There are several run-control cards that may be used with SCEPTRE. All transient runs must have a problem duration specified by the STOP TIME card. For this program the stop time is 1,000 nanoseconds. The type of integration subroutine used to solve the circuit can be specified at this point. If no specification is given, the subroutine giving the best compromise between speed and accuracy is used. The card RUN INITIAL CONDITIONS will include the initial condition solution as output information along with the transient solution. The rerun feature permits repeated solutions of the problem with revised parameters. In this case, three reruns are achieved as a result of the card RERUN DESCRIPTION (3) and then followed by the quantities to be changed for each rerun.

Here the input voltage expressed as a piecewise linear function is given for the three reruns. The same output quantities result from each rerun.

A terminate feature can also be applied if certain variables reach some prescribed limits. For example, if it were desired in the circuit of Fig. 8-24 that the analysis terminate when the voltage from base to emitter of T_2 exceeds zero volts, the card TERMINATE IF (VCET2.GT.0.) could be inserted before the stop time card. The run would then terminate if and when this condition is met. For this program the master run and rerun 1 are carried through to the 1,000-nanosecond duration. Reruns 2 and 3 are terminated earlier when the terminate condition is applied.

Another example from the SCEPTRE User's Manual [6] will demonstrate the temporary stored-model feature. The circuit to be analyzed is shown in Fig. 8-25a for a sinusoidal input of 1 mv amplitude and 100 kHz frequency. An equivalent circuit for the transistor is shown in Fig. 8-25b.

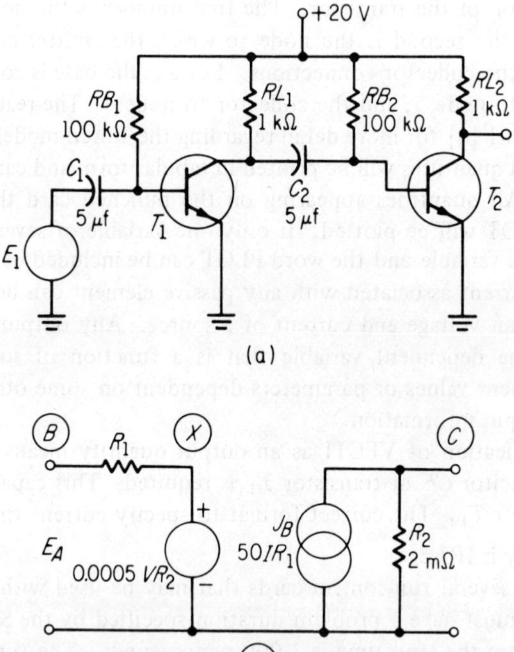

FIG. 8-25. (a) Amplifier circuit. (b) Transistor equivalent circuit.

Figure 8-26 shows the circuit prepared for SCEPTRE. It should be noted that for small-signal analysis the d-c source can be shorted. The program to analyze this circuit is given in Table 8-23.

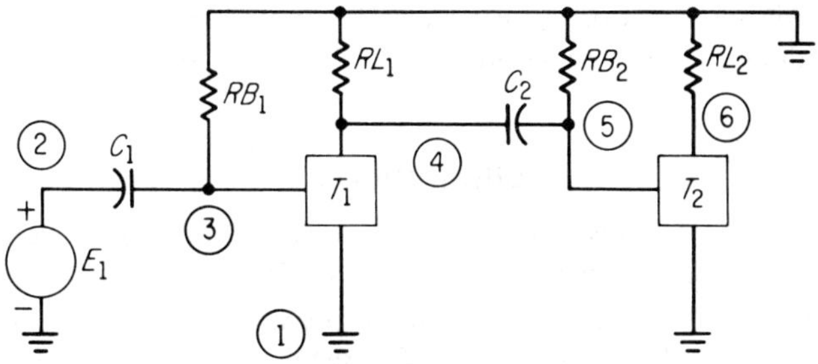

FIG. 8-26. Circuit prepared for SCEPTRE input.

For this program the transistor model is stored only until the program is executed. The source E_1 is a sinusoidal signal with period equal to 10,000 nanoseconds and is expressed in the form of an equation.

There are many other interesting features of SCEPTRE that cannot be covered here. The references at the end of this chapter provide more detail to the many program features. While SCEPTRE does not provide a-c analysis capabilities, it is a very useful program and one that is growing in usage.

Other specific programs have not been discussed in this text. This is not intended to indicate that programs such as CIRCUS, PANE, NET-1, NET-2,

TABLE 8-23. Program to Analyze Circuit of Fig. 8-26.

```
MODEL DESCRIPTION
MODEL SS1 (TEMP) (B-E-C)
ELEMENTS
EA, E-X = .0005*VR2
R2, C-E = 2000
R1, B-X = .3
JB, C-E = 50.*IR1
CIRCUIT DESCRIPTION
ELEMENTS
E1, 1-2 = (. 001*SIN ( . 000628*TIME))
C1, 2-3 = 5E6
C2, 4-5 = 5E6
RB1, 3-1 = 100
RL1, 1-4 = 1
RB2, 5-1 = 100
RL2, 1-6 = 1
T1, 3-1-4 = MODEL SS1
T2, 5-1-6 = MODEL SS1
OUTPUTS
VRL1, VRL2, VC1, VC2, PLOT
RUN CONTROLS
STOP TIME 30000
INTEGRATION ROUTINE = TRAP
END
```

and CORNAP are not worthy of note. The limited amount of space available in this text does not allow these programs to be considered.

BIBLIOGRAPHY

1. Comer, D. J., *Introduction to Semiconductor Circuit Design,* Addison-Wesley Publishing Company, Reading, Mass., 1968.
2. Gray, P. E., D. De Witt, A. R. Boothroyd, and J. F. Gibbons, *Physical Electronics and Circuit Models of Transistors,* SEEC Vol. 2, John Wiley & Sons, Inc., New York, 1964.
3. Jensen, R. W., and M. D. Lieberman, *IBM Electronic Circuit Analysis Program*, Prentice-Hall, Inc., Englewood Cliffs, N. J., 1968.
4. Sedore, S. R., "SCEPTRE: A Program for Automatic Network Analysis," *IBM Journal,* Vol. 11 (November, 1967), pp. 627-637.
5. Sedore, S. R., and J. R. Sents, "Trends in Transient Automatic Circuit Analysis Programs," IBM Publications No. 67-825-1791, IBM Corporation, Oswego, N. Y., 1967.
6. Sedore, S. R., J. R. Sents, and H. W. Mathers, *SCEPTRE SUPPORT, Vol. 1: Revised User's Manual,* Technical Report No. AFWL-TR-67-124, Vol. I, Air Force Weapons Laboratory, Kirtland AFB, New Mexico.
7. Warten, R. M., "Automatic Step Size Control for Runge-Kutta Integration" *IBM Journal*, Vol. 7, pp. 340-341 (October, 1963).
8. 1620 Electronic Circuit Analysis Program (ECAP), (1620-EE-02X) User's Manual, IBM Publications H20-0170-1, IBM Corporation, White Plains, N.Y.

PROBLEMS

8-1 Calculate the quiescent output voltage as R varies from 40 kΩ to 200 kΩ in 10-kΩ steps. Assume $\beta = 100$, $V_{BE(ON)} = 0.5$ V, and $\bar{r}_c = 40$ kΩ.

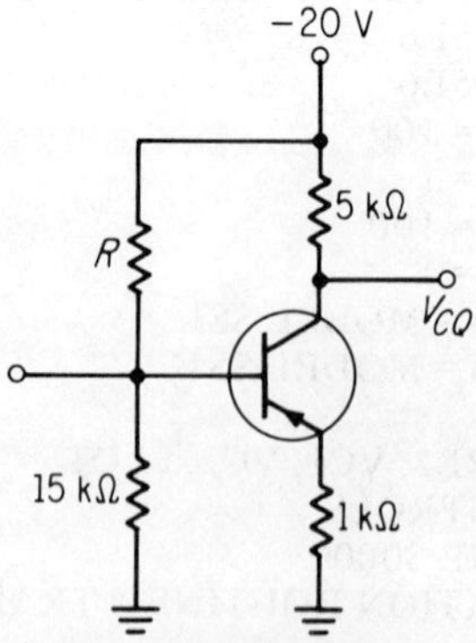

PROB. 8-1

8-2 Find V_{CQ} in the circuit of Prob. 8-1 for worst-case conditions when the nominal value of R is 100 kΩ. Resistor tolerances are 10 percent. Assume that β can vary from 50 to 180, $V_{BE(ON)}$ can vary from 0.4 V to 0.6 V, and $\bar{r}_c$ can vary from 30 kΩ to 50 kΩ.

8-3 Find the minimum input voltage that will saturate the transistor. Assume $\beta = 50$, $V_{BE(ON)} = 0.6$ V, and that $\bar{r}_c = \infty$. Use the d-c analysis program.

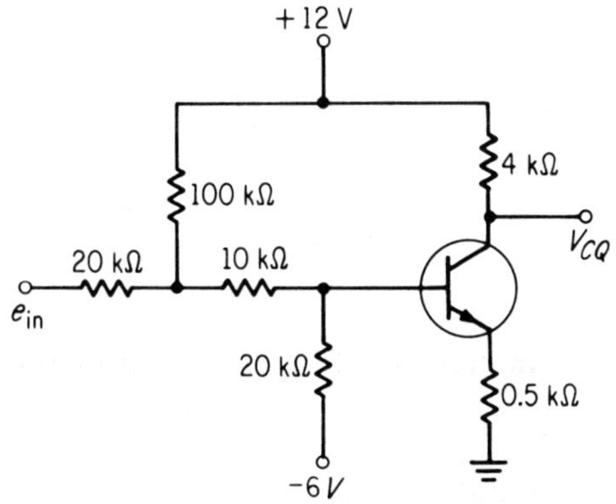

PROB. 8-3

8-4 Find the maximum input voltage that can be applied to cause V_{CQ} to be less than or equal to +10 V. Assume $\beta = 150$ and $V_{BE} = 0.4$ V. Use the d-c analysis program and the circuit of Prob. 8-3.

8-5 Select R to set I at (50 ± 10) μa.

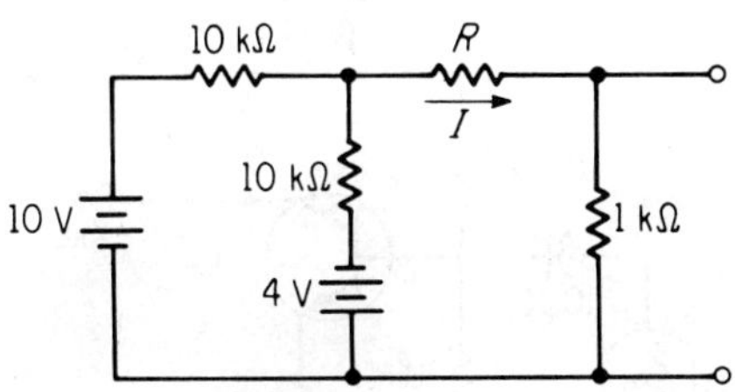

PROB. 8-5

8-6 Plot d-c power to the load and the output voltage as functions of load resistance as R_L varies from 100 Ω to 10 kΩ.

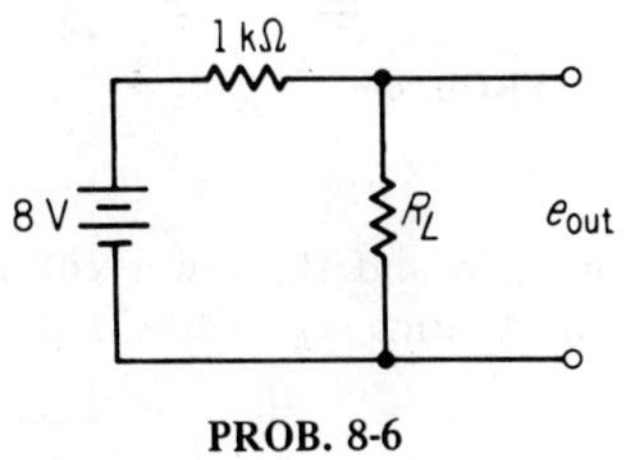

PROB. 8-6

8-7 Repeat Prob. 8-6 for the circuit shown. Assume $\beta = 100$, $V_{BE(ON)} = 0.5$ V, and that the transistor input impedance is negligible.

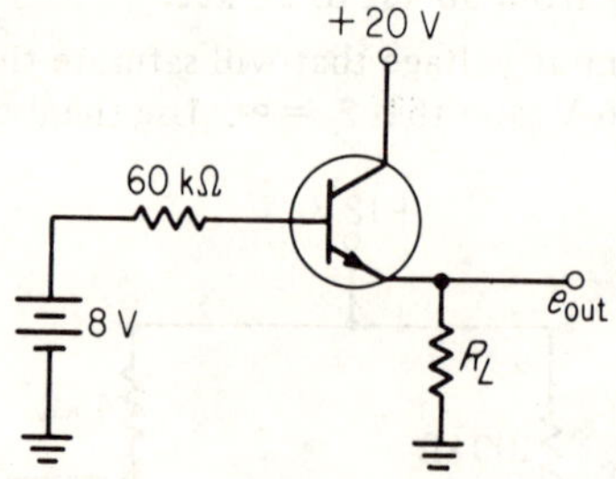

PROB. 8-7

8-8 When R_L is 2 kΩ in Prob. 8-7, find the sensitivity of e_{out} with respect to current gain β.

8-9 **(a)** Find the quiescent point of the circuit shown assuming $\beta = 100$, $V_{BE(ON)} = 0.6$ V, $r_{bb}' = 200\ \Omega$ and $r_c = \infty$.

(b) Using the same parameters and $r_{b'e}$ as calculated in (a), plot the frequency response of the voltage gain up to a frequency of 50 kHz.

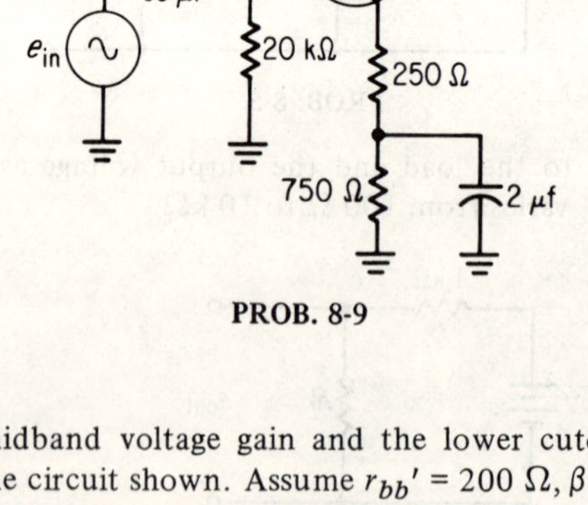

PROB. 8-9

8-10 Find the midband voltage gain and the lower cutoff frequency (−3-db point) for the circuit shown. Assume $r_{bb}' = 200\ \Omega$, $\beta = 80$, $r_{b'e} = 1{,}400\ \Omega$, and $r_c = \infty$.

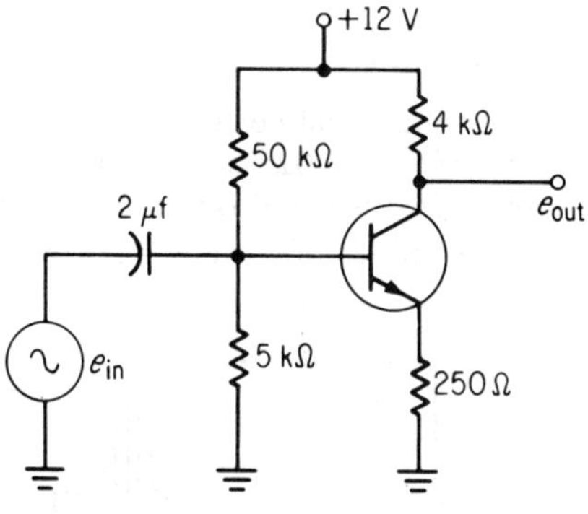

PROB. 8-10

8-11 Find the voltage gain for the circuit of Prob. 3-9.

8-12 Work Prob. 3-10 using ECAP.

8-13 Work Prob. 3-12 using ECAP.

8-14 Work Prob. 3-14 using ECAP.

8-15 Work Prob. 3-16 using ECAP.

8-16 Work Prob. 3-21 using ECAP.

Problems 8-17 to 8-29 may be worked manually, by ECAP, or by SCEPTRE whichever method is appropriate.

8-17 Calculate the quiescent collector currents in order to determine $r_{b'e}$ then

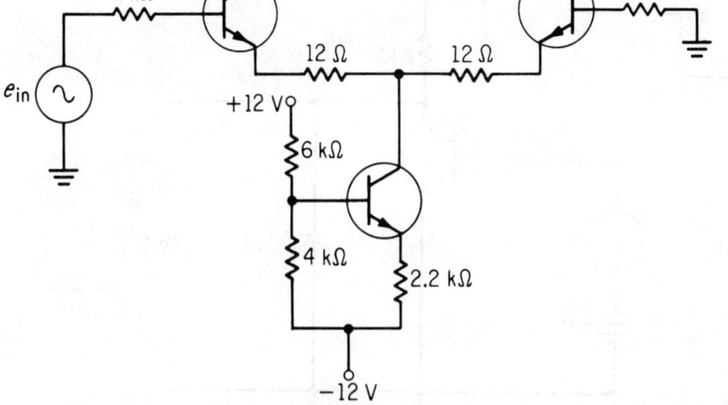

PROB. 8-17

calculate the ac voltage gain of the differential amplifier. Assume that $r_{bb}' = 200\ \Omega$, $V_{BE(ON)} = 0.5$ V, $\beta = 50$, and $r_c = \infty$ for all transistors.

8-18 Calculate the output voltage and power if the primary inductance is 0.4 h, the secondary inductance is 0.025 h, and the coefficient of coupling is 0.99995. The input signal is 0.3 V rms at 1 kHz and $\beta = 20$, $r_{bb}' + r_{b'e} = 40\ \Omega$, and the quiescent collector current is 100 ma.

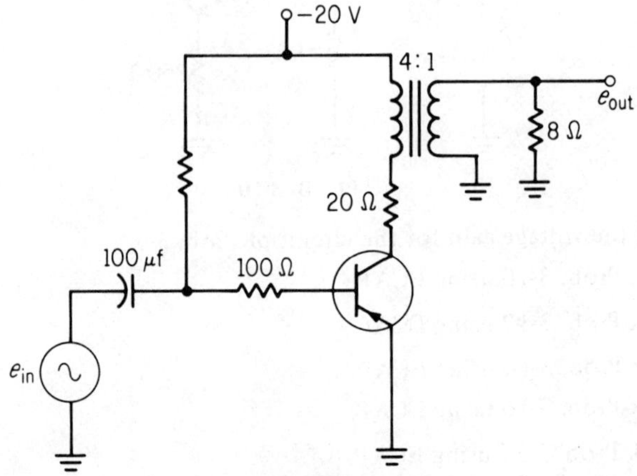

PROB. 8-18

8-19 Rework Prob. 8-18 if $e_{in} = 2$ V rms at 1 kHz. Are your results reasonable? Explain.

8-20 Plot the frequency response of the circuit of Fig. 8-5 if $R_s = 75\Omega$, $r_{bb}' = 50\ \Omega$, $r_{b'e} = 300\ \Omega$, $C_d = 50$ pf, $C_{b'c} = 2.5$ pf, $g_m = 0.2$ mho, and $R_L = 400\ \Omega$.

8-21 Plot the output voltage as a function of time. Assume that $r_c = 33\ k\Omega$, $\beta = 90$, $V_{BE(ON)} = 0.2$ V, and $r_b' = 400\ \Omega$.

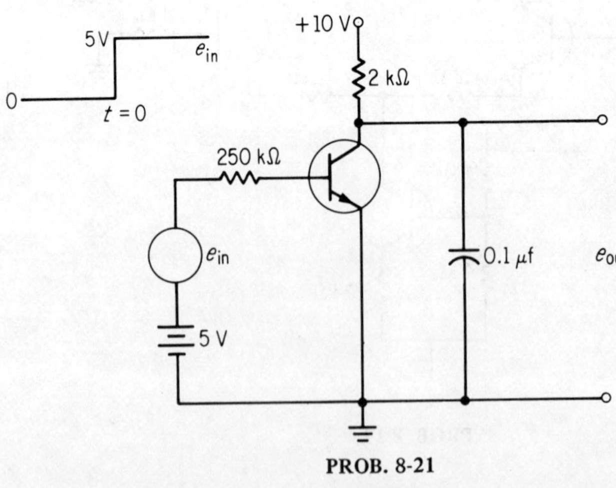

PROB. 8-21

8-22 Synthesize the nonrecurring ramp voltage shown. The total fall time must be less than 5 μs and the waveform generator must present an output impedance less than 100 Ω.

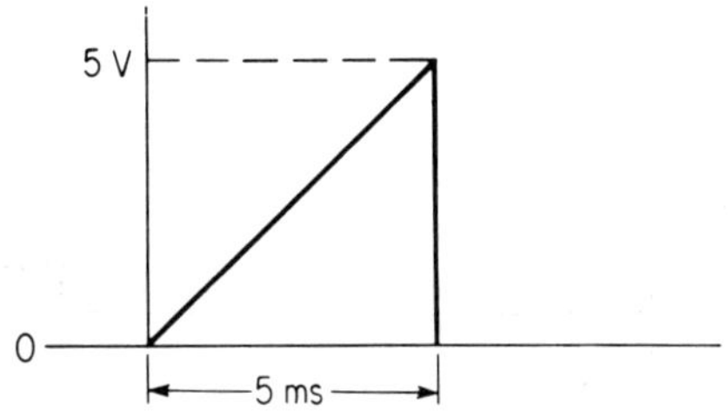

PROB. 8-22

8-23 Plot the output voltage of the circuit if the signal in Prob. 8-22 is applied. Assume that $r_c = 33$ kΩ, $\beta = 90$, $V_{BE(ON)} = 0.2$ V, and $r'_b = 400$ Ω.

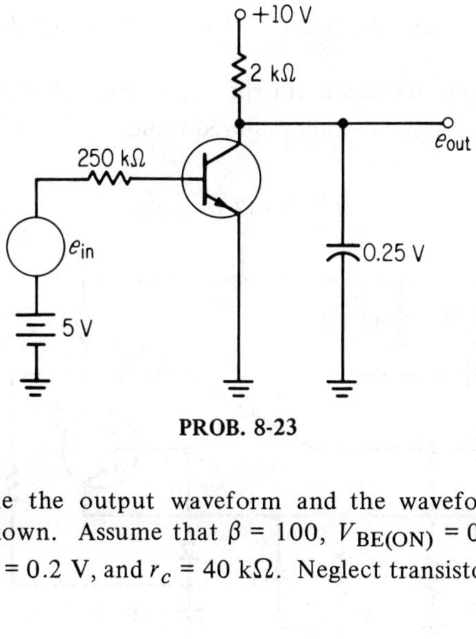

PROB. 8-23

8-24 Determine the output waveform and the waveform at the base of the circuit shown. Assume that $\beta = 100$, $V_{BE(ON)} = 0.3$ V, $r'_b = 1$ kΩ, $R_{cs} = 20$ Ω, $V_s = 0.2$ V, and $r_c = 40$ kΩ. Neglect transistor switching times.

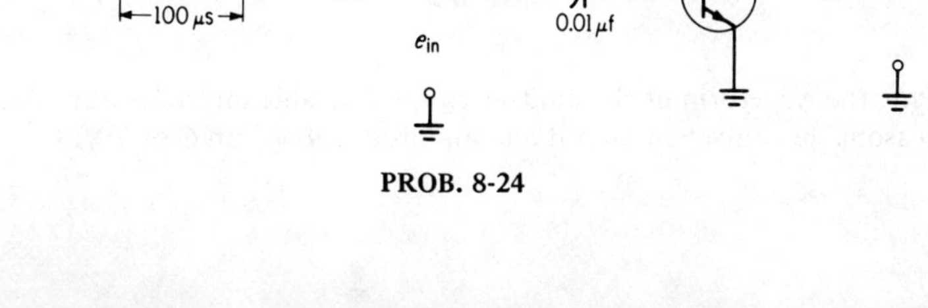

PROB. 8-24

8-25 Plot the output voltage of the Schmitt trigger circuit if $V_s = 0.2$ V, $V_{BE(ON)} = 0.6$ V, $\beta = 40$, $r_c = 30\text{k}\Omega$, $r_b' = 500\ \Omega$, and $R_{cs} = 10\ \Omega$. Neglect transistor switching times.

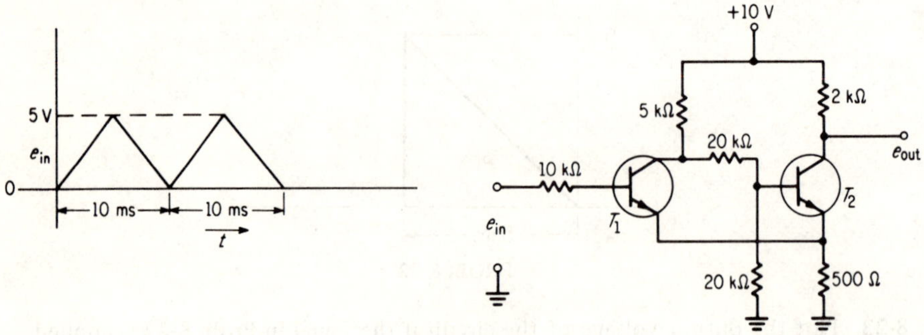

PROB. 8-25

8-26 Rework Prob. 8-25 if the 5-kΩ collector resistor of T_1 is changed to 2.5 kΩ.

8-27 Calculate the output waveform for the one-shot circuit if $V_{CE(sat)} = 0$ and $V_{BE(ON)} = 0.2$ V. Assume other required values.

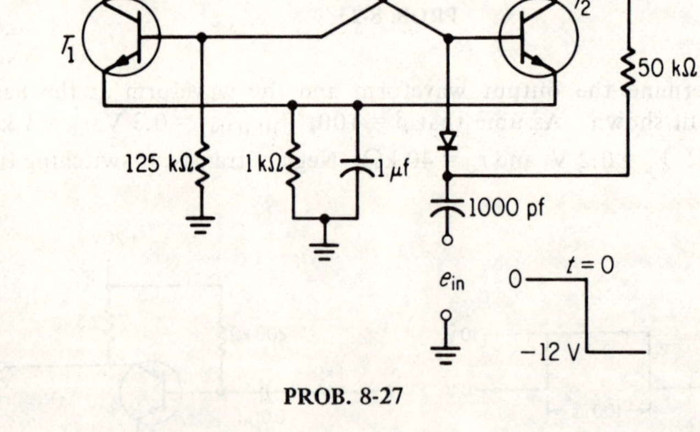

PROB. 8-27

8-28 Plot the waveform of the emitter coupled astable multivibrator. Assume a reasonable equivalent circuit and an initial voltage on C of 7 V.

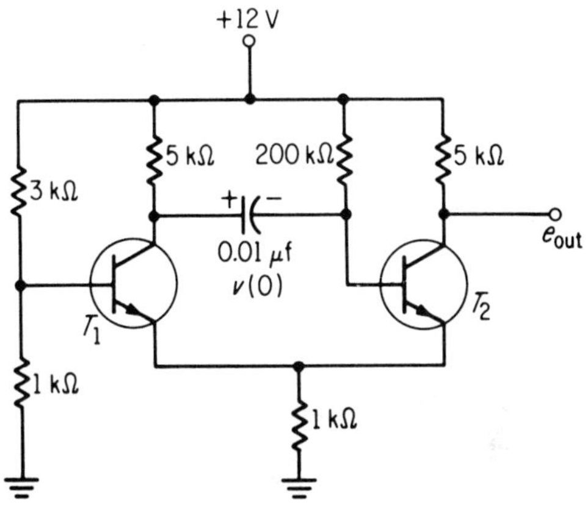

PROB. 8-28

8-29 Use reasonable parameter values for the three-region model of a silicon

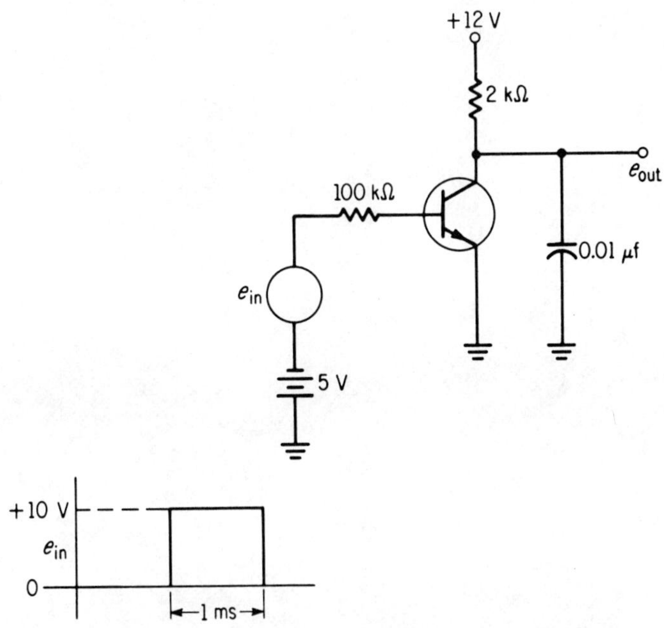

PROB. 8-29

transistor to evaluate the output waveform. Neglect transistor switching times.

APPENDIX

SUBROUTINES

This section lists various subroutines that were called by main programs throughout the text.

MPRD

There are two such subroutines used in this text. One is for use on the IBM 1132 while the other is used with the IBM 360/50. This subroutine is used to form the product of two matrices.

IBM 1132

```
      SUBROUTINE MPRD (A,B,R,N,M,MSA,MSB,L)
      DIMENSION A(1),B(1),R(1)
C        SPECIAL CASE FOR DIAGONAL BY DIAGONAL
      MS=MSA*10+MSB
      IF(MS-22) 30,10,30
   10 DO 20 I=1,N
   20 R(I)=A(I)*B(I)
      RETURN
C        ALL OTHER CASES
   30 IR=1
      DO 90 K=1,L
      DO 90 J=1,N
      R(IR)=0
      DO 80 I=1,M
      IF(MS) 40,60,40
   40 CALL LO C(J,I,IA,N,M,MSA)
      CALL LO C(I,K,IB,M,L,MSB)
      IF(IA) 50,80,50
   50 IF(IB) 70,80,70
   60 IA=N*(I-1)+J
      IB=M*(K-1)+I
   70 R(IR)=R(IR)+A(IA)*B(IB)
   80 CONTINUE
   90 IR=IR+1
      RETURN
      END
```

IBM 360/50

```
      SUBROUTINE MPRD(N,K,A,B,R)
C     THIS SUBROUTINE FORMS THE MATRIX PRODUCT AXB
C     WHERE A IS AN
C     NXK MATRIX AND B IS A KXM MATRIX
```

```
      DIMENSION A(N,K),B(K,M),R(N,M)
      DO 1 I=1,N
      DO 1 J=1,M
      R(I,J)=0.0
      DO1 L=1,K
    1 R(I,J)=R(I,J)+A(I,L)*B(L,J)
      RETURN
      END
```

MINV

IBM 1132

```
      SUBROUTINE MINV(A,N,D,L,M)
      DIMENSION A(1),L(1),M(1)
C        SEARCH FOR LARGEST ELEMENT
      D=1.0
      NK=-N
      DO 80 K=1,N
      NK=NK+N
      L(K)=K
      M(K)=K
      KK=NK+K
      BIGA=A(KK)
      DO 20 J=K,N
      IZ=N*(J-1)
      DO 20 I=K,N
      IJ=IZ+I
   10 IF( ABS(BIGA)- ABS(A(IJ))) 15,20,20
   15 BIGA=A(IJ)
      L(K)=I
      M(K)=J
   20 CONTINUE
C        INTERCHANGE ROWS
      J=L(K)
      IF(J-K) 35,35,25
   25 KI=K-N
      DO 30 I=1,N
      KI=KI+N
      HOLD=-A(KI)
      JI=KI-K+J
      A(KI)=A(JI)
   30 A(JI) = HOLD
C        INTERCHANGE COLUMNS
   35 I=M(K)
      IF(I-K) 45,45,38
   38 JP=N*(I-1)
      DO 40 J=1,N
      JK=NK+J
      JI=JP+J
      HOLD=-A(JK)
      A(JK)=A(JI)
   40 A(JI) =HOLD
C        DIVIDE COLUMN BY MINUS PIVOT (VALUE OF PIVOT
```

```
C           ELEMENT IS CONTAINED IN BIGA)
   45 IF(ABS(BIGA)-1.E-20)46,46,48
   46 D=0.0
      RETURN
   48 DO 55 I=1,N
      IF(I-K) 50,55,50
   50 IK=NK+I
      A(IK)=A(IK)/(-BIGA)
   55 CONTINUE
C           REDUCE MATRIX
      DO 65 I=1,N
      IK=NK+I
      HOLD=A(IK)
      IJ=I-N
      DO 65 J=1,N
      IJ=IJ+N
      IF(I-K) 60,65,60
   60 IF(J-K) 62,65,62
   62 KJ=IJ-I+K
      A(IJ)=HOLD*A(KJ)+A(IJ)
   65 CONTINUE
C           DIVIDE ROW BY PIVOT
      KJ=K-N
      DO 75 J=1,N
      KJ=KJ+N
      IF(J-K) 70,75,70
   70 A(KJ)=A(KJ)/BIGA
   75 CONTINUE
C           PRODUCT OF PIVOTS
      D=D*BIGA
C           REPLACE PIVOT BY RECIPROCAL
      A(KK)=1.0/BIGA
   80 CONTINUE
C           FINAL ROW AND COLUMN INTERCHANGE
      K=N
  100 K=(K-1)
      IF(K) 150,150,105
  105 I=L(K)
      IF(I-K) 120,120,108
  108 JQ=N*(K-1)
      JR=N*(I-1)
      DO 110 J=1,N
      JK=JQ+J
      HOLD=A(JK)
      JI=JR+J
      A(JK)=-A(JI)
  110 A(JI) =HOLD
  120 J=M(K)
      IF(J-K) 100,100,125
  125 KI=K-N
      DO 130 I=1,N
      KI=KI+N
```

```
      HOLD=A(KI)
      JI=KI-K+J
      A(KI)=-A(JI)
  130 A(JI) =HOLD
      GO TO 100
  150 RETURN
      END
```

GMPRD

IBM 1132

```
C       MATRIX PRODUCT
        SUBROUTINE GMPRD(A,B,R,N,M,L)
        DIMENSION A(1),B(1),R(1)
        IR=0
        IK=-M
        DO 10 K=1,L
        IK=IK+M
        DO 10 J=1,N
        IR=IR+1
        JI=J-N
        IB=IK
        R(IR)=0
        DO 10 I=1,M
        JI=JI+N
        IB=IB+1
     10 R(IR)=R(IR)+A(JI)*B(IB)
        RETURN
        END
```

INVR

IBM 1620

```
C       MATRIX INVERSION WITH ACCOMPANYING SOLUTION OF
C       LINEAR EQUATIONS
        SUBROUTINE INVR(A,N,B,M)
        DIMENSION IPIVOT(20),A(10,10),B(10,1),INDEX(20,2),
      1   PIVOT(20)
        EQUIVALENCE (IROW,JROW), (ICOLUM,JCOLUM), (AMAX,T,
      1 SWAP)
C       INITIALIZATION
     10 DETERM=1.0
     15 DO 20 J=1,N
     20 IPIVOT(J)=0
     30 DO 550 I=1,N
C          SEARCH FOR PIVOT ELEMENT
     40 AMAX=0.0
     45 DO 105 J=1,N
     50 IF (IPIVOT(J)-1) 60,105,60
     60 DO 100 K=1,N
     70 IF (IPIVOT(K)-1) 80,100,740
     80 IF(ABS(AMAX)-ABS(A(J,K)))85,100,100
     85 IROW=J
     90 ICOLUM=K
     95 AMAX=A(J,K)
```

```
  100 CONTINUE
  105 CONTINUE
  110 IPIVOT(ICOLUM)=IPIVOT(ICOLUM)+1
C         INTERCHANGE ROWS TO PUT PIVOT ELEMENT ON
C         DIAGONAL
  130 IF (IROW-ICOLUM) 140,260,140
  140 DETERM=-DETERM
  150 DO 200 L=1,N
  160 SWAP=A(IROW,L)
  170 A(IROW,L)=A(ICOLUM,L)
  200 A(ICOLUM,L)=SWAP
  205 IF(M) 260,260,210
  210 DO 250 L=1,M
  220 SWAP=B(IROW,L)
  230 B(IROW,L)=B(ICOLUM,L)
  250 B(ICOLUM,L)=SWAP
  260 INDEX(I,1)=IROW
  270 INDEX(I,2)=ICOLUM
  310 PIVOT(I)=A(ICOLUM,ICOLUM)
  320 DETERM=DETERM*PIVOT(I)
C         DIVIDE PIVOT ROW BY PIVOT ELEMENT
  330 A(ICOLUM,ICOLUM)=1.0
  340 DO 350 L=1,N
  350 A(ICOLUM,L)=A(ICOLUM,L)/PIVOT(I)
  355 IF(M) 380,380,360
  360 DO 370 L=1,M
  370 B(ICOLUM,L)=B(ICOLUM,L) /PIVOT(I)
C         REDUCE NON-PIVOT ROWS
  380 DO 550 L1=1,N
  390 IF(L1-ICOLUM) 400,550,400
  400 T=A(L1,ICOLUM)
  420 A(L1,ICOLUM)=0.0
  430 DO 450 L=1,N
  450 A(L1,L)=A(L1,L)-A(ICOLUM,L)*T
  455 IF(M) 550,550,460
  460 DO 500 L=1,M
  500 B(L1,L)=B(L1,L)-B(ICOLUM,L)*T
  550 CONTINUE
C         INTERCHANGE COLUMNS
  600 DO 710 I=1,N
  610 L=N+1-I
  620 IF (INDEX(L,1)-INDEX(L,2)) 630,710,630
  630 JROW=INDEX(L,1)
  640 JCOLUM=INDEX(L,2)
  650 DO 705 K=1,N
  660 SWAP=A(K,JROW)
  670 A(K,JROW)=A(K,JCOLUM)
  700 A(K,JCOLUM)=SWAP
  705 CONINTUE
  710 CONTINUE
  740 RETURN
      END
```

INDEX